Marat V. Markin
Real Analysis

Also of Interest

Elementary Functional Analysis
Marat V. Markin, 2018
ISBN 978-3-11-061391-9, e-ISBN (PDF) 978-3-11-061403-9,
e-ISBN (EPUB) 978-3-11-061409-1

Elementary Operator Theory
Marat V. Markin, 2019
ISBN 978-3-11-060096-4, e-ISBN (PDF) 978-3-11-060098-8,
e-ISBN (EPUB) 978-3-11-059888-9

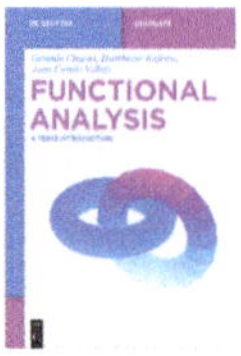

Functional Analysis. A Terse Introduction
Gerardo Chacón, Humberto Rafeiro, Juan Camilo Vallejo, 2016
ISBN 978-3-11-044191-8, e-ISBN (PDF) 978-3-11-044192-5,
e-ISBN (EPUB) 978-3-11-043364-7

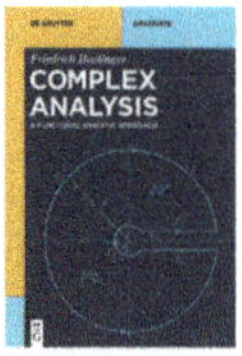

Complex Analysis. A Functional Analytic Approach
Friedrich Haslinger, 2017
ISBN 978-3-11-041723-4, e-ISBN (PDF) 978-3-11-041724-1,
e-ISBN (EPUB) 978-3-11-042615-1

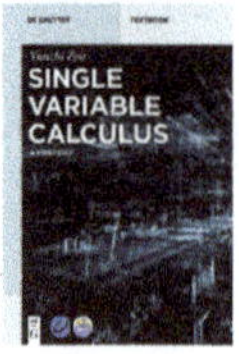

Single Variable Calculus. A First Step
Yunzhi Zou, 2018
ISBN 978-3-11-052462-8, e-ISBN (PDF) 978-3-11-052778-0,
e-ISBN (EPUB) 978-3-11-052785-8

Marat V. Markin

Real Analysis

Measure and Integration

DE GRUYTER

Mathematics Subject Classification 2010
28-01, 28A10, 28A12, 28A15, 28A20, 28A25

Author
Prof. Dr. Marat V. Markin
California State University, Fresno
Department of Mathematics
5245 North Backer Avenue
Fresno, CA 93740
USA
mmarkin@csufresno.edu

ISBN 978-3-11-060097-1
e-ISBN (PDF) 978-3-11-060099-5
e-ISBN (EPUB) 978-3-11-059882-7

Library of Congress Control Number: 2019931612

Bibliographic information published by the Deutsche Nationalbibliothek
The Deutsche Nationalbibliothek lists this publication in the Deutsche Nationalbibliografie; detailed bibliographic data are available on the Internet at http://dnb.dnb.de.

Cover image: Merrymoonmary / Getty Images
Typesetting: VTeX UAB, Lithuania
Printing and binding: CPI books GmbH, Leck

www.degruyter.com

With utmost appreciation to all my teachers.

Preface

The author discusses valueless measures in pointless spaces.
Paul Halmos

The Purpose of the Book and Targeted Audience

The book is intended as a text for a *one-semester* Master's level graduate course in real analysis with emphasis on the measure and integration theory to be taught within the existing constraints of the standard for the United States graduate curriculum (fifteen weeks with two seventy-five-minute lectures per week). Real analysis, being, as a rule, a core course in every graduate program in mathematics, is also of significant interest to a wider audience of STEM (science, technology, engineering, and mathematics) graduate students or advanced undergraduates with a solid background in proof-based intermediate analysis.

Book's Philosophy, Scope, and Specifics

The philosophy of the book, which makes it quite distinct from many existing texts on the subject, is based on treating the concepts of measure and integration starting with the most general abstract setting and then introducing and studying the Lebesgue measure and integration on the real line as an important particular case.

The book consists of *nine chapters* and an *appendix* taking the reader from the *basic set classes*, through *measures*, *outer measures* and the general procedure of *measure extension* described in the celebrated *Carathéodory's Extension Theorem* and applied to the construct of the *Lebesgue–Stieltjes measures* as a central particular case. It further treats *measurable functions* and *various types of convergence* of sequences of such based on the idea of measure, the treatment including the classical *Luzin's* and *Egorov's Theorem* as well as the *Lebesgue* and *Riesz Theorems*. Then the fundamentals of the abstract Lebesgue integration and the basic *limit theorems*, such as *Fatou's Lemma* and *Lebesgue's Dominated Convergence Theorem*, are furnished, the discourse culminating into the comparison of the Lebesgue and Riemann integrals and characterization of the Riemann integrable functions.

Chapter 1 outlines certain necessary preliminaries, including the fundamentals of metric spaces. The course is designed to be taught starting with Chapter 2, Chapter 1 being referred to whenever the need arises, for instance when dealing with the concepts of upper and lower limits of numeric or set sequences, using the properties of the inverse image operation, or defining the *Borel sets*.

Chapter 7 is dedicated to studying *convergence in p-norm*, *L_p spaces*, and the basics of *normed vector spaces*. It further demonstrates the deficiencies of the Riemann

https://doi.org/10.1515/9783110600995-201

integral relative to its Lebesgue counterpart and prepares the students for more advanced courses in functional analysis and operator theory.

Chapter 8 is dedicated to using the novel approach based on the Lebesgue measure and integration theory machinery to develop a better understanding of differentiation and extend the classical *total change formula* linking differentiation with integration to a substantially wider class of functions.

Chapter 9 on *signed measures* can be considered as a "bonus" chapter to be taught should the time constraints of a one-semester course permit.

The Appendix gives a concise treatise of the *Axiom of Choice*, its equivalents (the *Hausdorff Maximal Principle*, *Zorn's Lemma*, and *Zermello's Well-Ordering Principle*), and *ordered sets*, which is fundamental for proving the famed *Vitali Theorem* on the existence of a non-Lebesgue measurable set in $\mathbb{R}$.

Being designed as a text to be used in a classroom, the book constantly calls for the student's actively mastering the knowledge of the subject matter. There are problems at the end of each chapter, starting with Chapter 2 and totaling at 125. These problems are indispensable for understanding the material and moving forward. Many important statements, such as the *Approximation of Borel Sets Proposition* (Proposition 4.6) (Section 4.8, Problem 13), are given as problems and frequently referred to in the main body. There are also 358 exercises throughout the text, including Chapter 1 and the Appendix, which require of the student to prove or verify a statement or an example, fill in certain details in a proof, or provide an intermediate step or a counterexample. They are also an inherent part of the material. More difficult problems, such as Section 8.6, Problem 11, are marked with an asterisk, many problems and exercises are supplied with "existential" hints.

The book is generous on examples and contains numerous remarks accompanying definitions, examples, and statements to discuss certain subtleties, raise questions on whether the converse assertions are true, whenever appropriate, or whether the conditions are essential.

As amply demonstrated by experience, students tend to better remember statements by their names rather than by numbers. Thus, a distinctive feature of the book is that every theorem, proposition, corollary, and lemma, unless already possessing a name, is endowed with a descriptive one, making it easier to remember, which, in this author's humble opinion, is quite a bargain when the price for better understanding and retention of the material is a little clumsiness while making a longer reference. Each statement is referred to by its name and not just the number, e. g., the *Characterization of Riemann Integrability* (Theorem 6.10), as opposed to merely Theorem 6.10.

With no pretense on furnishing the history of the subject, the text provides certain dates and lists every related name as a footnote.

Acknowledgments

First and foremost, I wish to express my heartfelt gratitude to my mother, Svetlana A. Markina, for believing in me and all the moral support in the course of this undertaking.

My utmost appreciation goes to Mr. Edward Sichel, my pupil and graduate advisee, for his invaluable assistance while having tirelessly worked with me on proofreading the manuscript and making a number of insightful suggestions, which have contributed to its improvement.

I am very thankful to Dr. Przemyslaw Kajetanowicz (Department of Mathematics, CSU, Fresno) for his kind assistance with graphics.

My sincere acknowledgments are also due to the following associates of the *Walter de Gruyter GmbH*: Dr. Apostolos Damialis, Acquisitions Editor in Mathematics, for seeing value in my manuscript and making authors his highest priority, Ms. Nadja Schedensack, Project Editor in Mathematics and Physics, for her superb efficiency in managing all project related matters, as well as Ms. Ina Talandienė and Ms. Ieva Spudulytė, VTeX Book Production, for their expert editorial and LaTeX typesetting contributions.

Clovis, California, USA
December 2018–January 2019

Marat V. Markin

Contents

1 Preliminaries

In this chapter, we outline certain terminology, notations, and preliminary facts essential for our subsequent discourse.

1.1 Set Theoretic Basics

1.1.1 Some Terminology and Notations

- The logic *quantifiers* $\forall$, $\exists$, and $\exists!$ stand for *"for all,""there exist(s),"* and *"there exists a unique,"* respectively.
- $\mathbb{N} := \{1, 2, 3, \dots\}$ is the set of *natural numbers*.
- $\mathbb{Z} := \{0, \pm 1, \pm 2, \dots\}$ is the set of *integers*.
- $\mathbb{Q}$ is the set of *rational numbers*.
- $\mathbb{R}$ is the set of *real numbers*.
- $\mathbb{C}$ is the set of *complex numbers*.
- $\mathbb{Z}_+$, $\mathbb{Q}_+$, and $\mathbb{R}_+$ are the sets of nonnegative integers, rationals, and reals, respectively.
- $\overline{\mathbb{R}} := [-\infty, \infty]$ is the set of *extended real numbers* (*extended real line*).
- For $n \in \mathbb{N}$, $\mathbb{R}^n$ and $\mathbb{C}^n$ are the n-spaces of all *ordered n-tuples* of real and complex numbers, respectively.

Let X be a set. Henceforth, all sets are supposed to be subsets of X.
- $\mathscr{P}(X)$ is the *power set* of X, i. e., the collection of all subsets of X.
- 2^X is the set of all binary functions $f : X \to \{0, 1\}$ provided $X \neq \emptyset$.
- Sets $A, B \subseteq X$ with $A \cap B = \emptyset$ are called *disjoint*.
- Let I be a nonempty indexing set. The sets of a collection $\{A_i\}_{i \in I}$ of subsets of X are said to be *pairwise disjoint* if

$$A_i \cap A_j = \emptyset, \; i, j \in I, i \neq j.$$

- For $A, B \subseteq X$, $A \setminus B := \{x \in X \mid x \in A, \text{ but } x \notin B\}$ is the *difference* of A and B, in particular, $A^c := X \setminus A = \{x \in X \mid x \notin A\}$ is the *complement* of A and $A \setminus B = A \cap B^c$.
- For $A, B \subseteq X$, $A \triangle B := (A \setminus B) \cup (B \setminus A)$ is the *symmetric difference* of A and B.
- Let I be a nonempty indexing set and $\{A_i\}_{i \in I}$ be a collection of subsets of X. *De Morgan's laws* state

$$\left(\bigcup_{i \in I} A_i\right)^c = \bigcap_{i \in I} A_i^c \text{ and } \left(\bigcap_{i \in I} A_i\right)^c = \bigcup_{i \in I} A_i^c.$$

  More generally,

$$B \setminus \bigcup_{i \in I} A_i = \bigcap_{i \in I} B \setminus A_i \text{ and } B \setminus \bigcap_{i \in I} A_i = \bigcup_{i \in I} B \setminus A_i.$$

https://doi.org/10.1515/9783110600995-001

- The *Cartesian product* of sets $A_i \subseteq X$, $i = 1, \ldots, n$ ($n \in \mathbb{N}$),

$$A_1 \times \cdots \times A_n := \{(x_1, \ldots, x_n) \mid x_i \in A_i,\ i = 1, \ldots, n\}.$$

Definition 1.1 (Monotone Set Sequences). A sequence $(A_n)_{n\in\mathbb{N}}$ (another notation: $\{A_n\}_{n=1}^{\infty}$) of subsets of X is said to *increase* (to be *increasing*) if

$$A_n \subseteq A_{n+1},\ n \in \mathbb{N}.$$

We also say that $(A_n)_{n\in\mathbb{N}}$ increases to $\bigcup_{n=1}^{\infty} A_n$ and write

$$A_n \uparrow \bigcup_{i=1}^{\infty} A_i.$$

A sequence $(A_n)_{n\in\mathbb{N}}$ of subsets of X is said to *decrease* (to be *decreasing*) if

$$A_n \supseteq A_{n+1},\ n \in \mathbb{N}.$$

We also say that $(A_n)_{n\in\mathbb{N}}$ decreases to $\bigcap_{n=1}^{\infty} A_n$ and write

$$A_n \downarrow \bigcap_{i=1}^{\infty} A_i.$$

An increasing or decreasing sequence $(A_n)_{n\in\mathbb{N}}$ of subsets of X is called *monotone*.

Exercise 1.1. Give an example of:
(a) an increasing set sequence,
(b) a decreasing set sequence,
(c) a set sequence that is not monotone.

Definition 1.2 (Upper and Lower Limits of a Set Sequence). For an arbitrary sequence $(A_n)_{n\in\mathbb{N}}$ of subsets of X,
- *upper limit* or *limit superior* of $(A_n)_{n\in\mathbb{N}}$ is

$$\overline{\lim_{n\to\infty}} A_n := \bigcap_{n=1}^{\infty} \bigcup_{k=n}^{\infty} A_k,$$

another notation is $\limsup_{n\to\infty} A_n$;
- *lower limit* or *limit inferior* of $(A_n)_{n\in\mathbb{N}}$ is

$$\underline{\lim_{n\to\infty}} A_n := \bigcup_{n=1}^{\infty} \bigcap_{k=n}^{\infty} A_k,$$

another notation is $\liminf_{n\to\infty} A_n$.

Example 1.1. For $X = \mathbb{R}$ and

$$A_n := \begin{cases} [0,n], & n \in \mathbb{N} \text{ is odd}, \\ [-n,n] & n \in \mathbb{N} \text{ is even}, \end{cases}$$

$$\overline{\lim_{n\to\infty}} A_n = \mathbb{R} \text{ and } \underline{\lim_{n\to\infty}} A_n = [0,\infty)$$

Exercise 1.2. Verify.

Remarks 1.1.

- Upper and lower limits are well-defined and exist for any set sequence and, by the definition,

$$U_n := \bigcup_{k=n}^{\infty} A_k \downarrow \overline{\lim_{n\to\infty}} A_n, \; V_n := \bigcap_{k=n}^{\infty} A_k \uparrow \underline{\lim_{n\to\infty}} A_n.$$

- As follows from the definition,

$$\overline{\lim_{n\to\infty}} A_n = \{x \in X \mid x \in A_n \textit{ frequently}, \text{ i. e., for infinitely many } n \in \mathbb{N}\}$$

and

$$\underline{\lim_{n\to\infty}} A_n = \{x \in X \mid x \in A_n \textit{ eventually}, \text{ i. e., for } n \geq N(x) \text{ with some } N(x) \in \mathbb{N}\}.$$

Exercise 1.3. Explain.

This immediately implies the inclusion

$$\underline{\lim_{n\to\infty}} A_n \subseteq \overline{\lim_{n\to\infty}} A_n;$$

Definition 1.3 (Limit of a Set Sequence). If, for a sequence $(A_n)_{n\in\mathbb{N}}$ of subsets of X,

$$\underline{\lim_{n\to\infty}} A_n = \overline{\lim_{n\to\infty}} A_n,$$

we use the notation $\lim_{n\to\infty} A_n$ for the common value of the lower and upper limits and call this set the *limit* of $(A_n)_{n\in\mathbb{N}}$.

Remarks 1.2.

- As Example 1.1 shows, the limit of a set sequence need not exist.
- For an *increasing set sequence* $(A_n)_{n\in\mathbb{N}}$,

$$\lim_{n\to\infty} A_n = \bigcup_{n=1}^{\infty} A_n$$

and, for a *decreasing set sequence* $(A_n)_{n\in\mathbb{N}}$,

$$\lim_{n\to\infty} A_n = \bigcap_{n=1}^{\infty} A_n.$$

Exercise 1.4. Verify.

Example 1.2.

$$\lim_{n\to\infty}[0,n) = \bigcup_{n=1}^{\infty}[0,n) = [0,\infty) \text{ and } \lim_{n\to\infty}[n,\infty) = \bigcap_{n=1}^{\infty}[n,\infty) = \emptyset.$$

Exercise 1.5. Give an example of a nonmonotone set sequence $(A_n)_{n\in\mathbb{N}}$ for which $\lim_{n\to\infty} A_n$ exists.

1.1.2 Cardinality and Countability

Definition 1.4 (Similarity of Sets). Sets A and B are said to be *similar* if there exists a one-to-one correspondence (bijection) between them.

Notation. $A \sim B$.

Remark 1.3. Similarity is an *equivalence relation* (*reflexive*, *symmetric*, and *transitive*) on the power set $\mathscr{P}(X)$ of a nonempty set X.

Exercise 1.6. Verify.

Thus, in the context, we can use the term *"equivalence"* synonymously to *"similarity."*

Definition 1.5 (Cardinality). Equivalent sets are said to have the same number of elements or *cardinality*. Cardinality is a characteristic of an equivalence class of similar sets.

Notation. $\mathscr{P}(X) \ni A \mapsto |A|$.

Remark 1.4. Thus, $A \sim B$ *iff* $|A| = |B|$, i. e., two sets are equivalent iff they share the same cardinality.

Examples 1.3. 1. For a nonempty set X, $\mathscr{P}(X) \sim 2^X$.
2. $|\mathbb{N}| = |\mathbb{Z}| = |\mathbb{Q}| := \aleph_0$.
3. $|[0,1]| = |\mathbb{R}| = |\mathbb{C}| := \mathfrak{c}$.
   See, e. g., [6, 8].

Definition 1.6 (Domination). If sets A and B are such that A is equivalent to a subset of B, we write

$$A \preceq B$$

and say that B *dominates* A. If, in addition, $A \nsim B$, we write

$$A \prec B$$

and say that B *strictly dominates* A.

Remark 1.5. The relation $\preceq$ is a *partial order* (*reflexive*, *antisymmetric*, and *transitive*) on the power set $\mathscr{P}(X)$ of a nonempty set X (see Appendix A).

Exercise 1.7. Verify *reflexivity* and *transitivity*.

The *antisymmetry* of $\preceq$ is the subject of the following celebrated theorem.

Theorem 1.1 (Schröder–Bernstein Theorem). *If, for sets A and B, $A \preceq B$ and $B \preceq A$, then $A \sim B$.*[1]

For a proof, see, e. g., [6].

Remark 1.6. The set partial order $\preceq$ defines a partial order $\leq$ on the set of cardinals:

$$|A| \leq |B| \Leftrightarrow A \preceq B.$$

Thus, the *Schröder–Bernstein Theorem* can be equivalently reformulated in terms of cardinalities as follows:

If, for sets A and B, $|A| \leq |B|$ and $|B| \leq |A|$, then $|A| = |B|$.

Theorem 1.2 (Cantor's Theorem). *Every set X is strictly dominated by its power set $\mathscr{P}(X)$:*[2]

$$X \prec \mathscr{P}(X).$$

Equivalently,

$$|X| < |\mathscr{P}(X)|.$$

For a proof, see, e. g., [6].

In view of Examples 1.3, we obtain the following.

Corollary 1.1. *For a nonempty set X, $X \prec 2^X$, i. e., $|X| < |2^X|$.*

Definition 1.7 (Countable/Uncountable Set). A *countable set* is a set with the same *cardinality* as a subset of the set $\mathbb{N}$ of natural numbers, i. e., *equivalent* to a subset of $\mathbb{N}$.

A set, which is not *countable*, is called *uncountable*.

Remarks 1.7.

- A countable set A is either *finite*, i. e., equivalent to a set of the form $\{1, \dots, n\} \subset \mathbb{N}$ with some $n \in \mathbb{N}$, in which case, we say that A has n elements, or *countably infinite*, i. e., equivalent to the entire $\mathbb{N}$.

---

1 Ernst Schröder (1841–1902), Felix Bernstein (1878–1956).

2 Georg Cantor (1845–1918).

- For a finite set A of n elements ($n \in \mathbb{N}$),

$$|A| = |\{1, \dots, n\}| = n.$$

For a countably infinite set A,

$$|A| = |\mathbb{N}| = \aleph_0$$

(see Examples 1.3).
- In some sources, the term *"countable"* is used in the sense of *"countably infinite."* To avoid ambiguity, the term *"at most countable"* can be used when finite sets are included in consideration.

The subsequent statement immediately follows from *Cantor's Theorem* (Theorem 1.2).

Proposition 1.1 (Uncountable Sets). *The sets $\mathscr{P}(\mathbb{N})$ and $2^{\mathbb{N}}$ (the set of all binary sequences) are uncountable.*

Theorem 1.3 (Properties of Countable Sets).
(1) *Every infinite set contains a countably infinite subset (based on the Axiom of Choice (see Appendix* A)).
(2) *Any subset of a countable set is countable.*
(3) *The union of countably many countable sets is countable.*
(4) *The Cartesian product of finitely many countable sets is countable.*

Exercise 1.8. Prove that
(a) the set $\mathbb{Z}$ of all *integers* and the set of all *rational numbers* are countable;
(b) for any $n \in \mathbb{N}$, $\mathbb{Z}^n$ and $\mathbb{Q}^n$ are countable;
(c) the set of all *algebraic numbers* (the roots of polynomials with integer coefficients) is countable.

Subsequently, we also need the following useful result.

Proposition 1.2 (Cardinality of the Collection of Finite Subsets). *The cardinality of the collection of all finite subsets of an infinite set coincides with the cardinality of the set.*

For a proof, see, e. g., [8, 11, 18].

1.2 Terminology Related to Functions

Let X and Y be nonempty sets and $\emptyset \neq D \subseteq X$,

$$f : D \to Y.$$

- The set D is called the *domain (of definition)* of f.
- The *value* of f corresponding to an $x \in D$ is designated by $f(x)$.
- The set

$$\{f(x) \mid x \in D\}$$

of all values of f is called the *range* of f (also *codomain* or *target set*).
- For a set $A \subseteq D$, the set of values of f corresponding to all elements of A

$$f(A) := \{f(x) \mid x \in A\}$$

is called the *image* of A under the function f.
Thus, the *range* of f is the image $f(D)$ of the whole domain D.
- For a set $B \subseteq Y$, the set of all elements of the domain that map to the elements of B,

$$f^{-1}(B) := \{x \in D \mid f(x) \in B\}$$

is called the *inverse image* (or *preimage*) of B.

Example 1.4. For $X = Y := \mathbb{R}$ and $f(x) := x^2$ with $D := [-1, 2]$,
- $f([-1, 2]) = [0, 4]$ and $f([1, 2]) = [1, 4]$.
- $f^{-1}([-2, -1]) = \emptyset$, $f^{-1}([0, 1]) = [-1, 1]$, and $f^{-1}([1, 4]) = \{-1\} \cup [1, 2]$.

Theorem 1.4 (Properties of Inverse Image). *Let X and Y be nonempty sets and $\emptyset \neq D \subseteq X$*

$$f : D \to Y.$$

Then, for an arbitrary nonempty collection $\{B_i\}_{i \in I}$ of subsets of Y,
(1) $f^{-1}\left(\bigcup_{i \in I} B_i\right) = \bigcup_{i \in I} f^{-1}(B_i)$,
(2) $f^{-1}\left(\bigcap_{i \in I} B_i\right) = \bigcap_{i \in I} f^{-1}(B_i)$, *and*
(3) *for any $B_1, B_2 \subseteq Y$, $f^{-1}(B_1 \setminus B_2) = f^{-1}(B_1) \setminus f^{-1}(B_2)$,*

i. e., preimage preserves all set operations.

Exercise 1.9.
(a) Prove.
(b) Show that image preserves unions, i. e., for an arbitrary nonempty collection $\{A_i\}_{i \in I}$ of subsets of D,

$$f\left(\bigcup_{i \in I} A_i\right) = \bigcup_{i \in I} f(A_i),$$

and unions *only*. Give corresponding counterexamples for intersections and differences.

1.3 Upper and Lower Limits

Definition 1.8 (Upper and Lower Limits). Let $(x_n)_{n\in\mathbb{N}}$ (another notation is $\{x_n\}_{n=1}^{\infty}$) be a sequence of real numbers.

The *upper limit* or *limit superior* of $(x_n)_{n\in\mathbb{N}}$ is defined as follows:

$$\overline{\lim_{n\to\infty}} x_n := \lim_{n\to\infty} \sup_{k\geq n} x_k = \inf_{n\in\mathbb{N}} \sup_{k\geq n} x_k \in \overline{\mathbb{R}}.$$

The *lower limit* or *limit inferior* of $(x_n)_{n\in\mathbb{N}}$ is defined as follows:

$$\underline{\lim}_{n\to\infty} x_n := \lim_{n\to\infty} \inf_{k\geq n} x_k = \sup_{n\in\mathbb{N}} \inf_{k\geq n} x_k \in \overline{\mathbb{R}}.$$

Alternative notations are $\limsup_{n\to\infty} x_n$ and $\liminf_{n\to\infty} x_n$, respectively.

Example 1.5. For

$$x_n := \begin{cases} n, & n \in \mathbb{N} \text{ is odd}, \\ -1/n, & n \in \mathbb{N} \text{ is even}, \end{cases}$$

$$\overline{\lim_{n\to\infty}} x_n = \infty \text{ and } \underline{\lim}_{n\to\infty} x_n = 0.$$

Exercise 1.10.

(a) Verify.

(b) Explain why the upper and lower limits, unlike the regular limit, are guaranteed to exist for an arbitrary sequence of real numbers.

(c) Show that

$$\underline{\lim}_{n\to\infty} x_n \leq \overline{\lim_{n\to\infty}} x_n.$$

Proposition 1.3 (Characterization of Limit Existence). *For a sequence of real numbers $(x_n)_{n\in\mathbb{N}}$,*

$$\lim_{n\to\infty} x_n \in \overline{\mathbb{R}}$$

exists iff

$$\underline{\lim}_{n\to\infty} x_n = \overline{\lim_{n\to\infty}} x_n,$$

in which case

$$\lim_{n\to\infty} x_n = \underline{\lim}_{n\to\infty} x_n = \overline{\lim_{n\to\infty}} x_n.$$

1.4 Fundamentals of Metric Spaces

In this section, we furnish a brief discourse of *metric spaces*, i. e., sets endowed with a notion of *distance*, whose properties mimic those of the regular distance in three-dimensional space. Distance brings to life various topological notions such as *limit*, *continuity*, *openness*, *closedness*, *compactness*, and *denseness*, the geometric notion of *boundedness*, as well as the notions of *fundamentality* of sequences and *completeness*. We are to touch upon these concepts here.

1.4.1 Definition and Examples

Definition 1.9 (Metric Space). A *metric space* is a nonempty set X with a *metric* (or *distance function*), i. e., a mapping

$$\rho(\cdot,\cdot) : X \times X \to \mathbb{R}$$

subject to the following *metric axioms*:
1. $\rho(x,y) \geq 0$, $x,y \in X$. *Nonnegativity*
2. $\rho(x,y) = 0$ iff $x = y$. *Separation*
3. $\rho(x,y) = \rho(y,x)$, $x,y \in X$. *Symmetry*
4. $\rho(x,z) \leq \rho(x,y) + \rho(y,z)$, $x,y,z \in X$. *Triangle inequality*

For any fixed $x,y \in X$, the number $\rho(x,y)$ is called the *distance* of x from y, or from y to x, or between x and y.

Notation. (X,ρ).

Examples 1.6.
1. Any nonempty set X is a *metric space* relative to the *discrete metric*

$$X \ni x,y \mapsto \rho_d(x,y) := \begin{cases} 0 & \text{if } x = y, \\ 1 & \text{if } x \neq y. \end{cases}$$

2. The *real line* $\mathbb{R}$ or the *complex plane* $\mathbb{C}$ is a metric space relative to the regular distance function $\rho(x,y) := |x-y|$.
3. Let $n \in \mathbb{N}$ and $1 \leq p \leq \infty$. The real/complex *n-space*, $\mathbb{R}^n$ or $\mathbb{C}^n$, is a *metric space* relative to *p-metric*,

$$\rho_p(x,y) = \begin{cases} \left[\sum_{k=1}^n |x_k - y_k|^p\right]^{1/p} & \text{if } 1 \leq p < \infty, \\ \max_{1\leq k\leq n} |x_k - y_k| & \text{if } p = \infty, \end{cases}$$

where $x := (x_1,\dots,x_n)$ and $y := (y_1,\dots,y_n)$, designated by $l_p^{(n)}$ (real or complex, respectively).

Remarks 1.8.
- For $n = 1$, all these metrics coincide with $\rho(x,y) = |x - y|$.
- For $n = 2, 3$ and $p = 2$, we have the usual *Euclidean distance*.
- $(\mathbb{C}, \rho) = (\mathbb{R}^2, \rho_2)$.

4. Let $1 \le p \le \infty$. The set l_p of all real or complex sequences $(x_k)_{k\in\mathbb{N}}$ satisfying

$$\sum_{k=1}^{\infty} |x_k|^p < \infty \quad (1 \le p < \infty),$$

$$\sup_{k\in\mathbb{N}} |x_k| < \infty \qquad (p = \infty)$$

(*p-summable/bounded* sequences, respectively) is a *metric space* relative to *p-metric*

$$\rho_p(x,y) = \begin{cases} \left[\sum_{k=1}^{\infty} |x_k - y_k|^p\right]^{1/p} & \text{if } 1 \le p < \infty, \\ \sup_{k\in\mathbb{N}} |x_k - y_k| & \text{if } p = \infty, \end{cases}$$

where $x := (x_k)_{k\in\mathbb{N}}, y := (y_k)_{k\in\mathbb{N}} \in l_p$.

5. The set $C[a,b]$ of all real/complex-valued functions *continuous* on $[a,b]$ $(-\infty < a < b < \infty)$ is a metric space relative to *p-metric*

$$C[a,b] \ni f, g \mapsto \rho_p(f,g) = \begin{cases} \left[\int_a^b |f(x) - g(x)|^p \, dx\right]^{1/p} & \text{if } 1 \le p < \infty, \\ \max_{a\le x\le b} |f(x) - g(x)| & \text{if } p = \infty. \end{cases}$$

Exercise 1.11. Verify examples 3–5 for $p = 1$ and $p = \infty$ only.

1.4.2 Convergence

Definition 1.10 (Convergence and Limit of a Sequence). A sequence of points $(x_n)_{n\in\mathbb{N}}$ (another notation is $\{x_n\}_{n=1}^{\infty}$) in a *metric space* (X, ρ) is said to *converge* (to be *convergent*) to a point $x \in X$ if

$$\forall \varepsilon > 0 \; \exists N \in \mathbb{N} \; \forall n \ge N : \; \rho(x_n, x) < \varepsilon,$$

i. e.,

$$\lim_{n\to\infty} \rho(x_n, x) = 0 \; (\rho(x_n, x) \to 0, \; n \to \infty).$$

We write in this case

$$\lim_{n\to\infty} x_n = x \quad \text{or} \quad x_n \to x, \; n \to \infty$$

and say that x is the *limit* of $(x_n)_{n\in\mathbb{N}}$.

A sequence $(x_n)_{n\in\mathbb{N}}$ in a *metric space* (X, ρ) is called *convergent* if it converges to some $x \in X$ and *divergent* otherwise.

Theorem 1.5 (Uniqueness of Limit). *The limit of a convergent sequence $(x_n)_{n\in\mathbb{N}}$ in a metric space (X,ρ) is unique.*

Exercise 1.12. Prove.

Hint. Use the *triangle inequality*.

Theorem 1.6 (Characterization of Convergence). *A sequence $(x_n)_{n\in\mathbb{N}}$ in a metric space (X,ρ) converges to a point $x \in X$ iff every subsequence $(x_{n(k)})_{k\in\mathbb{N}}$ of $(x_n)_{n\in\mathbb{N}}$ contains a subsequence $(x_{n(k(j))})_{j\in\mathbb{N}}$ such that*

$$x_{n(k(j))} \to x,\ j \to \infty.$$

Exercise 1.13. Prove.

Hint. Prove the *"if"* part *by contrapositive*.

Examples 1.7.

1. A sequence $(x_n)_{n\in\mathbb{N}}$ is convergent in a discrete space (X,ρ_d) *iff* it is *eventually constant*, i. e.,

$$\exists N \in \mathbb{N}\, \forall n \geq N : x_n = x_N.$$

2. Convergence of a sequence in the space $l_p^{(n)}$ ($n \in \mathbb{N}$ and $1 \leq p \leq \infty$) is equivalent to *componentwise convergence*, i. e.,

$$(x_1^{(k)},\dots,x_n^{(k)}) \to (x_1,\dots,x_n),\ k \to \infty \Leftrightarrow \forall i = 1,\dots,n : x_i^{(k)} \to x_i,\ k \to \infty.$$

3. Convergence of a sequence in the space l_p $(1 \leq p \leq \infty)$

$$\left(x_n^{(k)}\right)_{n\in\mathbb{N}} =: x^{(k)} \to x := (x_n)_{n\in\mathbb{N}},\ k \to \infty,$$

   implies *termwise convergence*, i. e.,

$$\forall n \in \mathbb{N} : x_n^{(k)} \to x_n,\ k \to \infty,$$

   the converse not being true. Indeed, in l_p $(1 \leq p \leq \infty)$, the sequence $(e_k := (\delta_{nk})_{n\in\mathbb{N}})_{k\in\mathbb{N}}$, where δ_{nk} is the *Kronecker*[3] *delta*, converges to the zero sequence $0 := (0,0,0,\dots)$ *componentwise*, but *does not* converge.

4. Convergence in $(C[a,b],\rho_\infty)$ (see Examples 1.6) is *uniform convergence* on $[a,b]$, i. e.,

$$f_n \to f,\ n \to \infty,\ \text{in } (C[a,b],\rho_\infty)$$

   iff

$$\forall \varepsilon > 0\, \exists N \in \mathbb{N}\, \forall n \geq N\, \forall x \in [a,b] : |f_n(x) - f(x)| < \varepsilon.$$

3 Leopold Kronecker (1823–1891).

Uniform convergence on $[a,b]$ implies *pointwise convergence* on $[a,b]$, i. e.,

$$\forall \varepsilon > 0\, \forall x \in [a,b]\, \exists N \in \mathbb{N}\, \forall n \geq N : |f_n(x) - f(x)| < \varepsilon.$$

Exercise 1.14.
(a) Verify.
(b) Give an example showing that a function sequence $(f_n)_{n\in\mathbb{N}}$ in $C[a,b]$ pointwise convergent on $[a,b]$ need not converge on $[a,b]$ uniformly.

1.4.3 Completeness

The concept of *completeness* is a basic property of metric spaces underlying many important facts (see, e. g., [14]).

1.4.3.1 Cauchy/Fundamental Sequences

Definition 1.11 (Cauchy/Fundamental Sequence). A sequence $(x_n)_{n\in\mathbb{N}}$ in a metric space (X,ρ) is called a *Cauchy sequence*, or a *fundamental sequence*, if[4]

$$\forall\, \varepsilon > 0\, \exists N \in \mathbb{N}\, \forall\, m,n \geq N : \rho(x_m, x_n) < \varepsilon.$$

Remark 1.9. The latter is equivalent to

$$\rho(x_m, x_n) \to 0,\ m,n \to \infty,$$

or to

$$\sup_{k\in\mathbb{N}} \rho(x_{n+k}, x_n) \to 0,\ n \to \infty.$$

Examples 1.8.
1. A sequence is fundamental in a discrete space (X,ρ_d) *iff* it is *eventually constant*.
2. The sequence $(1/n)_{n\in\mathbb{N}}$ is *fundamental* in $\mathbb{R}$ and the sequence $(n)_{n\in\mathbb{N}}$ is not.
3. The sequence $(x_n := (1, 1/2, \dots, 1/n, 0, 0, \dots))_{n\in\mathbb{N}}$ is *fundamental* in l_p $(1 < p \leq \infty)$ but not in l_1.
4. The sequence $(e_n := (\delta_{nk})_{k\in\mathbb{N}})_{n\in\mathbb{N}}$, where δ_{nk} is the *Kronecker delta*, is not fundamental in l_p $(1 \leq p \leq \infty)$.

Exercise 1.15. Verify.

Remark 1.10. Every *convergent* sequence $(x_n)_{n\in\mathbb{N}}$ is *fundamental* but not vice versa, i. e., a fundamental sequence need not converge. For instance, the sequence $(1/n)_{n\in\mathbb{N}}$ is fundamental, but divergent in $\mathbb{R}\setminus\{0\}$ with the regular distance.

Exercise 1.16. Verify.

4 Augustin-Louis Cauchy (1789–1857).

1.4.3.2 Complete Metric Spaces

Definition 1.12 (Complete Metric Space). A metric space (X, ρ), in which every Cauchy/fundamental sequence converges, is called *complete* and *incomplete* otherwise.

Examples 1.9.
1. The spaces $\mathbb{R}$ and $\mathbb{C}$ are *complete* relative to the regular distance as is known from intermediate analysis courses.
2. The spaces $\mathbb{R} \setminus \{0\}$, $(0, 1)$, and $\mathbb{Q}$ are *incomplete* relative to the regular distance.
3. A *discrete* metric space (X, ρ_d) is *complete*.

   Exercise 1.17. Verify 2 and 3.

4. **Theorem 1.7** (Completeness of $(C[a,b], \rho_\infty)$). *The (real or complex) space $(C[a,b], \rho_\infty)$ $(-\infty < a < b < \infty)$ is complete.*

   See, e. g., [14].

1.4.4 Balls and Boundedness

Definition 1.13 (Balls and Spheres). Let (X, ρ) be a *metric space* and $r \geq 0$.
- The *open ball* of radius r centered at a point $x_0 \in X$ is the set
$$B(x_0, r) := \{x \in X \mid \rho(x, x_0) < r\}.$$
- The *closed ball* of radius r centered at a point $x_0 \in X$ is the set
$$\overline{B}(x_0, r) := \{x \in X \mid \rho(x, x_0) \leq r\}.$$
- The *sphere* of radius r centered at a point $x_0 \in X$ is the set
$$S(x_0, r) := \{x \in X \mid \rho(x, x_0) = r\} = \overline{B}(x_0, r) \setminus B(x_0, r).$$

Remarks 1.11.
- When contextually important to indicate which space the balls/spheres are considered in, the letter designating the space in question is added as a subscript. For example, for (X, ρ), we use the notation
$$B_X(x_0, r), \ \overline{B}_X(x_0, r), \text{ and } S_X(x_0, r), \ x_0 \in X, r \geq 0,$$
respectively.
- As is easily seen, for an arbitrary $x_0 \in X$,
$$B(x_0, 0) = \emptyset \text{ and } \overline{B}(x_0, 0) = S(x_0, 0) = \{x_0\} \quad (\textit{trivial} \text{ cases}).$$

Exercise 1.18.
(a) Explain the latter.
(b) Describe balls and spheres in $\mathbb{R}$ and $\mathbb{C}$ with the regular distance, give some examples.
(c) Sketch the *unit sphere* $S(0,1)$ in $(\mathbb{R}^2, \rho_1)$, $(\mathbb{R}^2, \rho_2)$, and $(\mathbb{R}^2, \rho_\infty)$.
(d) Describe balls and spheres in $(C[a,b], \rho_\infty)$.
(e) Let (X, ρ_d) be a *discrete* metric space and $x_0 \in X$ be arbitrary. Describe $B(x_0, r)$, $\overline{B}(x_0, r)$, and $S(x_0, r)$ for different values of $r \geq 0$.

Convergence can be equivalently redefined using ball terminology.

Definition 1.14 (Equivalent Definition of Convergence). A sequence of points $(x_n)_{n\in\mathbb{N}}$ in a *metric space* (X, ρ) is said to *converge* (to be *convergent*) to a point $x \in X$ if

$$\forall \varepsilon > 0 \, \exists N \in \mathbb{N} \, \forall n \geq N : \, x_n \in B(x_0, \varepsilon),$$

in which case we say that the sequence $(x_n)_{n\in\mathbb{N}}$ is *eventually* in the ε-ball $B(x, \varepsilon)$.

Definition 1.15 (Bounded Set). Let (X, ρ) be a *metric space*. A nonempty set $A \subseteq X$ is called *bounded* if

$$\operatorname{diam}(A) := \sup_{x,y\in A} \rho(x,y) < \infty.$$

The number $\operatorname{diam}(A)$ is called the *diameter* of A.

The *empty set* $\emptyset$ is regarded to be bounded with $\operatorname{diam}(\emptyset) := 0$.

Examples 1.10.
1. In a metric space (X, ρ), an open/closed ball of radius $r > 0$ is a bounded set of diameter *at most* $2r$.
2. In $(\mathbb{R}, \rho)$, the sets $(0, 1]$, $\{1/n\}_{n\in\mathbb{N}}$ are bounded and the sets $(-\infty, 1)$, $\{n^2\}_{n\in\mathbb{N}}$ are not.
3. In l_∞, the set $\{(x_n)_{n\in\mathbb{N}} \mid |x_n| \leq 1, \ n \in \mathbb{N}\}$ is bounded and, in l_p $(1 \leq p < \infty)$, it is not.
4. In $(C[0,1], \rho_\infty)$, the power set $\{x^n\}_{n\in\mathbb{Z}_+}$ is bounded and, in $(C[0,2], \rho_\infty)$, it is not.

Exercise 1.19.
(a) Verify.
(b) Show that a set A is *bounded* iff it is contained in some (closed) ball, i. e.,

$$\exists x \in X \, \exists r \geq 0 : \, A \subseteq \overline{B}(x, r).$$

(c) Describe all bounded sets in a *discrete* metric space (X, ρ_d).
(d) Give an example of a metric space (X, ρ), in which, for a ball $\overline{B}(x, r)$ with some $x \in X$ and $r > 0$,

$$\operatorname{diam}(\overline{B}(x, r)) < 2r.$$

Definition 1.16 (Bounded Function). Let T be a nonempty set and (X,ρ) be a *metric space*. A function $f : T \to X$ is called *bounded* if the *set of its values* $f(T)$ is bounded in (X,ρ).

Remark 1.12. As a particular case, for $T = \mathbb{N}$, we obtain the definition of a *bounded sequence*.

Theorem 1.8 (Boundedness of Fundamental Sequences). *Each fundamental sequence $(x_n)_{n\in\mathbb{N}}$ in a metric space (X,ρ) is bounded.*

Exercise 1.20.
(a) Prove.
(b) Give an example showing that a bounded sequence $(x_n)_{n\in\mathbb{N}}$ in a metric space (X,ρ) need not be fundamental.

Corollary 1.2 (Boundedness of Convergent Sequences). *Each convergent sequence $(x_n)_{n\in\mathbb{N}}$ in a metric space (X,ρ) is bounded.*

1.4.5 Interior Points, Open Sets

Definition 1.17 (Interior Point). Let (X,ρ) be a metric space. A point $x \in X$ is called an *interior point* of a nonempty set $A \subseteq X$ if A contains a nontrivial open ball centered at x, i. e.,

$$\exists r > 0 : B(x,r) \subseteq A.$$

Examples 1.11.
1. In an arbitrary metric space (X,ρ), any point $x \in X$ is, obviously, an interior point of an open ball $B(x,r)$ or a closed ball $\overline{B}(x,r)$ with an arbitrary $r > 0$.
2. For the set $[0,1)$ in $\mathbb{R}$ with the regular distance, the points $0 < x < 1$ are interior and the point $x = 0$ is not.
3. A singleton $\{x\}$ in $\mathbb{R}$ with the regular distance, has no interior points.

Exercise 1.21. Verify.

Definition 1.18 (Interior of a Set). The *interior* of a nonempty set A in a metric space (X,ρ) is the set of all interior points of A.

Notation. $\operatorname{int}(A)$.

Remark 1.13. Thus,

$$\operatorname{int}(A) \subseteq A.$$

As the prior examples demonstrate, this inclusion can be *proper* and $\operatorname{int}(A)$ can be *empty*.

Definition 1.19 (Open Set). A *nonempty set* A in a metric space (X,ρ) is called *open* if each point of A is its *interior point*, i. e., $A = \text{int}(A)$.

Remark 1.14. The *empty set* $\emptyset$ is regarded to be open and the whole space X is trivially open as well.

Exercise 1.22.
(a) Verify that, in $\mathbb{R}$ with the regular distance, the intervals of the form (a,∞), $(-\infty,b)$, and (a,b) $(-\infty < a < b < \infty)$ are open sets.
(b) Prove that, in a *metric space* (X,ρ), an *open ball* $B(x_0,r)$ $(x_0 \in X, r \geq 0)$ is an *open set*.
(c) Describe all open sets in a *discrete* metric space (X,ρ_d).

Theorem 1.9 (Properties of Open Sets). *The open sets in a metric space (X,ρ) have the following properties:*
(1) *$\emptyset$ and X are open sets;*
(2) *an arbitrary union of open sets is an open set;*
(3) *an arbitrary finite intersection of open sets is an open set.*

Exercise 1.23.
(a) Prove.
(b) Give an example showing that an *infinite intersection* of open sets need not be open.

Definition 1.20 (Metric Topology). The collection $\mathscr{G}$ of all open sets in a metric space (X,ρ) is called the *metric topology* on X generated by metric ρ.

See, e. g., [18, 21, 25].

1.4.6 Limit Points, Closed Sets

Definition 1.21 (Limit Point, Derived Set). Let (X,ρ) be a metric space. A point $x \in X$ is called a *limit point* (also an *accumulation point* or a *cluster point*) of a set A in X if every open ball centered at x contains a point of A *distinct* from x, i. e.,

$$\forall r > 0 : \ B(x,r) \cap (A \setminus \{x\}) \neq \emptyset.$$

The set A' of all limit points of A is called the *derived set* of A.

Example 1.12. In $\mathbb{R}$, $[0,1)' = [0,1]$, $\mathbb{Z}' = \emptyset$, $\mathbb{Q}' = \mathbb{R}$.

Exercise 1.24. Verify.

Remarks 1.15.

- A *limit point* x of a set A need not belong to A. It may even happen that none of them does, i. e.,

$$A' \subseteq A^c.$$

- Each open ball centered at a limit point x of a set A in a metric space (X, ρ) contains *infinitely many* points of A distinct from x.
- To have a *limit point*, a set A in a metric space (X, ρ) must necessarily be *nonempty* and even *infinite*. However, an infinite set need not have limit points.

Exercise 1.25.

(a) Verify and give corresponding examples.
(b) Describe the situation in a *discrete* metric space (X, ρ_d).
(c) Give examples showing that an *interior point* of a set need not be its *limit point* and vise versa.

Definition 1.22 (Closure of a Set). The *closure* $\overline{A}$ of a set A in a metric space (X, ρ) is the set consisting of all points, which are either points of A or limit points of A, i. e.,

$$\overline{A} := A \cup A'.$$

Example 1.13. In $\mathbb{R}$, $\overline{[0,1)} = [0,1]$, $\overline{\mathbb{Z}} = \mathbb{Z}$, $\overline{\mathbb{Q}} = \mathbb{R}$.

Exercise 1.26. Verify (see Example 1.12).

Remarks 1.16.

- Obviously $\emptyset' = \emptyset$, and hence, $\overline{\emptyset} = \emptyset$.
- We always have the inclusion

$$A \subseteq \overline{A},$$

which may be *proper*.
- A point $x \in \overline{A}$ *iff* every nontrivial open ball centered at x contains a point of A (not necessarily *distinct* from x).

Exercise 1.27. Verify and give a corresponding example.

Definition 1.23 (Closed Set). Let (X, ρ) be a metric space. A set A in X is called *closed* if it contains all its limit points, i. e., $A' \subseteq A$, and hence, $A = \overline{A}$.

Remarks 1.17.

- The whole space X is trivially closed.
- Also closed are the sets with no limit points, in particular, *finite sets*, including the empty set $\emptyset$.

- A set in a metric space (X,ρ), which is simultaneously closed and open is called *clopen*. There are always at least two (*trivial*) clopen sets: $\emptyset$ and X. However, there can exist nontrivial ones.

Exercise 1.28.
(a) Verify that in $\mathbb{R}$ with the regular distance the intervals of the form $[a,\infty)$, $(-\infty,b]$, and $[a,b]$ $(-\infty < a < b < \infty)$ are closed sets.
(b) Verify that the sets $(0,1)$ and $\{2\}$ are *clopen* in the metric space $(0,1) \cup \{2\}$ with the regular distance.
(c) Describe all closed sets in a *discrete* metric space (X,ρ_d).

Theorem 1.10 (Characterizations of Closed Sets). *For an arbitrary nonempty set $A \subseteq X$ in a metric space (X,ρ), the following statements are equivalent:*
1. *The set A is closed in (X,ρ).*
2. *The complement A^c of the set A is an open set in (X,ρ).*
3. *(Sequential Characterization) For any sequence $(x_n)_{n\in\mathbb{N}}$ in A convergent in (X,ρ), $\lim_{n\to\infty} x_n \in A$, i. e., the set A contains the limits of all its convergent sequences.*

Theorem 1.11 (Properties of Closed Sets). *The closed sets in a metric space (X,ρ) have the following properties:*
(1) *$\emptyset$ and X are closed sets;*
(2) *an arbitrary intersection of closed sets is a closed set;*
(3) *a finite union of closed sets is a closed set.*

Exercise 1.29.
(a) Prove.
(b) Give an example showing that an *infinite union* of closed sets need not be closed.

1.4.7 Dense Sets and Separable Spaces

Definition 1.24 (Dense Set). A set A in a metric space (X,ρ) is called *dense* if $\overline{A} = X$.

Example 1.14. The set $\mathbb{Q}$ of the rational numbers is dense in $\mathbb{R}$ (see Example 1.13).

Theorem 1.12 (Sequential Characterization of Dense Sets). *A set A is dense in a metric space (X,ρ) iff*

$$\forall x \in X\ \exists\, (x_n)_{n\in\mathbb{N}} \subseteq A : x_n \to x,\ n \to \infty.$$

Definition 1.25 (Separable Metric Space). A metric space (X,ρ) containing a *countable dense subset* is a called *separable*.

Remark 1.18. Any countable metric space is, obviously, separable. However, as the following examples show, a metric space need not be countable to be separable.

Examples 1.15.
1. The spaces $l_p^{(n)}$ are *separable* for ($n \in \mathbb{N}$, $1 \le p \le \infty$), which includes the cases of $\mathbb{R}$ and $\mathbb{C}$ with the regular distances.
   Indeed, as a countable dense set here, one can consider that of all ordered n-tuples with (real or complex) rational components.
2. The spaces l_p are *separable* for $1 \le p < \infty$.
   Indeed, as a countable dense set here, one can consider that of all eventually zero sequences with real or complex rational terms.
3. The space l_∞ is *not* separable (see, e. g., [14]).
4. The space $(C[a,b], \rho_\infty)$ $(-\infty < a < b < \infty)$ is separable, which follows from *Weierstrass Approximation Theorem* (see, e. g., [14]) when we consider as a countable dense set that of all polynomials with rational coefficients.

Exercise 1.30. When is a *discrete* metric space (X, ρ) separable?

Open sets in separable metric spaces allow the following description.

Proposition 1.4 (Open Sets in Separable Spaces). *Every open set in a separable metric space is a countable a union of open balls.*

See, e. g., [5, 10, 18].

For in $\mathbb{R}$ with the regular distance, we have the following useful one:

Proposition 1.5 (Open Sets of the Real Line). *Every open set in $\mathbb{R}$ is a countable union of pairwise disjoint open intervals.*

1.4.8 Compactness

Definition 1.26 (Cover, Subcover, Open Cover). A collection $\mathscr{C} = \{C_i\}_{i \in I}$ of subsets of a nonempty set X is said to be a *cover* of a set $A \subseteq X$, or *to cover* A, if

$$A \subseteq \bigcup_{i \in I} C_i. \tag{1.1}$$

A subcollection $\mathscr{C}'$ of a cover $\mathscr{C}$ of A, which is also a cover of A, is called a *subcover* of $\mathscr{C}$.

If (X, ρ) is a metric space, a cover of a set $A \subseteq X$ consisting of open sets is called an *open cover* of A.

Remark 1.19. In particular, when $A = X$, (1.1) acquires the form

$$X = \bigcup_{i \in I} C_i.$$

Examples 1.16.
1. The collection $\{[n, n+1)\}_{n\in\mathbb{Z}}$ is a *cover* for $\mathbb{R}$.
2. The collection $\{(n, n+1)\}_{n\in\mathbb{Z}}$ is *not a cover* for $\mathbb{Z}$.
3. The collection of all concentric open balls in a metric space (X,ρ) centered at a fixed point $x \in X$

$$\{B(x,r) \mid r > 0\}$$

is an *open cover* of X, the subcollection

$$\{B(x,n) \mid n \in \mathbb{N}\}$$

being its countable *subcover.*
4. Let A be a *dense set* in a metric space (X,ρ). For any $\varepsilon > 0$, the collection of ε-balls

$$\{B(x,\varepsilon) \mid x \in A\},$$

is an open cover of X.
5. Let $\{r_n\}_{n\in\mathbb{N}}$ be a countably infinite subset of $\mathbb{R}$. The collection of intervals

$$\left\{\left[r_n - 1/2^{n+1}, r_n + 1/2^{n+1}\right] \,\middle|\, n \in \mathbb{N}\right\}$$

does not cover $\mathbb{R}$. This is true even when the set is *dense* in $\mathbb{R}$, as is the case, e. g., for $\mathbb{Q}$.

Exercise 1.31. Verify.

Definition 1.27 (Compactness). A set A is said to be *compact* in a metric space (X,ρ) if each open cover $\mathscr{O}$ of A contains a finite subcover $\mathscr{O}'$.

A metric space (X,ρ) is called *compact* if the set X is compact in (X,ρ).

Definition 1.28 (Sequential Compactness). A *nonempty set* A in a metric space (X,ρ) is said to be *sequentially compact* in (X,ρ) if every sequence $(x_n)_{n\in\mathbb{N}}$ in A has a subsequence $(x_{n(k)})_{k\in\mathbb{N}}$ convergent to a limit in A.

A metric space (X,ρ) is called *sequentially compact* if the set X is sequentially compact in (X,ρ).

Remarks 1.20.
- Compactness in the sense of the prior definition is also called compactness in the *Heine*[5]*–Borel*[6] *sense.*

---

5 Heinrich Heine (1821–1881).
6 Émile Borel (1871–1956).

- Sequential compactness is also called compactness in the *Bolzano*[7]*–Weierstrass*[8] *sense*.
- The notions of compactness and sequential compactness in a metric space are equivalent.

Examples 1.17.

1. *Finite sets*, including the empty set $\emptyset$, are *compact* in an arbitrary metric space (X,ρ) and only finite sets are compact in a *discrete* metric space (X,ρ_d).
2. The sets $[0,\infty)$, $(0,1]$, and $\{1/n\}_{n\in\mathbb{N}}$ are *not compact* in $\mathbb{R}$ and the set $\{0\}\cup\{1/n\}_{n\in\mathbb{N}}$ is.
3. The set $E := \{e_n := (\delta_{nk})_{k\in\mathbb{N}}\}_{n\in\mathbb{N}}$, where δ_{nk} is the *Kronecker delta*, is *closed* and *bounded*, but *not compact* in l_∞ since its open cover by 1/2-balls

$$\{B(e_n,1/2)\}_{n\in\mathbb{N}},$$

has no finite subcover.
The same example works in l_p $(1 \le p < \infty)$.

Exercise 1.32. Verify.

Theorem 1.13 (Heine–Borel Theorem). *A set A is compact in $\mathbb{R}^n$ $(n \in \mathbb{N})$ with the Euclidean metric iff it is closed and bounded.*

See, e. g., [14].

Example 1.18. Thus, any closed and bounded interval $[a,b]$ $(-\infty < a < b < \infty)$ is a *compact set* in $\mathbb{R}$.

Theorem 1.14 (Properties of Compact Sets). *The compact sets in a metric space (X,ρ) have the following properties:*

(1) *a compact set is necessarily bounded, but not vice versa;*
(2) *a compact set is necessarily closed, but not vice versa;*
(3) *a closed subset of a compact set is compact, in particular, a closed set in a compact metric space (X,ρ) is compact;*
(4) *an arbitrary intersection of compact sets is compact;*
(5) *a finite union of compact sets is compact.*

See, e. g., [14].

The following classical result is a direct implication of the equivalence of different forms of compactness.

7 Bernard Bolzano (1781–1848).
8 Karl Weierstrass (1815–1897).

Theorem 1.15 (Bolzano–Weierstrass Theorem). *Each bounded sequence of real or complex numbers contains a convergent subsequence.*

1.4.9 Continuity

Definition 1.29 (Continuity of a Function). Let (X,ρ) and (Y,σ) be metric spaces. A function $f : X \to Y$ is called *continuous* at a point $x_0 \in X$ if

$$\forall \varepsilon > 0 \; \exists \delta > 0 \; \forall x \in X \text{ with } \rho(x, x_0) < \delta : \; \sigma(f(x), f(x_0)) < \varepsilon.$$

A function $f : X \to Y$ is called *continuous* on X if it is continuous at every point of X. The set of all such functions is designated as $C(X, Y)$ and we write $f \in C(X, Y)$.

Remarks 1.21.

- When X and Y are subsets of $\mathbb{R}$ with the regular distance, we obtain the familiar calculus (ε, δ)-definitions.
- When $Y = \mathbb{R}$ or $Y = \mathbb{C}$, the shorter notation $C(X)$ is used.

Definition 1.30 (Equivalent Definition of Continuity). Let (X,ρ) and (Y,σ) be metric spaces. A function $f : X \to Y$ is called *continuous* at a point $x_0 \in X$ if

$$\forall \varepsilon > 0 \; \exists \delta > 0 : \; f(B_X(x_0, \delta)) \subseteq B_Y(f(x_0), \varepsilon).$$

It is often convenient to describe continuity in terms of sequences.

Theorem 1.16 (Sequential Characterization of Local Continuity). *Let (X,ρ) and (Y,σ) be metric spaces. A function $f : X \to Y$ is continuous at a point $x_0 \in X$ iff, for each sequence $(x_n)_{n\in\mathbb{N}}$ in X such that*

$$\lim_{n\to\infty} x_n = x_0 \textit{ in } (X, \rho),$$

we have:

$$\lim_{n\to\infty} f(x_n) = f(x_0) \textit{ in } (Y, \sigma).$$

Exercise 1.33. Prove.

Hint. The *necessity* is proved directly. Prove the *sufficiency* by *contrapositive*.

Theorem 1.17 (Properties of Numeric Continuous Functions). *Let (X,ρ) be a metric space and $Y = \mathbb{R}$ or $Y = \mathbb{C}$ with the regular distance.*

If f and g are continuous at a point $x_0 \in X$, then

(1) $\forall c \in \mathbb{R}$ (or $c \in \mathbb{C}$), *cf is continuous at x_0;*
(2) *$f + g$ is continuous at x_0;*

(3) *$f \cdot g$ is continuous at x_0;*
(4) *provided $g(x_0) \neq 0$, $\frac{f}{g}$ is continuous at x_0.*

Theorem 1.18 (Continuity of Composition). *Let (X,ρ), (Y,σ), and (Z,τ), $f : X \to Y$ and $g : Y \to Z$.*

If for some $x_0 \in X$ f is continuous at x_0 and g is continuous at $y_0 = f(x_0)$, then the composition $g(f(x))$ is continuous at x_0.

Exercise 1.34. Prove Theorems 1.17 and 1.18 using the *sequential approach.*

Remark 1.22. The statements of Theorems 1.17 and 1.18 are naturally carried over to functions continuous on the whole space (X,ρ).

The following characterization of continuity is sometimes given as its definition (see, e. g., [18, 21]).

Theorem 1.19 (Characterization of Continuity). *Let (X,ρ) and (Y,σ) be metric spaces. A function $f : X \to Y$ is continuous on X iff, for each open set A' in (Y,σ) its preimage $f^{-1}(A')$ is open in (X,ρ), or equivalently,*

$$f^{-1}(\mathscr{G}') := \left\{ f^{-1}(A') \,\middle|\, A \in \mathscr{G}' \right\} \subseteq \mathscr{G},$$

where $\mathscr{G}'$ is the metric topology on Y generated by metric σ and $\mathscr{G}$ is the metric topology on X generated by metric ρ.

2 Basic Set Classes

In this chapter, we introduce and discuss basic set classes needed for our subsequent discourse.

Henceforth, X is regarded to be a *nonempty set*.

2.1 Semirings, Semi-algebras

Definition 2.1 (Semiring, Semi-algebra). Let X be a nonempty set. A *semiring* (of sets) on X is a nonempty collection $\mathscr{S}$ of subsets of X such that:
(1) $\emptyset \in \mathscr{S}$;
(2) if $A_1, \dots, A_n \in \mathscr{S}$ ($n \in \mathbb{N}$), then $\bigcap_{i=1}^{n} A_i \in \mathscr{S}$ (*closedness under finite intersections*);
(3) if $A, B \in \mathscr{S}$, then

$$A \setminus B = \bigcup_{i=1}^{n} C_i$$

with some $n \in \mathbb{N}$ and pairwise disjoint $C_i \in \mathscr{S}$, $i = 1, \dots, n$.

A *semi-algebra* $\mathscr{S}$ (of sets) on X is a semiring on X such that $X \in \mathscr{S}$.

Remarks 2.1.
- Condition (2) can be replaced with the following equivalent one:

$$\text{if } A, B \in \mathscr{R}, \text{ then } A \cap B \in \mathscr{R}.$$

- Condition (3) makes condition (1) redundant.

Examples 2.1.
1. On an arbitrary nonempty set X, its power set $\mathscr{P}(X)$ is a *semi-algebra*.
2. On an arbitrary nonempty set X, a finite collection $\mathscr{S} := \{A_1, \dots, A_n\}$ ($n \in \mathbb{N}$) of pairwise disjoint subsets, which includes $\emptyset$, is a *semiring*.
   In particular, $\{\emptyset, X\}$ and $\{\emptyset\}$ are *semirings* on X, the former being also a *semi-algebra*.
   For $X :- \{0, 1, 2\}$

$$\mathscr{S} := \{\emptyset, \{0\}, \{1, 2\}\}$$

   is a *semiring* on X.
3. On $\mathbb{R}$,

$$\mathscr{S}_1 := \{(a, b] \mid -\infty < a < b < \infty\} \cup \{\emptyset\}$$

   and

$$\mathscr{S}_2 := \{[a, b) \mid -\infty < a < b < \infty\} \cup \{\emptyset\}$$

   are *semirings* but *not* semi-algebras.

https://doi.org/10.1515/9783110600995-002

4. The set collections

$$\mathscr{S}_1 := \{(a_1, b_1] \times (a_2, b_2] \mid -\infty < a_i < b_i < \infty,\ i = 1, 2\} \cup \{\emptyset\}$$

and

$$\mathscr{S}_2 := \{[a_1, b_1) \times [a_2, b_2) \mid -\infty < a_i < b_i < \infty,\ i = 1, 2\} \cup \{\emptyset\}$$

are *semirings* but *not* semi-algebras on $\mathbb{R}^2$ (see the *Product Semiring Proposition* (Proposition 2.1), Section 2.6, Problem 2).

5. The set collections

$$\mathscr{S}_1 := \{(c, d] \mid a \le c < d \le b\} \cup \{\emptyset\}$$

and

$$\mathscr{S}_2 := \{[c, d) \mid a \le c < d \le b\} \cup \{\emptyset\}$$

are *semi-algebras* on $(a, b]$ and $[a, b)$ $(-\infty < a < b < \infty)$, respectively.

6. The set collections

$$\mathscr{S}_1 := \{(c_1, d_1] \times (c_2, d_2] \mid a_i \le c_i < d_i \le b_i,\ i = 1, 2\} \cup \{\emptyset\}$$

and

$$\mathscr{S}_2 := \{[c_1, d_1) \times [c_2, d_2) \mid a_i \le c_i < d_i \le b_i,\ i = 1, 2\} \cup \{\emptyset\}$$

are *semi-algebras* on $(a_1, b_1] \times (a_2, b_2]$ and $[a_1, b_1) \times [a_2, b_2)$ $(-\infty < a_1 < b_1 < \infty$, $-\infty < a_2 < b_2 < \infty)$, respectively (see the *Product Semiring Proposition* (Proposition 2.1), Section 2.6, Problem 2).

7. On $\mathbb{R}$,

$$\mathscr{C} := \{(a, b) \mid -\infty < a < b < \infty\} \cup \{\emptyset\}$$

is *not* a semiring.

Exercise 2.1.
(a) Verify.
(b) Give an example showing that a semiring need not be closed under finite unions.

2.2 Rings, Algebras

Definition 2.2 (Ring, Algebra). Let X be a nonempty set. A *ring* (of sets) on X is a nonempty collection $\mathscr{R}$ of subsets of X such that

(1) $\emptyset \in \mathscr{R}$;
(2) if $A_1, \dots, A_n \in \mathscr{R}$ $(n \in \mathbb{N})$, then $\bigcup_{i=1}^n A_i \in \mathscr{R}$ (*closedness under finite unions*);
(3) if $A, B \in \mathscr{R}$, then $A \setminus B \in \mathscr{R}$ (*closedness under set differences*).

An *algebra* $\mathscr{A}$ (of sets) on X is a ring on X such that $X \in \mathscr{A}$.

Remarks 2.2.
- Condition (2) can be replaced with the following equivalent one:

$$\text{if } A, B \in \mathscr{R}, \text{ then } A \cup B \in \mathscr{R}.$$

- Condition (3) makes condition (1) redundant (cf. Remark 2.1).
- As follows from conditions (2) and (3) by *De Morgan's laws*, a ring $\mathscr{R}$ is *closed under finite intersections*, i. e.,

$$\text{if } A_1, \dots, A_n \in \mathscr{R} \ (n \in \mathbb{N}), \text{ then } \bigcap_{i=1}^n A_i \in \mathscr{R}.$$

- As follows from condition (3), an algebra $\mathscr{A}$ is *closed under complements*, i. e.,

$$\text{if } A \in \mathscr{A}, \text{ then } A^c \in \mathscr{A}.$$

- Each ring/algebra is a semiring/semi-algebra, respectively, but not vice versa.

Exercise 2.2. Explain, verify.

Examples 2.2.
1. On an arbitrary nonempty set X,
   (a) the power set $\mathscr{P}(X)$ and $\{\emptyset, X\}$ are *algebras* and $\{\emptyset\}$ is a *ring* but *not* an algebra;
   (b) for a finite collection $\mathscr{C} := \{A_1, \dots, A_n\}$ $(n \in \mathbb{N})$ of pairwise disjoint subsets, the collection $\mathscr{R}$ of all finite unions of A_i, $i = 1, \dots, n$, including $\emptyset$ $(\bigcup_{i \in \emptyset} A_i = \emptyset)$, is a *ring* on X consisting of 2^n elements.
   With $\mathscr{C}$ being a *partition* of X, i. e., in addition,

$$\bigcup_{i=1}^n A_i = X,$$

   $\mathscr{R}$ is an *algebra* on X.
   In particular, for $X := \{0, 1, 2\}$ and the semiring

$$\mathscr{S} := \{\emptyset, \{0\}, \{1, 2\}\}$$

   (see Examples 2.1) forming a partition of X,

$$\mathscr{A} := \{\emptyset, \{0\}, \{1, 2\}, X\}$$

is an *algebra* on X; for $X := [a, b]$ $(-\infty < a < b < \infty)$ and

$$\mathscr{C} : \{[a, (a+b)/2], ((a+b)/2, b]\}$$

forming a partition of X,

$$\mathscr{A} := \{\emptyset, [a, (a+b)/2], ((a+b)/2, b], X\}$$

is an *algebra* on X.

2. The set collections

$$\mathscr{S}_1 := \{(a, b] \mid -\infty < a < b < \infty\} \cup \{\emptyset\}$$

and

$$\mathscr{S}_2 := \{[a, b) \mid -\infty < a < b < \infty\} \cup \{\emptyset\}$$

are *semirings* but *not* rings on $\mathbb{R}$ and the set collections

$$\mathscr{R}_1 := \left\{ \bigcup_{i=1}^{n} (a_i, b_i] \,\middle|\, n \in \mathbb{N},\ -\infty < a_i < b_i < \infty,\ i = 1, \dots, n \right\} \cup \{\emptyset\}$$

and

$$\mathscr{R}_2 := \left\{ \bigcup_{i=1}^{n} [a_i, b_i) \,\middle|\, n \in \mathbb{N},\ -\infty < a_i < b_i < \infty,\ i = 1, \dots, n \right\} \cup \{\emptyset\},$$

are *rings* but *not* algebras on $\mathbb{R}$.

3. The set collections

$$\mathscr{S}_1 := \{(c, d] \mid a \le c < d \le b\} \cup \{\emptyset\}$$

and

$$\mathscr{S}_2 := \{[c, d) \mid a \le c < d \le b\} \cup \{\emptyset\}$$

are *semi-algebras* but *not* algebras on $(a, b]$ and $[a, b)$ $(-\infty < a < b < \infty)$, respectively, and the set collections

$$\mathscr{A}_1 := \left\{ \bigcup_{i=1}^{n} (c_i, d_i] \,\middle|\, n \in \mathbb{N},\ a \le c_i < d_i \le b,\ i = 1, \dots, n \right\} \cup \{\emptyset\}$$

and

$$\mathscr{A}_2 := \left\{ \bigcup_{i=1}^{n} [c_i, d_i) \,\middle|\, n \in \mathbb{N},\ a \le c_i < d_i \le b,\ i = 1, \dots, n \right\} \cup \{\emptyset\}$$

are *algebras* on $(a, b]$ and $[a, b)$ $(-\infty < a < b < \infty)$, respectively.

4. The set collections

$$\mathscr{S}_1 := \{(c_1, d_1] \times (c_2, d_2] \mid a_i \le c_i < d_i \le b_i,\ i = 1, 2\} \cup \{\emptyset\}$$

and

$$\mathscr{S}_2 := \{[c_1, d_1) \times [c_2, d_2) \mid a_i \le c_i < d_i \le b_i,\ i = 1, 2\} \cup \{\emptyset\}$$

are *semi-algebras* on $(a_1, b_1] \times (a_2, b_2]$ and $[a_1, b_1) \times [a_2, b_2)$ $(-\infty < a_1 < b_1 < \infty,$ $-\infty < a_2 < b_2 < \infty)$, respectively, and the collections

$$\mathscr{A}_1 := \left\{ \bigcup_{i=1}^{n} (c_1^{(i)}, d_1^{(i)}] \times (c_2^{(i)}, d_2^{(i)}] \;\middle|\; a_1 \le c_1^{(i)} < d_1^{(i)} \le b_1,\ a_2 \le c_2^{(i)} < d_2^{(i)} \le b_2,\ i = 1, \dots, n \right\} \cup \{\emptyset\}$$

and

$$\mathscr{A}_2 := \left\{ \bigcup_{i=1}^{n} [c_1^{(i)}, d_1^{(i)}) \times [c_2^{(i)}, d_2^{(i)}) \;\middle|\; a_1 \le c_1^{(i)} < d_1^{(i)} \le b_1,\ a_2 \le c_2^{(i)} < d_2^{(i)} \le b_2,\ i = 1, \dots, n \right\} \cup \{\emptyset\}$$

are *algebras* on $(a_1, b_1] \times (a_2, b_2]$ and $[a_1, b_1) \times [a_2, b_2)$ $(-\infty < a_1 < b_1 < \infty,$ $-\infty < a_2 < b_2 < \infty)$, respectively.

An algebra can be equivalently defined as follows:

Definition 2.3 (Algebra: Alternative Definition). Let X be a nonempty set. An *algebra* (of sets) on X is a nonempty collection $\mathscr{A}$ of subsets of X such that:
(1) $\emptyset, X \in \mathscr{A}$;
(2) if $A_1, \dots, A_n \in \mathscr{A}$ $(n \in \mathbb{N})$, then $\bigcup_{i=1}^{n} A_i \in \mathscr{A}$ (*closedness under finite unions*);
(3) if $A \in \mathscr{A}$, then $A^c \in \mathscr{A}$ (*closedness under complements*).

Exercise 2.3. Verify the equivalence of Definitions 2.2 and 2.3.

2.3 σ-Rings, σ-Algebras

Definition 2.4 (σ-Ring, σ-Algebra). Let X be a nonempty set. A *σ-ring* (of sets) on X is a nonempty collection $\mathscr{R}$ of subsets of X such that:
(1) $\emptyset \in \mathscr{R}$;
(2) if $(A_n)_{n \in \mathbb{N}}$ is a sequence in $\mathscr{R}$, then $\bigcup_{i=1}^{\infty} A_i \in \mathscr{R}$ (*closedness under countable unions*);
(3) if $A, B \in \mathscr{R}$, then $A \setminus B \in \mathscr{R}$ (*closedness under set differences*).

A *σ-algebra* Σ (of sets) on X is a σ-ring on X such that $X \in \Sigma$.

Remarks 2.3.

- Condition (3) makes condition (1) redundant (cf. Remarks 2.2).
- As follows from conditions (2) and (3) by *De Morgan's laws*, a σ-ring $\mathscr{R}$ is *closed under countable intersections*, i. e.,

$$\text{if } (A_n)_{n\in\mathbb{N}} \text{ is a sequence in } \mathscr{R}, \text{ then } \bigcap_{i=1}^{\infty} A_i \in \mathscr{R}$$

  (cf. Remarks 2.2).
- Each σ-ring/σ-algebra is a ring/algebra, respectively, but not vice versa.
- Hence, in particular, each σ-algebra Σ is *closed under complements*, i. e.,

$$\text{if } A \in \Sigma, \text{ then } A^c \in \Sigma.$$

- Each ring/algebra consisting of finitely many elements is a σ-ring/σ-algebra, respectively.

Exercise 2.4. Verify.

Examples 2.3.

1. All finite algebras from Examples 2.2 are *σ-algebras*.
2. On an arbitrary nonempty set X,
   (a) the power set $\mathscr{P}(X)$ and $\{\emptyset, X\}$ are *σ-algebras* and $\{\emptyset\}$ is a *σ-ring* but *not* a σ-algebra;
   (b)
   $$\Sigma := \{A \subseteq X \mid A \text{ or } A^c \text{ is } \textit{countable}\}$$
   is a *σ-algebra*;
   (c) provided X is *infinite*,
   $$\mathscr{A} := \{A \subseteq X \mid A \text{ or } A^c \text{ is } \textit{finite}\}$$
   is an *algebra*, but *not* a σ-algebra and
   $$\mathscr{R} := \{A \subseteq X \mid A \text{ is } \textit{finite}\}$$
   is a *ring*, but *not* a σ-ring on X;
   (d) provided X is *countably infinite*, if a countably infinite collection $\{A_n\}_{n\in\mathbb{N}}$ is a *partition* of X, i. e., the sets A_i, $i \in \mathbb{N}$, are *pairwise disjoint* and
   $$\bigcup_{i=1}^{\infty} A_i = X,$$
   the collection Σ of all countable unions of A_i, $i \in \mathbb{N}$, including $\emptyset$ ($\bigcup_{i\in\emptyset} A_i = \emptyset$), is a *$\sigma$-algebra* of cardinality $2^{\aleph_0} = \mathfrak{c}$ (cf. Section 2.6, Problem 8).

3. An example of a σ-ring, which is not a σ-algebra, can be obtained by taking an arbitrary σ-algebra Σ on a nonempty set X, e. g., $\mathscr{P}(X)$, and considering it on the set $X' := X \cup \{x'\}$, where $x' \notin X$. For instance, $\mathscr{P}([0,1))$ is a σ-algebra on $[0,1)$ but only a σ-ring on $[0,1]$.

Exercise 2.5. Verify.

A σ-algebra can be equivalently defined as follows.

Definition 2.5 (σ-Algebra: Alternative Definition). Let X be a nonempty set. A *σ-algebra* (of sets) on X is a nonempty collection Σ of subsets of X such that:
(1) $\emptyset, X \in \Sigma$;
(2) for any sequence $(A_n)_{n\in\mathbb{N}}$ in Σ, $\bigcup_{i=1}^{\infty} A_i \in \Sigma$ (*closedness under countable unions*);
(3) if $A \in \Sigma$, then $A^c \in \Sigma$ (*closedness under complements*).

Exercise 2.6. Verify the equivalence of Definitions 2.4 and 2.5.

2.4 Monotone Classes

Definition 2.6 (Monotone Class). Let X be a nonempty set. A *monotone class* (of sets) on X is a nonempty collection $\mathscr{M}$ of subsets of X such that, if $(A_n)_{n\in\mathbb{N}}$ is a monotone sequence in $\mathscr{M}$, then $\lim_{n\to\infty} A_n \in \mathscr{M}$ (*closedness under the limits of monotone set sequences*).

Remarks 2.4.
- As follows from the definition, a σ-ring, in particular a σ-algebra, is a monotone class.

  Exercise 2.7. Verify.

  This fact readily provides many examples of monotone classes.
- However, as the following example shows, a monotone class need not even be a semiring.

Examples 2.4.
1. On a nonempty *finite* set X, any nonempty set collection $\mathscr{C}$ is a *monotone class*.
2. On $\mathbb{R}$,

$$\mathscr{M} := \{[m,n] \mid m,n \in \mathbb{Z},\ m < n\} \cup \{(-\infty,n] \mid n \in \mathbb{Z}\} \cup \{[n,\infty) \mid n \in \mathbb{Z}\} \cup \{\emptyset\} \cup \{\mathbb{R}\}$$

   is a *monotone class*.

Exercise 2.8.
(a) Verify.
(b) Show that $\mathscr{M}$ from the prior example is not a semiring.
(c) Would $\mathscr{M}$ from the prior example be a monotone class if $m, n \in \mathbb{R}$?

Theorem 2.1 (Monotone Ring Theorem). *A monotone ring is a σ-ring.*

Proof. Let $\mathscr{R}$ be simultaneously a ring and a monotone class on a nonempty set X.

Then $\mathscr{R}$ immediately meets condition (3) of the definition of a σ-ring (see Definition 2.4). It remains to verify condition (2) of Definition 2.4, i. e., the closedness of $\mathscr{R}$ under countable unions.

For an arbitrary sequence $(A_n)_{n\in\mathbb{N}}$ in $\mathscr{R}$, consider the set sequence

$$B_n := \bigcup_{i=1}^{n} A_i, \quad n \in \mathbb{N}.$$

Since $\mathscr{R}$ is a ring

$$B_n \in \mathscr{R}, \quad n \in \mathbb{N}.$$

The sequence $\{B_n\}_{n=1}^{\infty}$ is *increasing* since

$$B_n = \bigcup_{i=1}^{n} A_i \subseteq \bigcup_{i=1}^{n+1} A_i = B_{n+1}, \quad n \in \mathbb{N},$$

which, since $\mathscr{R}$ is a monotone class, implies that

$$\mathscr{R} \ni \lim_{n\to\infty} B_n = \bigcup_{i=1}^{\infty} B_i = \bigcup_{i=1}^{\infty} A_i.$$

Hence, $\mathscr{R}$ is a σ-ring. □

As an important particular case, we obtain the following.

Corollary 2.1 (Monotone Algebra Theorem). *A monotone algebra is a σ-algebra.*

2.5 Generated Set Classes

Here, we introduce the important notion of the ring, algebra, σ-ring, σ-algebra, and monotone class "generated" by a nonempty collection $\mathscr{C}$ of subsets of a nonempty set X. To proceed, we need the following lemma.

2.5.1 Intersection Lemma

Lemma 2.1 (Intersection Lemma). *The intersection of an arbitrary nonempty collection of rings, algebras, σ-rings, σ-algebras, or monotone classes on a nonempty set X is a ring, algebra, σ-ring, σ-algebra, and monotone class on X, respectively.*

Proof. Let $\{\mathscr{R}_i\}_{i\in I}$ be a nonempty collection of *rings* on X.

Observe that

$$\bigcap_{i\in I}\mathscr{R}_i \neq \emptyset$$

since $\emptyset \in \mathscr{R}_i$ for each $i \in I$.

For any $A, B \in \bigcap_{i\in I}\mathscr{R}_i$,

$$A, B \in \mathscr{R}_i,\ i \in I,$$

which, with $\mathscr{R}_i$, $i \in I$, being a ring, implies that

$$A \cup B, A \setminus B \in \mathscr{R}_i,\ i \in I,$$

and hence,

$$A \cup B, A \setminus B \in \bigcap_{i\in I}\mathscr{R}_i.$$

Thus, $\bigcap_{i\in I}\mathscr{R}_i$ is a *ring* on X.

The statement for a nonempty collection $\{\mathscr{A}_i\}_{i\in I}$ of *algebras* on X follows immediately with one more condition to verify:

$$X \in \bigcap_{i\in I}\mathscr{A}_i,$$

which is trivially true since

$$X \in \mathscr{A}_i,\ i \in I.$$

□

Exercise 2.9. Prove the statement for *σ-rings*, *σ-algebras*, and *monotone classes*.

Remark 2.5. The analogue of the prior lemma does not hold for a nonempty collection of *semirings*.

For example, on $X := \{0, 1, 2\}$, the following set collections:

$$\mathscr{C}_1 := \{\emptyset, \{0\}, \{1, 2\}, \{0, 1, 2\}\} \text{ and } \mathscr{C}_2 := \{\emptyset, \{0\}, \{1\}, \{2\}, \{0, 1, 2\}\}$$

are *semirings* ($\mathscr{C}_1$ is even an *algebra* on X (see Examples 2.2)), but their intersection

$$\mathscr{C}_1 \cap \mathscr{C}_2 = \{\emptyset, \{0\}, \{0, 1, 2\}\}$$

is not.

Exercise 2.10. Verify (cf. Section 2.6, Problem 9).

2.5.2 Generated Set Classes

Definition 2.7 (Generated Set Classes). Let $\mathscr{C}$ be a nonempty collection of subsets of a nonempty set X.

- $r(\mathscr{C}) := \bigcap_{\mathscr{R} \text{ is a ring}, \mathscr{C} \subseteq \mathscr{R}} \mathscr{R}$ is the *smallest ring containing* $\mathscr{C}$ called the *ring generated by* $\mathscr{C}$.
- $a(\mathscr{C}) := \bigcap_{\mathscr{A} \text{ is an algebra}, \mathscr{C} \subseteq \mathscr{A}} \mathscr{A}$ is the *smallest algebra containing* $\mathscr{C}$ called the *algebra generated by* $\mathscr{C}$.
- $\sigma r(\mathscr{C}) := \bigcap_{\mathscr{R} \text{ is a } \sigma\text{-ring}, \mathscr{C} \subseteq \mathscr{R}} \mathscr{R}$ is the *smallest σ-ring containing* $\mathscr{C}$ called the *σ-ring generated by* $\mathscr{C}$.
- $\sigma a(\mathscr{C}) := \bigcap_{\Sigma \text{ is a } \sigma\text{-algebra}, \mathscr{C} \subseteq \Sigma} \Sigma$ is the *smallest σ-algebra containing* $\mathscr{C}$ called the *σ-algebra generated by* $\mathscr{C}$.
- $m(\mathscr{C}) := \bigcap_{\mathscr{M} \text{ is a monotone class}, \mathscr{C} \subseteq \mathscr{M}} \mathscr{M}$ is the *smallest monotone class containing* $\mathscr{C}$ called the *monotone class generated by* $\mathscr{C}$.

Exercise 2.11.
(a) Explain why the generated classes are well-defined.
(b) If $X \in \mathscr{C}$, $r(\mathscr{C}) = a(\mathscr{C})$ and $\sigma r(\mathscr{C}) = \sigma a(\mathscr{C})$.
(c) If $\mathscr{C}_1 \subset \mathscr{C}_2$,

$$r(\mathscr{C}_1) \subseteq r(\mathscr{C}_2),\ a(\mathscr{C}_1) \subseteq a(\mathscr{C}_2),\ \sigma r(\mathscr{C}_1) \subseteq \sigma r(\mathscr{C}_2),\ \sigma a(\mathscr{C}_1) \subseteq \sigma a(\mathscr{C}_2).$$

(d) Give an example showing that the notion of the semiring generated by a nonempty collection $\mathscr{C}$ of subsets of a nonempty set X, i. e., the *smallest semiring on X containing* $\mathscr{C}$, is not well-defined (see Remark 2.5 and Section 2.6, Problem 9).

Examples 2.5. For a nonempty set X,
1. $r(\{\emptyset\}) = \sigma r(\{\emptyset\}) = m(\{\emptyset\}) = \{\emptyset\}$;
2. $r(\{X\}) = a(\{X\}) = \sigma r(\{X\}) = \sigma a(\{X\}) = \{\emptyset, X\}$ and $m(\{X\}) = \{X\}$.

Exercise 2.12. Verify.

Theorem 2.2 (Generated Ring Theorem). *Let $\mathscr{S}$ be a semiring on a nonempty set X. Then*

$$r(\mathscr{S}) := \left\{ \bigcup_{i=1}^{n} A_i \,\middle|\, n \in \mathbb{N},\ A_i \in \mathscr{S},\ i = 1, \dots, n \right\}, \tag{2.1}$$

i. e., the collection of all finite unions of sets of a semiring $\mathscr{S}$ on a nonempty set X is the ring generated by $\mathscr{S}$.

Proof. Let

$$\mathscr{C} := \left\{ \bigcup_{i=1}^{n} A_i \,\middle|\, n \in \mathbb{N},\ A_i \in \mathscr{S},\ i = 1, \dots, n \right\}.$$

Then

$$\mathscr{S} \subseteq \mathscr{C} \subseteq r(\mathscr{S}). \tag{2.2}$$

Exercise 2.13. Explain.

Let us show that $\mathscr{C}$ is a *ring* on X.
Indeed, for arbitrary

$$A = \bigcup_{i=1}^{m} A_i, \ m \in \mathbb{N}, \ A_i \in \mathscr{S}, \ i = 1, \ldots, m,$$

and

$$B = \bigcup_{j=1}^{n} B_j \in \mathscr{C}, \ n \in \mathbb{N}, \ B_j \in \mathscr{S}, \ j = 1, \ldots, n,$$

by the very definition of $\mathscr{C}$,

$$A \cup B \in \mathscr{C}.$$

Further, by *De Morgan's laws*,

$$A \setminus B = \bigcup_{i=1}^{m} A_i \setminus \bigcup_{j=1}^{n} B_j = \bigcup_{i=1}^{m} \left(A_i \setminus \bigcup_{j=1}^{n} B_j \right) = \bigcup_{i=1}^{m} \bigcap_{j=1}^{n} A_i \setminus B_j.$$

Since $\mathscr{S}$ is a *semiring*, without loss of generality, we can regard that the sets A_i, $i = 1, \ldots, m$, are *pairwise disjoint*, and so are B_j, $j = 1, \ldots, n$ (see Section 2.6, Problem 10).
Furthermore, for all $i = 1, \ldots, m$, $j = 1, \ldots, n$,

$$A_i \setminus B_j = \bigcup_{k=1}^{l(i,j)} C_k^{(i,j)}$$

with some $l(i,j) \in \mathbb{N}$ and *pairwise disjoint* $C_k^{(i,j)} \in \mathscr{S}$, $k = 1, \ldots, l(i,j)$.
Hence,

$$A \setminus B = \bigcup_{i=1}^{m} \bigcap_{j=1}^{n} A_i \setminus B_j = \bigcup_{i=1}^{m} \bigcap_{j=1}^{n} \bigcup_{k=1}^{l(i,j)} C_k^{(i,j)} \in \mathscr{C}.$$

Exercise 2.14. Explain.

Thus, we conclude that $\mathscr{C}$ is a ring on X, which in view of $\mathscr{S} \subseteq \mathscr{C}$ implies that

$$r(\mathscr{S}) \subseteq \mathscr{C}. \tag{2.3}$$

Inclusions (2.2) and (2.3) jointly imply that

$$\mathscr{C} = r(\mathscr{S}),$$

which completes the proof. □

Remarks 2.6.
- In particular, if $X \in \mathscr{S}$,

$$r(\mathscr{S}) = a(\mathscr{S})$$

(see Exercise 2.11).
- As is noted in the above proof, without loss of generality, the sets $A_1, \dots, A_n$ ($n \in \mathbb{N}$) in definition (2.1) of the *Generated Ring Theorem* (Theorem 2.2) can be regarded to be *pairwise disjoint* (see Section 2.6, Problem 10), i. e., the *Generated Ring Theorem* (Theorem 2.2) can be equivalently restated as follows.

Theorem 2.3 (Generated Ring Theorem). *The collection of all finite unions of pairwise disjoint sets of a semiring $\mathscr{S}$ on a nonempty set X is the ring generated by $\mathscr{S}$.*

Examples 2.6.
1. On an arbitrary nonempty set X, a finite collection $\mathscr{S} := \{A_1, \dots, A_n\}$ ($n \in \mathbb{N}$) of pairwise disjoint subsets, which includes $\emptyset$, is a *semiring* (see Examples 2.1), and hence, by the *Generated Ring Theorem* (Theorem 2.2), the collection of all finite unions of sets of $\mathscr{S}$ is the ring $r(\mathscr{S})$ generated by $\mathscr{S}$.
   In particular, when $\mathscr{S}$ is a *partition* of X (see Examples 2.2), $r(\mathscr{S})$ is the algebra generated by $\mathscr{S}$, i. e., in this case, $r(\mathscr{S}) = a(\mathscr{S})$.
2. By the *Generated Ring Theorem* (Theorem 2.2),

$$\mathscr{R}_1 := \left\{ \bigcup_{i=1}^{n} (a_i, b_i] \,\middle|\, n \in \mathbb{N},\ -\infty < a_i < b_i < \infty,\ i = 1, \dots, n \right\} \cup \{\emptyset\} = r(\mathscr{S}_1)$$

and

$$\mathscr{R}_2 := \left\{ \bigcup_{i=1}^{n} [a_i, b_i) \,\middle|\, n \in \mathbb{N},\ -\infty < a_i < b_i < \infty,\ i = 1, \dots, n \right\} \cup \{\emptyset\} = r(\mathscr{S}_2),$$

where

$$\mathscr{S}_1 := \{(a, b] \mid -\infty < a < b < \infty\} \cup \{\emptyset\}$$

and

$$\mathscr{S}_2 := \{[a, b) \mid -\infty < a < b < \infty\} \cup \{\emptyset\}$$

(see Examples 2.2).
3. By the *Generated Ring Theorem* (Theorem 2.2),

$$\mathscr{A}_1 := \left\{ \bigcup_{i=1}^{n} (c_i, d_i] \,\middle|\, n \in \mathbb{N},\ a \le c_i < d_i \le b,\ i = 1, \dots, n \right\} \cup \{\emptyset\} = a(\mathscr{S}_1)$$

and

$$\mathscr{A}_2 := \left\{ \bigcup_{i=1}^{n} [c_i, d_i) \,\middle|\, n \in \mathbb{N},\ a \le c_i < d_i \le b,\ i = 1, \dots, n \right\} \cup \{\emptyset\} = a(\mathscr{S}_2),$$

where

$$\mathscr{S}_1 := \{(c,d] \mid a \leq c < d \leq b\} \cup \{\emptyset\}$$

and

$$\mathscr{S}_2 := \{[c,d) \mid a \leq c < d \leq b\} \cup \{\emptyset\}$$

$(-\infty < a < b < \infty)$ (see Examples 2.2).

4. By the *Generated Ring Theorem* (Theorem 2.2),

$$\mathscr{A}_1 := \left\{ \bigcup_{i=1}^{n} (c_1^{(i)}, d_1^{(i)}] \times (c_2^{(i)}, d_2^{(i)}] \;\middle|\; a_1 \leq c_1^{(i)} < d_1^{(i)} \leq b_1,\ a_2 \leq c_2^{(i)} < d_2^{(i)} \leq b_2, \right.$$
$$\left. i = 1, \dots, n \right\} \cup \{\emptyset\} = a(\mathscr{S}_1)$$

and

$$\mathscr{A}_2 := \left\{ \bigcup_{i=1}^{n} [c_1^{(i)}, d_1^{(i)}) \times [c_2^{(i)}, d_2^{(i)}) \;\middle|\; a_1 \leq c_1^{(i)} < d_1^{(i)} \leq b_1,\ a_2 \leq c_2^{(i)} < d_2^{(i)} \leq b_2, \right.$$
$$\left. i = 1, \dots, n \right\} \cup \{\emptyset\} = a(\mathscr{S}_2),$$

where

$$\mathscr{S}_1 := \{(c_1, d_1] \times (c_2, d_2] \mid a_i \leq c_i < d_i \leq b_i,\ i = 1, 2\} \cup \{\emptyset\}$$

and

$$\mathscr{S}_2 := \{[c_1, d_1) \times [c_2, d_2) \mid a_i \leq c_i < d_i \leq b_i,\ i = 1, 2\} \cup \{\emptyset\}$$

$(-\infty < a_1 < b_1 < \infty,\ -\infty < a_2 < b_2 < \infty)$ (see Examples 2.2).

Theorem 2.4 (Monotone Class Theorem). *Let $\mathscr{R}$ be a ring on a nonempty set X. Then*

$$\sigma r(\mathscr{R}) = m(\mathscr{R}).$$

Proof. Since $\sigma r(\mathscr{R})$ is a also monotone class containing $\mathscr{R}$ (see Remarks 2.4), we have the inclusion

$$m(\mathscr{R}) \subseteq \sigma r(\mathscr{R}). \tag{2.4}$$

Let us show that $m(\mathscr{R})$ is a *ring*.

For a fixed $A \in m(\mathscr{R})$, consider

$$\mathscr{L}(A) := \{C \subseteq X \mid A \cup C, A \setminus C, C \setminus A \in m(\mathscr{R})\}.$$

Whenever $A \in \mathscr{R}$,

$$\mathscr{R} \subseteq \mathscr{L}(A). \tag{2.5}$$

Indeed, since $\mathscr{R}$ is a ring, for each $C \in \mathscr{R}$,

$$A \cup C, A \setminus C, C \setminus A \in \mathscr{R} \subseteq m(\mathscr{R}).$$

Further, for each $A \in m(\mathscr{R})$, $\mathscr{L}(A)$ is a *monotone class*. Indeed, for an arbitrary increasing sequence $(C_n)_{n\in\mathbb{N}}$ in $\mathscr{L}(A)$,

$$A \cup C_n, A \setminus C_n, C_n \setminus A \in m(\mathscr{R}),\ n \in \mathbb{N},$$

and, as is easily seen, the sequences $(A \cup C_n)_{n\in\mathbb{N}}$ and $(C_n \setminus A)_{n\in\mathbb{N}}$ are *increasing*, whereas the sequence $(A \setminus C_n)_{n\in\mathbb{N}}$ is *decreasing* in $m(\mathscr{R})$.

Exercise 2.15. Verify.

Since, $m(\mathscr{R})$ is a monotone class,

$$m(\mathscr{R}) \ni \lim_{n\to\infty} (A \cup C_n) = \bigcup_{n=1}^{\infty} (A \cup C_n) = A \cup \bigcup_{n=1}^{\infty} C_n,$$

$$m(\mathscr{R}) \ni \lim_{n\to\infty} (C_n \setminus A) = \bigcup_{n=1}^{\infty} (C_n \setminus A) = \bigcup_{n=1}^{\infty} C_n \setminus A,$$

and, in view of *De Morgan's laws*,

$$m(\mathscr{R}) \ni \lim_{n\to\infty} (A \setminus C_n) = \bigcap_{n=1}^{\infty} (A \setminus C_n) = A \setminus \bigcup_{n=1}^{\infty} C_n.$$

Hence, by the definition of $\mathscr{L}(A)$ (see (2.5)),

$$\bigcup_{n=1}^{\infty} C_n \in \mathscr{L}(A).$$

The case of a decreasing sequence $(C_n)_{n\in\mathbb{N}}$ in $\mathscr{L}(A)$ is considered in the same fashion.

Exercise 2.16. Consider.

Thus, for each $A \in \mathscr{R}$, $\mathscr{L}(A)$ is a monotone class containing $\mathscr{R}$, and hence, we have the inclusion

$$m(\mathscr{R}) \subseteq \mathscr{L}(A),$$

which implies that, for any $A \in \mathscr{R}$ and $C \in m(\mathscr{R})$,

$$A \cup C, A \setminus C, C \setminus A \in m(\mathscr{R}).$$

Whence, we conclude that, for any $C \in m(\mathscr{R})$,

$$\mathscr{R} \subseteq \mathscr{L}(C).$$

With $\mathscr{L}(C)$ being a monotone class, the latter implies that, for each $C \in m(\mathscr{R})$,

$$m(\mathscr{R}) \subseteq \mathscr{L}(C).$$

Hence, for any $C_1, C_2 \in m(\mathscr{R})$,

$$C_1 \cup C_2, C_1 \setminus C_2, C_2 \setminus C_1 \in m(\mathscr{R}),$$

which proves that $m(\mathscr{R})$ is a *ring* on X indeed.

By the *Monotone Ring Theorem* (Theorem 2.1), we infer that $m(\mathscr{R})$ is a *σ-ring* on X, and hence,

$$\sigma r(\mathscr{R}) \subseteq m(\mathscr{R}). \tag{2.6}$$

Inclusions (2.4) and (2.6) imply that

$$\sigma r(\mathscr{R}) = m(\mathscr{R}),$$

which completes the proof. □

As an important particular case, we obtain the following.

Corollary 2.2. *Let $\mathscr{A}$ be an algebra on a nonempty set X. Then*

$$\sigma a(\mathscr{A}) = m(\mathscr{A}).$$

2.5.3 Borel Sets

Definition 2.8 (Borel σ-Algebra). Let (X, ρ) be a metric space and $\mathscr{G}$ be the *metric topology* generated by metric ρ (see Definition 1.20).

The *σ-algebra of Borel sets* or the *Borel σ-algebra* on X is the σ-algebra generated by the metric topology $\mathscr{G}$:[1]

$$\mathscr{B}(X) := \sigma a(\mathscr{G}).$$

Exercise 2.17.

(a) Let (X, ρ) be a metric space and $\mathscr{F}$ be the collection of all *closed* sets in (X, ρ). Prove that

1 Émile Borel (1871–1956).

(i) $\mathscr{B}(X) = \sigma a(\mathscr{F})$;
(ii) provided (X,ρ) is *separable*,

$$\mathscr{B}(X) = \sigma a\left(\{B(x,r) \mid x \in X,\ r > 0\}\right) = \sigma a\left(\{\overline{B}(x,r) \mid x \in X,\ r > 0\}\right)$$

(see *Open Sets in Separable Spaces Proposition* (Proposition 1.4));
(iii) each *countable set*, in particular each singleton, is a Borel set.

(b) Prove that:
(i) $\mathbb{N}, \mathbb{Z}, \mathbb{Q}, \mathbb{R} \setminus \mathbb{Q} \in \mathscr{B}(\mathbb{R})$;
(ii) $(a,b], [a,b) \in \mathscr{B}(\mathbb{R})$ $(-\infty < a < b < \infty)$;
(iii)

$$\begin{aligned}\mathscr{B}(\mathbb{R}) &= \sigma a\left(\{(a,b) \mid -\infty < a < b < \infty\}\right) = \sigma a\left(\{(a,b] \mid -\infty < a < b < \infty\}\right)\\ &= \sigma a\left(\{[a,b) \mid -\infty < a < b < \infty\}\right) = \sigma a\left(\{[a,b] \mid -\infty < a < b < \infty\}\right)\\ &= \sigma a\left(\{(-\infty,b) \mid b \in \mathbb{R}\}\right) = \sigma a\left(\{(a,\infty) \mid a \in \mathbb{R}\}\right)\\ &= \sigma a\left(\{(-\infty,b] \mid b \in \mathbb{R}\}\right) = \sigma a\left(\{[a,\infty) \mid a \in \mathbb{R}\}\right);\end{aligned}$$

(iv)

$$\begin{aligned}\mathscr{B}(\mathbb{R}) &= \sigma a\left(\{(a,b) \mid -\infty < a < b < \infty,\ a,b \in \mathbb{Q}\}\right)\\ &= \sigma a\left(\{(a,b] \mid -\infty < a < b < \infty,\ a,b \in \mathbb{Q}\}\right)\\ &= \sigma a\left(\{[a,b) \mid -\infty < a < b < \infty,\ a,b \in \mathbb{Q}\}\right)\\ &= \sigma a\left(\{[a,b] \mid -\infty < a < b < \infty,\ a,b \in \mathbb{Q}\}\right)\\ &= \sigma a\left(\{(-\infty,b) \mid b \in \mathbb{Q}\}\right) = \sigma a\left(\{(a,\infty) \mid a \in \mathbb{Q}\}\right)\\ &= \sigma a\left(\{(-\infty,b] \mid b \in \mathbb{Q}\}\right) = \sigma a\left(\{[a,\infty) \mid a \in \mathbb{Q}\}\right).\end{aligned}$$

Remark 2.7. If (X,ρ) is a *separable* metric space, $|\mathscr{B}(X)| \leq \mathfrak{c}$, in particular, $|\mathscr{B}(\mathbb{R}^n)| = \mathfrak{c}$ $(n \in \mathbb{N})$ (see, e. g., [3]). This immediately implies that there exist subsets of $\mathbb{R}^n$ $(n \in \mathbb{N})$ that are not Borel.

2.6 Problems

1. (a) Show that, for any $l > 0$,

$$\mathscr{S}_l := \{(a,b] \mid -\infty < a < b < \infty,\ b - a \leq l\} \cup \{\emptyset\}$$

is a semiring on $\mathbb{R}$.

(b) Show that

$$\mathscr{S}_{l_2} \subseteq \mathscr{S}_{l_1},\ 0 < l_2 < l_1.$$

(c) Find $\bigcap_{l>0} \mathscr{S}_l$.

2. Prove

Proposition 2.1 (Product Semiring). *Let $\mathscr{S}_i$ be a semiring on a nonempty set X_i, $i = 1, 2$. Then*

$$\mathscr{S}_1 \times \mathscr{S}_2 := \{A_1 \times A_2 \mid A_i \in \mathscr{S}_i,\ i = 1, 2\}$$

is a semiring on $X_1 \times X_2$.

3. Prove

Proposition 2.2 (Ring Characterization). *A nonempty collection $\mathscr{C}$ of subsets of a nonempty set X is a ring iff $\mathscr{C}$ is a semiring on X which is closed under finite unions.*

4. Show that each ring $\mathscr{R}$ is closed under symmetric differences, i. e., if $A, B \in \mathscr{R}$, then $A \triangle B \in \mathscr{R}$.
5. Let $\mathscr{R}_i$ be a ring on a nonempty set X_i, $i = 1, 2$. As follows from the *Product Semiring Proposition* (Proposition 2.1) (see Problem 2),

$$\mathscr{R}_1 \times \mathscr{R}_2 := \{A_1 \times A_2 \mid A_i \in \mathscr{S}_i,\ i = 1, 2\}$$

is a semiring on $X_1 \times X_2$. Give an example showing that $\mathscr{R}_1 \times \mathscr{R}_2$ need not be a ring.
6. If $\mathscr{R}$ is a σ-ring on a nonempty set X, then, for any set sequence $(A_n)_{n\in\mathbb{N}}$ in $\mathscr{R}$,
   (a) $\overline{\lim}_{n\to\infty} A_n \in \mathscr{R}$ and $\underline{\lim}_{n\to\infty} A_n \in \mathscr{R}$;
   (b) provided $\lim_{n\to\infty} A_n$ exists (see Definition 1.3), $\lim_{n\to\infty} A_n \in \mathscr{R}$.
7. Prove that the collection of all symmetric sets in $\mathbb{R}^2$:

$$\Sigma := \left\{A \subseteq \mathbb{R}^2 \,\middle|\, \forall\,(x, y) \in A :\ (-x, -y) \in A\right\} \cup \{\emptyset\}$$

is a *σ-algebra* on $\mathbb{R}^2$.
8. * Does there exist a *countably infinite* σ-algebra?
9. Let $\mathscr{C} := \{(-\infty, b] \mid b \in \mathbb{R}\}$.
   (a) Show that $\mathscr{C}$ is not a semiring on $\mathbb{R}$.
   (b) Show that

$$\mathscr{S} := \{(a, b] \mid -\infty \le a < b < \infty\} \cup \{\emptyset\}$$

   is a *semiring* on $\mathbb{R}$ containing $\mathscr{C}$ but *not* the smallest one.
   (c) Is there the smallest semiring on $\mathbb{R}$ containing $\mathscr{C}$ (see Problem 1)?
10. Let $\mathscr{S}$ be a *semiring* on a nonempty set X. Show that, without loss of generality, in a finite union

$$\bigcup_{i=1}^{n} A_i,\ n \in \mathbb{N},\ A_i \in \mathscr{S}, i = 1, \dots, n,$$

the sets A_i, $i = 1, \dots, n$, can be regarded to be *pairwise disjoint* (see the proof of the *Generated Ring Theorem* (Theorem 2.2) and Remarks 2.6).

Hint. Use induction.

11. Let $\mathscr{C}_1$ and $\mathscr{C}_2$ be nonempty collections of subsets of a nonempty set X.
    (a) Prove that, if $\mathscr{C}_1 \subseteq \mathscr{C}_2 \subseteq r(\mathscr{C}_1)$, then $r(\mathscr{C}_1) = r(\mathscr{C}_2)$.
    (b) Show that the analogous statements hold for the generated algebras, σ-rings, σ-algebras, and monotone classes.
12. Let $\mathscr{C}$ be a nonempty collection of subsets of a nonempty set X and an arbitrary set $B \subseteq X$ be fixed. Prove that
    (a) $r(\mathscr{C} \cap B) = r(\mathscr{C}) \cap B$, where

$$\mathscr{C} \cap B := \{C \cap B \mid C \in \mathscr{C}\};$$

    (b) $\sigma r(\mathscr{C} \cap B) = \sigma r(\mathscr{C}) \cap B$.

    Hint. For (a), show that $\mathscr{C} \cap B \subseteq r(\mathscr{C}) \cap B$ and $r(\mathscr{C}) \cap B$ is a ring on X.

13. Let $\mathscr{C} = \{A_1, \dots, A_n\}$ ($n \in \mathbb{N}$) be a collection of n subsets of a nonempty set X. Prove that
    (a) $a(\mathscr{C})$ consists of at most 2^{2^n} sets;
    (b) $\sigma a(\mathscr{C}) = a(\mathscr{C})$.

    Hint. For (a), consider all sets of the form

$$\hat{A}_1 \cap \dots \cap \hat{A}_n,$$

    where $\hat{A}_i = A_i$ or $\hat{A}_i = A_i^c$, $i = 1, \dots, n$.

3 Measures

The concept of *measure* of a set as an estimate of its size generalizes the familiar and intuitive notions of conventional length, area, and volume and is focal for our entire discourse.

3.1 Set Functions

Here, on a nonempty collection $\mathscr{C}$ of subsets of a nonempty set X, we consider set functions

$$\mu : \mathscr{C} \to \overline{R} := [-\infty, \infty]$$

subject to certain conditions.

Remarks 3.1. Before we proceed, we agree upon the following natural order and arithmetic rules involving infinity:

- for any $a \in \mathbb{R}$,

$$-\infty < a < \infty;$$

- for any $a \in \mathbb{R}$,

$$a + \infty := \infty =: \infty + a, \ a + (-\infty) := -\infty =: -\infty + a;$$

- the expressions $-\infty + \infty$ and $\infty + (-\infty)$ being indeterminate, we do not consider set functions that assume both $-\infty$ and ∞ values. Thus, either

$$\mu : \mathscr{C} \to (-\infty, \infty]$$

or

$$\mu : \mathscr{C} \to [-\infty, \infty).$$

Definition 3.1 (Various Properties of Set Functions). Let $\mathscr{C}$ be a nonempty collection of subsets of a nonempty set X. A function $\mu : \mathscr{C} \to (-\infty, \infty]$ is called:

(1) *nonnegative* if $\mu : \mathscr{C} \to [0, \infty]$, i. e., for each $A \in \mathscr{C}$, $\mu(A) \geq 0$;
(2) *finitely subadditive* (or *subadditive*), if for any $A_1, \ldots, A_n \in \mathscr{C}$ ($n \in \mathbb{N}$) with $\bigcup_{i=1}^n A_i \in \mathscr{C}$,

$$\mu\left(\bigcup_{i=1}^n A_i\right) \leq \sum_{i=1}^n \mu(A_i);$$

(3) *countably subadditive* (or *σ-subadditive*), if for any sequence $(A_n)_{n\in\mathbb{N}}$ in $\mathscr{C}$ with $\bigcup_{i=1}^\infty A_i \in \mathscr{C}$,

https://doi.org/10.1515/9783110600995-003

$$\mu\left(\bigcup_{i=1}^{\infty} A_i\right) \le \sum_{i=1}^{\infty} \mu(A_i);$$

(4) *finitely additive* (or *additive*), if for any pairwise disjoint $A_1, \dots, A_n \in \mathscr{C}$ $(n \in \mathbb{N})$ with $\bigcup_{i=1}^{n} A_i \in \mathscr{C}$,

$$\mu\left(\bigcup_{i=1}^{n} A_i\right) = \sum_{i=1}^{n} \mu(A_i);$$

(5) *countably additive* (or *σ-additive*), if for any pairwise disjoint sequence $(A_n)_{n\in\mathbb{N}}$ in $\mathscr{C}$ with $\bigcup_{i=1}^{\infty} A_i \in \mathscr{C}$,

$$\mu\left(\bigcup_{i=1}^{\infty} A_i\right) = \sum_{i=1}^{\infty} \mu(A_i);$$

(6) *monotone*, if for any $A, B \in \mathscr{C}$ with $A \subseteq B$, $\mu(A) \le \mu(B)$;
(7) *finite*, if for each $A \in \mathscr{C}$, $\mu(A) < \infty$;
(8) *countably finite* (or *σ-finite*), if there is a sequence $(A_n)_{n\in\mathbb{N}}$ in $\mathscr{C}$ with $\mu(A_n) < \infty$, $n \in \mathbb{N}$, such that

$$\bigcup_{i=1}^{\infty} A_i = X.$$

Example 3.1. The *length* μ is a *nonnegative*, *subadditive*, *additive*, *monotone*, and *finite* function on the semirings

$$\mathscr{S}_1 := \{(a,b] \mid -\infty < a < b < \infty\} \cup \{\emptyset\}$$

and

$$\mathscr{S}_2 := \{[a,b) \mid -\infty < a < b < \infty\} \cup \{\emptyset\}$$

(see Examples 2.1) with $\mu([a,b)) := b - a =: \mu((a,b])$, $\mu(\emptyset) := 0$ and is a *nonnegative*, *subadditive*, *additive*, *monotone*, and *σ-finite* function on the set collections

$$\mathscr{S}_1 \cup \{\mathbb{R}\} \text{ and } \mathscr{S}_2 \cup \{\mathbb{R}\}$$

with $\mu(\mathbb{R}) := \infty$.

Exercise 3.1.
(a) Verify.
(b) Let $\mathscr{C}$ be a nonempty collection of subsets of a nonempty set X and $\mu : \mathscr{C} \to (-\infty, \infty]$. Show that:
  (i) if $\emptyset \in \mathscr{C}$, $\mu(A) < \infty$ for at least one set $A \in \mathscr{C}$, and μ is *additive*, then $\mu(\emptyset) = 0$;
  (ii) if $\emptyset \in \mathscr{C}$, $\mu(A) < \infty$ for at least one set $A \in \mathscr{C}$, and μ is *σ-additive*, then $\mu(\emptyset) = 0$ and μ is *additive*.

3.2 Measure

3.2.1 Definition and Examples

Definition 3.2 (Measure). A *measure* is a nonnegative σ-additive function on a semiring.

Remark 3.2. In particular, a nonnegative σ-additive function on a ring, an algebra, a σ-ring, or a σ-algebra is a *measure* (see Remarks 2.2 and 2.3).

Examples 3.2.

1. Immediate trivial examples of a measure on an arbitrary semiring $\mathscr{S}$ on a nonempty set X are:
   (a) the *zero measure*
   $$\mathscr{S} \ni A \mapsto \mu(A) := 0,$$
   (b) the *infinite measure*
   $$\mathscr{S} \ni A \mapsto \mu(A) := \infty,$$
   (c) the *almost infinite measure*
   $$\mathscr{S} \ni A \mapsto \mu(A) := \begin{cases} \infty & \text{if } A \neq \emptyset, \\ 0 & \text{if } A = \emptyset. \end{cases}$$
2. For an arbitrary nonempty set X,
   (a)
   $$\mathscr{P}(X) \ni A \mapsto \mu(A) := \begin{cases} \text{number of elements in } A & \text{if } A \text{ is } \textit{finite}, \\ \infty & \text{if } A \text{ is } \textit{infinite} \end{cases}$$
   is a *measure* on $\mathscr{P}(X)$ called the *counting measure*;
   (b) for a fixed $x \in X$,
   $$\mathscr{P}(X) \ni A \mapsto \delta_x(A) := \begin{cases} 1 & \text{if } x \in A, \\ 0 & \text{if } x \notin A \end{cases}$$
   is a *measure* on $\mathscr{P}(X)$, called the *unit point mass measure at* x;
   (c) for fixed distinct $x_1, \dots, x_n \in X$ and $a_1, \dots, a_n \in [0, \infty)$ ($n \in \mathbb{N}$),
   $$\mathscr{P}(X) \ni A \mapsto \mu(A) := \sum_{i=1}^{n} a_i \delta_{x_i}(A) = \sum_{i \in \mathbb{N}: x_i \in A} a_i$$
   is a *measure* on $\mathscr{P}(X)$, which can be called a *mass distribution measure over the set* $\{x_1, \dots, x_n\}$;

(d) provided X is *infinite*, for *countably infinite* $\{x_n\}_{n\in\mathbb{N}} \subseteq X$ and a sequence $(a_n)_{n\in\mathbb{N}} \subset [0,\infty)$,

$$\mathscr{P}(X) \ni A \mapsto \mu(A) := \sum_{i=1}^{\infty} a_i \delta_{x_i}(A) = \sum_{i\in\mathbb{N}:\, x_i\in A} a_i,$$

is a *measure* on $\mathscr{P}(X)$, which can be called a *mass distribution measure over the set* $\{x_n\}_{n\in\mathbb{N}}$.

Exercise 3.2. Verify.

Remark 3.3. For $X := \mathbb{N}$, the *counting measure* on $\mathscr{P}(\mathbb{N})$ coincides with the *unit mass distribution measure* over $\mathbb{N}$:

$$\mathscr{P}(\mathbb{N}) \ni A \mapsto \mu(A) := \sum_{i=1}^{\infty} \delta_{x_i}(A) = \sum_{i\in\mathbb{N}:\, x_i\in A} a_i \text{ with } a_n = 1,\ n \in \mathbb{N}.$$

Exercise 3.3. Verify.

More meaningful examples of measures are forthcoming.

3.2.2 Properties of Measure

The subsequent major theorem establishes certain immediate properties of a measure on a ring.

Theorem 3.1 (Properties of Measure). *Let μ be a measure on a ring $\mathscr{R}$. Then:*
(1) *provided μ is not infinite, $\mu(\emptyset) = 0$;*
(2) *μ is additive on $\mathscr{R}$;*
(3) *μ is monotone on $\mathscr{R}$;*
(4) *if $A, B \in \mathscr{R}$ with $A \subseteq B$ and $\mu(A) < \infty$,*

$$\mu(B \setminus A) = \mu(B) - \mu(A);$$

(5) *if $A, B \in \mathscr{R}$ and $\mu(A) < \infty$ or $\mu(B) < \infty$,*

$$\mu(A \cup B) = \mu(A) + \mu(B) - \mu(A \cap B);$$

(6) *μ is subadditive;*
(7) *μ is σ-subadditive.*

Proof. Properties (1) and (2) (*additivity*) follow from a more general statement (see Exercise 3.1 (b)).

Property (3) (*monotonicity*) follows from the *additivity* and *nonnegativity* of μ. Indeed, for $A, B \in \mathscr{R}$ with $A \subseteq B$, since $\mathscr{R}$ is a ring,

$$B \setminus A \in \mathscr{R}$$

and, by the additivity and nonnegativity of μ,

$$\mu(B) = \mu(A \cup (B \setminus A)) = \mu(A) + \mu(B \setminus A) \geq \mu(A).$$

Let $A, B \in \mathscr{R}$ with $A \subseteq B$ and $\mu(A) < \infty$, by the additivity of μ,

$$\mu(B) = \mu(A) + \mu(B \setminus A),$$

and hence, the finite value $\mu(A)$ can be subtracted through, yielding

$$\mu(B \setminus A) = \mu(B) - \mu(A),$$

which shows that property (4) holds as well.

Let $A, B \in \mathscr{R}$ with $\mu(A) < \infty$ or $\mu(B) < \infty$. Considering that $\mathscr{R}$ is a ring,

$$B \setminus (A \cap B) \in \mathscr{R}.$$

Since

$$A \cup (B \setminus (A \cap B)) = A \cup B \text{ and } A \cap (B \setminus (A \cap B)) = \emptyset,$$

by the additivity of μ, we have

$$\mu(A \cup B) = \mu\left(A \cup (B \setminus (A \cap B))\right) = \mu(A) + \mu\left(B \setminus (A \cap B)\right). \tag{3.1}$$

By the *monotonicity* of μ,

$$\mu(A \cap B) \leq \min\left[\mu(A), \mu(B)\right] < \infty,$$

and hence, by (4), we infer that

$$\mu\left(B \setminus (A \cap B)\right) = \mu(B) - \mu(A \cap B). \tag{3.2}$$

From (3.1) and (3.2), we infer that

$$\mu(A \cup B) = \mu(A) + \mu(B) - \mu(A \cap B),$$

which completes the proof of property (5).

For any $A_1, \dots, A_n \in \mathscr{R}$ ($n \in \mathbb{N}$), let

$$B_i := A_i \setminus \bigcup_{j=1}^{i-1} A_j, \; i = 1, \dots, n,$$

with $\bigcup_{j=1}^{0} A_j := \emptyset$, i. e., $B_1 := A_1$.

Since $\mathscr{R}$ is a ring, $B_i \in \mathscr{R}$, $i = 1, \dots, n$.
Also,

$$\bigcup_{i=1}^{n} B_i = \bigcup_{i=1}^{n} A_i \in \mathscr{R}$$

and the sets B_i, $i = 1, \dots, n$, are *pairwise disjoint.*

Exercise 3.4. Verify.

Since

$$B_i \subseteq A_i, \; i = 1, \dots, n,$$

by the *monotonicity* of μ,

$$\mu(B_i) \le \mu(A_i), \; i = 1, \dots, n.$$

In view of this, by the *additivity* of μ,

$$\mu\left(\bigcup_{i=1}^{n} A_i\right) = \mu\left(\bigcup_{i=1}^{n} B_i\right) = \sum_{i=1}^{n} \mu(B_i) \le \sum_{i=1}^{n} \mu(A_i),$$

which completes the proof of property (6) (*subadditivity*).

For any sequence $(A_n)_{n\in\mathbb{N}}$ in $\mathscr{R}$ with $\bigcup_{i=1}^{\infty} A_i \in \mathscr{R}$, similarly to the proof of (6), we define the sequence

$$B_n := A_n \setminus \bigcup_{j=1}^{n-1} A_j \subseteq A_n, \; n \in \mathbb{N},$$

of pairwise disjoint sets in $\mathscr{R}$ such that

$$\bigcup_{i=1}^{\infty} B_i = \bigcup_{i=1}^{\infty} A_i \in \mathscr{R}$$

and by the *σ-additivity* and *monotonicity* of μ, we infer that

$$\mu\left(\bigcup_{i=1}^{\infty} A_i\right) = \mu\left(\bigcup_{i=1}^{\infty} B_i\right) = \sum_{i=1}^{\infty} \mu(B_i) \le \sum_{i=1}^{\infty} \mu(A_i).$$

Thus, property (7) (*σ-subadditivity*) holds as well. □

Remarks 3.4.

- In particular, the *Properties of Measure* (Theorem 3.1) hold for a measure on an algebra, a σ-ring, or a σ-algebra (see Remarks 2.2 and 2.3).

- As follows from Exercise 3.1 (b) and the proof of the prior theorem, properties (1) and (3)–(6) hold for a *noninfinite*, *nonnegative*, and *additive* function μ on a ring $\mathscr{R}$ and do not require from μ the property of σ-additivity. In particular, such a function μ is *monotone* and *subadditive*.
- By the property of *monotonicity*, the finiteness of a nonnegative additive function μ, in particular a measure, on an algebra $\mathscr{A}$ is equivalent to $\mu(X) < \infty$.

Exercise 3.5. Explain.

3.2.3 Continuity of Measure

The following is another important property of a measure on a ring.

Theorem 3.2 (Continuity of Measure). *Let μ be a measure on a ring $\mathscr{R}$. Then:*
(1) *for any increasing sequence $(A_n)_{n\in\mathbb{N}}$ in $\mathscr{R}$ with $\lim_{n\to\infty} A_n = \bigcup_{i=1}^{\infty} A_i \in \mathscr{R}$,*

$$\mu(\lim_{n\to\infty} A_n) = \lim_{n\to\infty} \mu(A_n) \quad \textit{(continuity from below)};$$

(2) *for any decreasing sequence $(A_n)_{n\in\mathbb{N}}$ in $\mathscr{R}$ with $\mu(A_1) < \infty$ and $\lim_{n\to\infty} A_n = \bigcap_{i=1}^{\infty} A_i \in \mathscr{R}$,*

$$\mu(\lim_{n\to\infty} A_n) = \lim_{n\to\infty} \mu(A_n) \quad \textit{(continuity from above)}.$$

Proof. Let $(A_n)_{n\in\mathbb{N}}$ be an increasing sequence in $\mathscr{R}$ with $\lim_{n\to\infty} A_n = \bigcup_{i=1}^{\infty} \in \mathscr{R}$.
There are two possibilities:
- $\exists N \in \mathbb{N} : \mu(A_N) = \infty$, or
- $\forall n \in \mathbb{N} : \mu(A_n) < \infty$.

Suppose that

$$\exists N \in \mathbb{N} : \mu(A_N) = \infty.$$

Then, since, for any $n \geq N$,

$$A_N \subseteq A_n \subseteq \bigcup_{i=1}^{\infty} A_i,$$

by the *monotonicity* of μ,

$$\mu(A_n) = \mu\left(\bigcup_{i=1}^{\infty} A_i\right) = \infty, \; n \geq N,$$

and hence,

$$\lim_{n\to\infty} \mu(A_n) = \infty = \mu\left(\bigcup_{i=1}^{\infty} A_i\right).$$

Now, suppose that

$$\forall n \in \mathbb{N} : \mu(A_n) < \infty. \tag{3.3}$$

The sequence

$$B_n := A_n \setminus A_{n-1} \in \mathscr{R},\ n \in \mathbb{N},$$

with $A_0 := \emptyset$ is *pairwise disjoint* and

$$\bigcup_{i=1}^{\infty} B_i = \bigcup_{i=1}^{\infty} A_i.$$

Exercise 3.6. Verify both statements.

Since

$$B_n \subseteq A_n,\ n \in \mathbb{N},$$

in view of (3.3), by the *Properties of Measure* (Theorem 3.1),

$$\mu(B_n) = \mu(A_n) - \mu(A_{n-1}),\ n \in \mathbb{N},$$

and hence, by the *σ-additivity* of μ,

$$\mu\left(\bigcup_{i=1}^{\infty} A_i\right) = \mu\left(\bigcup_{i=1}^{\infty} B_i\right) = \sum_{i=1}^{\infty} \mu(B_i) = \lim_{n\to\infty} \sum_{i=1}^{n} [\mu(A_n) - \mu(A_{n-1})] = \lim_{n\to\infty} \mu(A_n),$$

which completes the proof of the continuity from below.

Let $(A_n)_{n\in\mathbb{N}}$ be a decreasing sequence in $\mathscr{R}$ with $\mu(A_1) < \infty$ and $\lim_{n\to\infty} A_n = \bigcap_{i=1}^{\infty} A_i \in \mathscr{R}$.

Then $(A_1 \setminus A_n)_{n\in\mathbb{N}}$ is an *increasing* sequence in $\mathscr{R}$, and hence, by *De Morgan's laws* and the proven *continuity from below*,

$$\mu\left(A_1 \setminus \bigcap_{i=1}^{\infty} A_i\right) = \mu\left(\bigcup_{i=1}^{\infty} A_1 \setminus A_i\right) = \lim_{n\to\infty} \mu(A_1 \setminus A_n).$$

Where, in view of $\mu(A_1) < \infty$, by the *Properties of Measure* (Theorem 3.1),

$$\mu(A_1) - \mu\left(\bigcap_{i=1}^{\infty} A_i\right) = \lim_{n\to\infty} [\mu(A_1) - \mu(A_n)],\ n \in \mathbb{N},$$

which implies

$$\mu\left(\bigcap_{i=1}^{\infty} A_i\right) = \lim_{n\to\infty} \mu(A_n)$$

completing the proof of the continuity from above and of the theorem. □

Remarks 3.5.

- As is easily seen, the requirement of $\mu(A_1) < \infty$ in the *continuity of measure from above* can be replaced with a more general one:

$$\exists N \in \mathbb{N} : \ \mu(A_N) < \infty,$$

  i. e., the measure $\mu(A_n)$, $n \in \mathbb{N}$, is *eventually finite.*
- The foregoing requirement is essential and cannot be dropped.

Exercise 3.7. Explain and give a corresponding example.

3.2.4 More Examples of Measures

3.2.4.1 Jordan Measure

The *Jordan*[1] *measure* is an extension of the familiar notions of length, area, and volume to a lager collection of sets and to higher dimensions. For a set in $\mathbb{R}^n$ ($n \in \mathbb{N}$) to be Jordan measurable, i. e., such, to which Jordan measure can be meaningfully assigned, it must, in a sense, be "permissible," which, in particular, includes its being *bounded.* For instance, such subsets of $\mathbb{R}$ as $\mathbb{N}$, $\mathbb{Q}$, $[0, \infty)$, and $[0, 1] \cap \mathbb{Q}$ are not Jordan measurable. The *Lebesgue*[2] *measure* further extends the concept of measurability of a set in $\mathbb{R}^n$ to a still larger collection of sets making, in particular, all the foregoing sets to be Lebesgue measurable. As we see below, the Jordan measurable sets form a *ring* whereas the Lebesgue measurable sets constitute a *σ-algebra.*

Definition 3.3 (Partitions of $\mathbb{R}^n$). The *mth-order partition* of the n-space $\mathbb{R}^n$ ($m \in \mathbb{Z}_+$, $n \in \mathbb{N}$) is the following collection:

$$\pi_n^{(m)} := \left\{ Q^{(m)}(k_1, \dots, k_n) \,\middle|\, k_i \in \mathbb{Z},\ i = 1, \dots, n \right\}$$

of the pairwise disjoint *blocks*

$$Q^{(m)}(k_1, \dots, k_n) := \left\{ (x_1, \dots, x_n) \in \mathbb{R}^n \,\middle|\, k_i \in \mathbb{Z},\ \frac{k_i}{2^m} < x_i \le \frac{k_i + 1}{2^m},\ i = 1, \dots, n \right\}$$

with

$$\mathbb{R}^n = \bigcup_{k_i \in \mathbb{Z},\, i=1,\dots,n} Q^{(m)}(k_1, \dots, k_n).$$

Exercise 3.8. Describe the 0th-order and 1st-order partitions of $\mathbb{R}$ and $\mathbb{R}^2$.

Proposition 3.1 (Properties of Partitions). *Let $n \in \mathbb{N}$.*

(1) *For each $m \in \mathbb{Z}_+$, the $(m+1)$th-order partition of $\mathbb{R}^n$ is obtained from the blocks of the mth-order partition of $\mathbb{R}^n$ via partitioning the latter into 2^n congruent blocks;*

---

1 Camille Jordan (1838–1922).

2 Henri Lebesgue (1875–1941).

(2) *The diameter of each block of the mth-order partition of* $\mathbb{R}^n$ *in* $(\mathbb{R}^n, \rho_2)$ *is*

$$\sqrt{n}2^{-m}.$$

(3) *The mth-order partition of* $\mathbb{R}^n$ *naturally generates mth-order partitions of* $\mathbb{R}^k$ *for* $k = 1, \dots, n-1$.

(4) *Let A be a bounded set in* $(\mathbb{R}^n, \rho_2)$ $(n \in \mathbb{R})$. *Then, for each* $m \in \mathbb{Z}_+$, *the collection of blocks* $Q^{(m)}(k_1, \dots, k_n)$ *in the mth-order partition of* $\mathbb{R}^n$, *which have at least one point in common with A is finite.*

Exercise 3.9. Verify.

Definition 3.4 (Measure of a Partition Block). Let $m \in \mathbb{Z}_+$, $n \in \mathbb{N}$. The *measure* (or *volume*) of a partition block $Q := \prod_{i=1}^n (a_i, b_i] \in \pi_n^{(m)}$ is

$$\mu(Q) := \prod_{i=1}^{n} (b_i - a_i) = 2^{-mn}.$$

Remark 3.6. For $n = 1, 2, 3$, the prior definition is consistent with that of *length*, *area*, and *volume*, respectively.

Definition 3.5 (Measure of Finite Union of Partition Blocks). Let $m \in \mathbb{Z}_+$, $n \in \mathbb{N}$, and $Q_1, \dots, Q_k \in \pi_n^{(m)}$ $(k \in \mathbb{N})$ be distinct partition blocks. The *measure* of their union $\bigcup_{i=1}^k Q_i$ is naturally defined as follows:

$$\mu\left(\bigcup_{i=1}^{k} Q_i\right) := \sum_{i=1}^{k} \mu(Q_i).$$

In order to define Jordan measurability and measure for a *bounded* set A in $(\mathbb{R}^n, \rho_2)$, some groundwork is to be done first.

Let A be a *bounded* set in $(\mathbb{R}^n, \rho_2)$ $(n \in \mathbb{N})$. For each $m \in \mathbb{Z}_+$, let

$$A_{(m)} := \bigcup_{Q \in \pi_n^{(m)},\, Q \subseteq A} Q, \quad A^{(m)} := \bigcup_{Q \in \pi_n^{(m)},\, Q \cap A \neq \emptyset} Q,$$

and

$$\Delta A_{(m)} := A^{(m)} \setminus A_{(m)} = \bigcup_{Q \in \pi_n^{(m)},\, Q \cap A \neq \emptyset,\, Q \cap A^c \neq \emptyset} Q$$

(see Figure 3.1).

Remarks 3.7.

- If there does not exist a single block $Q \in \pi_n^{(m)}$ such that $Q \subseteq A$, e. g., when A is a singleton, $A_{(m)} := \emptyset$.
- If there does not exist a single block $Q \in \pi_n^{(m)}$ such that $Q \cap A \neq \emptyset$, which happens *iff* $A = \emptyset$, $A^{(m)} := \emptyset$.

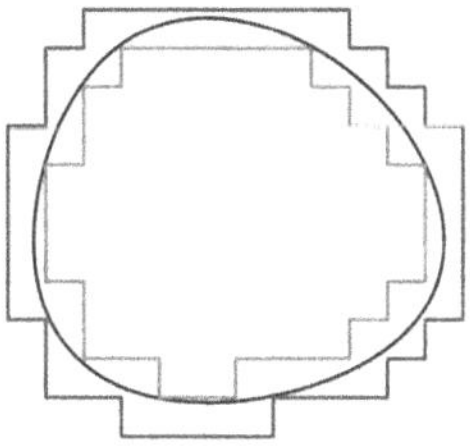

Figure 3.1: Geometric illustration for $n = 2$.

The measures of $A_{(m)}$, $A^{(m)}$, and $\Delta A_{(m)}$ are defined in the sense of Definition 3.5, with

$$\mu(\emptyset) := 0,$$

and, in view of

$$A_{(m)} \subseteq A^{(m)}, \; m \in \mathbb{Z}_+,$$

we have

$$\mu(A_{(m)}) \le \mu(A^{(m)}) \text{ and } \mu(\Delta A_{(m)}) = \mu(A^{(m)}) - \mu(A_{(m)}), \; m \in \mathbb{Z}_+.$$

As follows from the *Properties of Partitions* (Proposition 3.1 (1)) and the definition,

$$A_{(m)} \subseteq A_{(m+1)} \subseteq A^{(m+1)} \subseteq A^{(m)}, \; m \in \mathbb{Z}_+,$$

where

$$0 \le \mu(A_{(m)}) \le \mu(A_{(m+1)}) \le \mu(A^{(m+1)}) \le \mu(A^{(m)}), \; m \in \mathbb{Z}_+.$$

Hence, $(\mu(A_{(m)}))_{m\in\mathbb{Z}_+}$ is an *increasing sequence* bounded above by $\mu(A^{(0)})$ and $(\mu(A^{(m)}))_{m\in\mathbb{Z}_+}$ is a *decreasing sequence* bounded below by $\mu(A_{(0)})$, which, by the *Monotone Convergence Theorem*, implies that both sequences converge and

$$\lim_{m\to\infty} \mu(A_{(m)}) = \sup_{m\in\mathbb{Z}_+} \mu(A_{(m)}) \text{ and } \lim_{m\to\infty} \mu(A^{(m)}) = \inf_{m\in\mathbb{Z}_+} \mu(A^{(m)})$$

and makes the following concepts to be well-defined.

Definition 3.6 (Inner and Outer Measures of a Bounded Set). The *inner measure* of a bounded set A in $(\mathbb{R}^n, \rho_2)$ $(n \in \mathbb{N})$ is

$$\mu_*(A) := \lim_{m\to\infty} \mu(A_{(m)}) = \sup_{m\in\mathbb{Z}_+} \mu(A_{(m)}).$$

The *outer measure* of a bounded set A in $(\mathbb{R}^n, \rho_2)$ is

$$\mu^*(A) := \lim_{m\to\infty} \mu(A^{(m)}) = \inf_{m\in\mathbb{Z}_+} \mu(A^{(m)}).$$

Remark 3.8. $\mu_*(A) \le \mu^*(A)$.

Exercise 3.10. Explain.

Definition 3.7 (Jordan Measurability and Measure of a Bounded Set). A bounded set A in $(\mathbb{R}^n, \rho_2)$ $(n \in \mathbb{N})$ is called *Jordan measurable* if

$$\mu_*(A) = \mu^*(A),$$

in which case the number

$$\mu(A) := \mu_*(A) = \mu^*(A)$$

is called the *(n-dimensional) Jordan measure* of A.

Examples 3.3.

1. A partition block $Q \in \pi_n^{(m)}$ $(m \in \mathbb{Z}_+,\ n \in \mathbb{N})$ is Jordan measurable and its Jordan measure is precisely the measure in the sense of Definition 3.4.
2. A block $Q := \prod_{i=1}^n (a_i, b_i] \subset \mathbb{R}^n$ $(n \in \mathbb{N},\ -\infty < a_i < b_i < \infty,\ i = 1, \dots, n)$ is *Jordan measurable* and
$$\mu\left(\prod_{i=1}^n (a_i, b_i]\right) = \prod_{i=1}^n (b_i - a_i).$$
3. A *singleton* $\{x\}$ in $\mathbb{R}^n$ $(n \in \mathbb{N})$ is Jordan measurable and
$$\mu(\{x\}) = 0.$$
4. The set
$$T := \left\{(x_1, x_2) \in \mathbb{R}^2 \,\middle|\, x_1 \ge 0,\ x_2 \ge 0,\ x_1 + x_2 \le 1\right\}$$
is Jordan measurable and
$$\mu(T) = 1/2.$$
5. The sets
$$F_1 := [0, 1] \cap \mathbb{Q}$$
and
$$F_2 := [0, 1]^2 \cap \mathbb{Q}^2$$
are *not* Jordan measurable since
$$\mu_*(F_1) = \mu_*(F_2) = 0 \ne 1 = \mu^*(F_2) = \mu^*(F_1).$$
6. *Unbounded* sets in $(\mathbb{R}^n, \rho_2)$ $(n \in \mathbb{N})$ are *not* Jordan measurable.

Exercise 3.11. Verify 1–5 (for 2, the case of $n = 1$ only).

The following statement provides a characterization of Jordan measurability.

Proposition 3.2 (Characterization of Jordan Measurability). *A bounded set A in $(\mathbb{R}^n, \rho_2)$ $(n \in \mathbb{N})$ is Jordan measurable iff*

$$\lim_{m\to\infty} \mu(\Delta A_{(m)}) = 0.$$

Exercise 3.12. Prove.

Theorem 3.3 (Jordan Ring). *The collection $\mathscr{J}_n$ of all Jordan measurable sets in $\mathbb{R}^n$ $(n \in \mathbb{N})$ is a ring on $\mathbb{R}^n$.*

Proof. For any $A, B \in \mathscr{J}_n$. Let us show that, for each $m \in \mathbb{Z}_+$,

$$\Delta(A \cup B)_{(m)} \subseteq \Delta A_{(m)} \cup \Delta B_{(m)} \tag{3.4}$$

and

$$\Delta(A \setminus B)_{(m)} \subseteq \Delta A_{(m)} \cup \Delta B_{(m)}. \tag{3.5}$$

Indeed, for each $m \in \mathbb{Z}_+$ and any partition block $Q \in \pi_n^{(m)}$ with $Q \subseteq \Delta(A \cup B)_{(m)}$, there exist

$$x \in Q \cap (A \cup B) = (Q \cap A) \cup (Q \cap B)$$

and, in view of *De Morgan's laws*,

$$y \in Q \cap (A \cup B)^c = Q \cap (A^c \cap B^c) = (Q \cap A^c) \cap (Q \cap B^c).$$

If $x \in Q \cap A$, then since $y \in Q \cap A^c$,

$$Q \subseteq \Delta A_{(m)}.$$

If $x \in Q \cap B$, then since $y \in Q \cap B^c$,

$$Q \subseteq \Delta B_{(m)}.$$

Thus, in both cases,

$$Q \subseteq \Delta A_{(m)} \cup \Delta B_{(m)},$$

which proves inclusion (3.4).

Further, for each $m \in \mathbb{Z}_+$ and any partition block $Q \in \pi_n^{(m)}$ with $Q \subseteq \Delta(A \setminus B)_{(m)}$, there exist

$$x \in Q \cap (A \setminus B) = Q \cap (A \cap B^c) = (Q \cap A) \cap (Q \cap B^c)$$

and, in view of *De Morgan's laws*,

$$y \in Q \cap (A \setminus B)^c = Q \cap (A \cap B^c)^c = Q \cap (A^c \cup B) = (Q \cap A^c) \cup (Q \cap B).$$

If $y \in Q \cap A^c$, then since $x \in Q \cap A$,

$$Q \subseteq \Delta A_{(m)}.$$

If $y \in Q \cap B$, then since $x \in Q \cap B^c$,

$$Q \subseteq \Delta B_{(m)}.$$

Thus, in both cases,

$$Q \subseteq \Delta A_{(m)} \cup \Delta B_{(m)},$$

which proves inclusion (3.5).

By inclusions (3.4) and (3.5), for each $m \in \mathbb{Z}_+$, we have

$$0 \leq \mu(\Delta(A \cup B)_{(m)}) \leq \mu(\Delta A_{(m)}) + \mu(\Delta B_{(m)})$$

and

$$0 \leq \mu(\Delta(A \setminus B)_{(m)}) \leq \mu(\Delta A_{(m)}) + \mu(\Delta B_{(m)}).$$

Where by the *Characterization of Jordan Measurability* (Proposition 3.2) applied to A and B and the *Squeeze Theorem*, we infer that

$$\lim_{m\to\infty} \mu(\Delta(A \cup B)_{(m)}) = \lim_{m\to\infty} \mu(\Delta(A \setminus B)_{(m)}) = 0.$$

Now, applying the *Characterization of Jordan Measurability* (Proposition 3.2) to $A \cup B$ and $A \setminus B$, we conclude that

$$A \cup B, A \setminus B \in \mathscr{J}_n,$$

which completes the proof. □

Theorem 3.4 (Jordan Measure). *Let $n \in \mathbb{N}$. The set function*

$$\mathscr{J}_n \ni A \mapsto \mu(A) \tag{3.6}$$

is a measure (i. e., nonnegative and σ-additive) on the ring $\mathscr{J}_n$ of Jordan measurable sets in $\mathbb{R}^n$ called the Jordan measure.

Proof. The set function μ defined by (3.6) is, obviously, *nonnegative* and *finite* on $\mathscr{J}_n$ (see Definition 3.7).

It is also *subadditive*. Indeed, for arbitrary $A, B \in \mathscr{J}_n$ and any $m \in \mathbb{Z}_+$,

$$(A \cup B)_{(m)} \subseteq A \cup B \subseteq A^{(m)} \cup B^{(m)},$$

and hence,

$$\mu((A \cup B)_{(m)}) \le \mu(A^{(m)}) + \mu(B^{(m)}).$$

The latter, by the definition of the Jordan measure of a set (Definition 3.7), implies that

$$\mu(A \cup B) = \lim_{m\to\infty} \mu((A \cup B)_{(m)}) \le \lim_{m\to\infty} \mu(A^{(m)}) + \lim_{m\to\infty} \mu(B^{(m)}) = \mu(A) + \mu(B).$$

Now, to prove that μ is *additive*, it suffices to show that, for arbitrary *disjoint* $A, B \in \mathscr{J}_n$,

$$\mu(A \cup B) \ge \mu(A) + \mu(B).$$

Since, for any $m \in \mathbb{Z}_+$,

$$A_{(m)} \subseteq A,\ B_{(m)} \subseteq B, \text{ and } A_{(m)} \cup B_{(m)} \subseteq (A \cup B)_{(m)},$$

the disjointness of A and B implies *disjointness* for $A_{(m)}$ and $B_{(m)}$, $m \in \mathbb{Z}_+$, and hence,

$$\mu(A_{(m)}) + \mu(B_{(m)}) = \mu(A_{(m)} \cup B_{(m)}) \le \mu((A \cup B)_{(m)}),\ m \in \mathbb{Z}_+.$$

Where by the definition of the Jordan measure of a set (Definition 3.7),

$$\mu(A) + \mu(B) = \lim_{m\to\infty} \mu(A_{(m)}) + \lim_{m\to\infty} \mu(B_{(m)}) \le \lim_{m\to\infty} \mu((A \cup B)_{(m)}) = \mu(A \cup B).$$

Thus, μ is *additive* on $\mathscr{J}_n$.

Being a *finite*, *nonnegative*, and *additive* function on a ring, the Jordan measure possesses properties (1) and (3)–(6) of the *Properties of the Measure Theorem* (Theorem 3.1) (see Remarks 3.4), and hence, in particular, is *monotone*.

By the *Characterization of Measure on a Ring* (Proposition 3.3) (see Section 3.3, Problem 3) to prove the *σ-additivity* for μ, it suffices to show that μ is *σ-subadditive*.

Let $(A_k)_{k\in\mathbb{N}}$ be a sequence in $\mathscr{J}_n$ such that $\bigcup_{i=1}^{\infty} A_i \in \mathscr{J}_n$ and let $\varepsilon > 0$ be arbitrary. By the construct of the Jordan measure of a set, for each $k \in \mathbb{N}$, there exist a *closed* set $F_k \in \mathscr{J}_n$ and *open* set $G_k \in \mathscr{J}_n$ and such that

$$F_k \subseteq A_k \subseteq G_k \text{ and } \mu(G_k) - \mu(F_k) < \frac{\varepsilon}{2^k}$$

(see the *Approximation of Jordan Measurable Sets Proposition* (Proposition 3.6), Section 3.3, Problem 11), and hence, by the *monotonicity* of μ,

$$\mu(G_k) < \mu(F_k) + \frac{\varepsilon}{2^k} \le \mu(A_k) + \frac{\varepsilon}{2^k},\ k \in \mathbb{N}. \tag{3.7}$$

Similarly, there exist a *closed* set $F \in \mathscr{I}_n$ and *open* set $G \in \mathscr{I}_n$ and such that

$$F \subseteq \bigcup_{i=1}^{\infty} A_i \subseteq G \text{ and } \mu(G) - \mu(F) < \varepsilon,$$

and hence, by the *monotonicity* of μ,

$$\mu\left(\bigcup_{i=1}^{\infty} A_i\right) \leq \mu(G) < \mu(F) + \varepsilon. \tag{3.8}$$

Since

$$F \subseteq \bigcup_{i=1}^{\infty} A_i \subseteq \bigcup_{i=1}^{\infty} G_i,$$

the sets of the sequence $(G_n)_{n\in\mathbb{N}}$ form an *open cover* of the *closed* and *bounded*, and hence, by the *Heine–Borel Theorem* (Theorem 1.13), *compact* in $(\mathbb{R}^n, \rho_2)$ set F. Therefore, there is a *finite subcover*, i. e.,

$$\exists\, N \in \mathbb{N} : F \subseteq \bigcup_{k=1}^{N} G_k.$$

Where by the *monotonicity* and *subadditivity* of μ, we infer that

$$\mu(F) \leq \mu\left(\bigcup_{k=1}^{N} G_k\right) \leq \sum_{k=1}^{N} \mu(G_k),$$

which, in view of (3.7), implies

$$\mu(F) < \sum_{k=1}^{N}\left[\mu(A_k) + \frac{\varepsilon}{2^k}\right] < \sum_{k=1}^{\infty} \mu(A_k) + \sum_{k=1}^{\infty} \frac{\varepsilon}{2^k} = \sum_{k=1}^{\infty} \mu(A_k) + \varepsilon.$$

By (3.8),

$$\mu(A) < \mu(F) + \varepsilon < \sum_{k=1}^{\infty} \mu(A_k) + 2\varepsilon.$$

Since $\varepsilon > 0$ is arbitrary, passing to the limit as $\varepsilon \to 0+$, we arrive at

$$\mu(A) \leq \sum_{k=1}^{\infty} \mu(A_k),$$

and hence, μ is *σ-subadditive*.

Being *nonnegative*, *additive*, and *σ-subadditive* on the ring $\mathscr{I}_n$, by the *Characterization of Measure on a Ring* (Proposition 3.3) (see Section 3.3, Problem 3), μ is a *measure* on $\mathscr{I}_n$. □

Examples 3.4.

1. A *finite set* F in $\mathbb{R}^n$ ($n \in \mathbb{N}$) is Jordan measurable and
$$\mu(F) = 0.$$
2. A *countably infinite* set $C := \{x_n\}_{n\in\mathbb{N}}$ in $\mathbb{R}^n$ ($n \in \mathbb{N}$) such that there exists
$$\lim_{n\to\infty} x_n = x \in \mathbb{R}^n$$
is Jordan measurable and
$$\mu(C) = 0.$$
In particular, this applies to the set $\{1/n\}_{n\in\mathbb{N}}$.

Exercise 3.13.

(a) Verify.

(b) Give two examples (bounded and unbounded) showing that the countable union of Jordan measurable sets need not be Jordan measurable, i. e., that $\mathscr{J}_n$ is not a σ-ring.

Corollary 3.1. *Let*
$$\mathscr{S}_1 := \{(a,b] \mid -\infty < a < b < \infty\} \cup \{\emptyset\}.$$
Then
$$\mu(\emptyset) := 0, \ \mu((a,b]) := b - a, \ (a,b] \in \mathscr{S}_1,$$
is a measure on $\mathscr{S}_1$.

Corollary 3.2. *Let*
$$\mathscr{S}_2 := \{(a_1,b_1] \times (a_2,b_2] \mid -\infty < a_i < b_i < \infty, \ i = 1,2\} \cup \{\emptyset\}.$$
Then
$$\mu(\emptyset) := 0, \ \mu((a_1,b_1] \times (a_2,b_2]) := (b_1 - a_1)(b_2 - a_2), \ (a_1,b_1] \times (a_2,b_2] \in \mathscr{S}_2,$$
is a measure on $\mathscr{S}_2$.

Exercise 3.14. Prove the two corollaries (cf. Examples 3.3 and Example 3.1).

3.2.4.2 Lebesgue–Stieltjes Measures on Interval Semiring

Here, we introduce Lebesgue–Stieltjes[3] measures, for the time being, on the interval semiring
$$\mathscr{S}_1 := \{(a,b] \mid -\infty < a < b < \infty\} \cup \{\emptyset\}$$

3 Thomas Joannes Stieltjes (1856–1894).

(see Examples 2.1) as follows:

$$\lambda_F(\emptyset) := 0,\ \lambda_F((a,b]) := F(b) - F(a),\ (a,b] \in \mathscr{S}_1, \tag{3.9}$$

where a function $F : \mathbb{R} \to \mathbb{R}$ is *increasing*:

$$F(a) \le F(b),\ -\infty < a < b < \infty$$

and *right-continuous*:

$$F(x_0) = \lim_{x \to x_0+} F(x),\ x_0 \in \mathbb{R},$$

function.

Exercise 3.15. Give three examples of such a function F.

Theorem 3.5 (Lebesgue–Stieltjes Measures on Interval Semiring). *For an increasing and right-continuous function $F : \mathbb{R} \to \mathbb{R}$, the set function λ_F defined by (3.9) is a measure on the interval semiring $\mathscr{S}_1$ called the Lebesgue–Stieltjes measure on $\mathscr{S}_1$ associated with the function F.*

Proof. Since the function F is *increasing*, λ_F is *nonnegative*. As is easily verified, λ_F is also *additive* and *subadditive*.

Exercise 3.16. Verify.

To prove that λ_F is *σ-additive*, consider an arbitrary *pairwise disjoint* sequence $((a_n, b_n])_{n\in\mathbb{N}}$ in $\mathscr{S}_1$ such that

$$\bigcup_{i=1}^{\infty}(a_i, b_i] = (a,b] \in \mathscr{S}_1.$$

Since $\mathscr{S}_1$ is a semiring, it can be shown inductively that, for each $N \in \mathbb{N}$,

$$(a,b] \setminus \bigcup_{i=1}^{N}(a_i, b_i] = \bigcup_{k=1}^{l(N)} C_k^{(N)}$$

with some $l(N) \in \mathbb{N}$ and *pairwise disjoint* $C_k^{(N)} \in \mathscr{S}_1$, $k = 1, \dots, l(N)$.

Exercise 3.17. Show.

Where, by the *additivity* of λ_F, for each $N \in \mathbb{N}$,

$$\lambda_F((a,b]) = \sum_{i=1}^{N} \lambda_F((a_i, b_i]) + \sum_{k=1}^{l(N)} \lambda_F\left(C_k^{(N)}\right),$$

which implies that

$$\lambda_F((a,b]) \ge \sum_{i=1}^{N} \lambda_F((a_i, b_i]),\ N \in \mathbb{N},$$

where, passing to the limit as $N \to \infty$, we arrive at

$$\lambda_F((a,b]) \geq \sum_{i=1}^{\infty} \lambda_F((a_i,b_i]). \tag{3.10}$$

Since the function F is *right-continuous*,

$$\forall \varepsilon > 0 \; \exists a' \in (a,b): \; F(a') - F(a) < \varepsilon,$$

and hence,

$$\lambda_F((a,b]) - \lambda_F((a',b]) = F(b) - F(a) - [F(b) - F(a')] = F(a') - F(a) < \varepsilon; \tag{3.11}$$

also,

$$\forall n \in \mathbb{N} \; \exists b'_n > b_n: \; F(b'_n) - F(b_n) < \frac{\varepsilon}{2^n},$$

and hence, for each $N \in \mathbb{N}$,

$$\begin{aligned} \lambda_F((a_n,b'_n]) - \lambda_F((a_n,b_n]) &= F(b'_n) - F(a_n) - [F(b_n) - F(a_n)] \\ &= F(b'_n) - F(b_n) < \frac{\varepsilon}{2^n}. \end{aligned} \tag{3.12}$$

Since

$$[a',b] \subset (a,b] = \bigcup_{i=1}^{\infty}(a_i,b_i] \subseteq \bigcup_{i=1}^{\infty}(a_i,b'_i),$$

the collection $\{(a_n,b'_n)\}_{n\in\mathbb{N}}$ is an *open cover* of the *compact* in $\mathbb{R}$ set $[a',b]$, and hence, there is a *finite subcover* $\{(a_n,b'_n)\}_{n=1,\dots,N}$ ($N \in \mathbb{N}$), which implies that

$$(a',b] \subset [a',b] \subseteq \bigcup_{i=1}^{N}(a_i,b'_i) \subseteq \bigcup_{i=1}^{N}(a_i,b'_i]$$

(cf. the proof of Theorem 3.4).

By the *subadditivity* of λ_F,

$$\lambda_F((a',b]) \leq \sum_{i=1}^{N} \lambda_F((a_i,b'_i]) \leq \sum_{i=1}^{\infty} \lambda_F((a_i,b'_i]).$$

Where, by (3.11) and (3.12),

$$\lambda_F((a,b]) \leq \lambda_F((a',b]) + \varepsilon \leq \sum_{i=1}^{\infty}\left[\lambda_F((a_i,b_i]) + \frac{\varepsilon}{2^i}\right] + \varepsilon = \sum_{i=1}^{\infty} \lambda_F((a_i,b_i]) + 2\varepsilon.$$

Since $\varepsilon > 0$ is arbitrary, passing to the limit as $\varepsilon \to 0+$, we arrive at

$$\lambda_F((a,b]) \le \sum_{i=1}^{\infty} \lambda_F((a_i,b_i]). \tag{3.13}$$

Inequalities (3.10) and (3.13) jointly imply that

$$\lambda_F((a,b]) = \sum_{i=1}^{\infty} \lambda_F((a_i,b_i]),$$

which shows that λ_F is *σ-additive* on $\mathscr{S}_1$ and completes the proof. □

Remarks 3.9.

- Of special importance is the particular case of

$$F(x) := x, \ x \in \mathbb{R},$$

with the associated measure

$$\lambda(\emptyset) := 0, \ \lambda((a,b]) := b - a, \ (a,b] \in \mathscr{S}_1,$$

called the *Lebesgue measure* on the interval semiring $\mathscr{S}_1$, which coincides with the Jordan measure on $\mathscr{S}_1$ (see Corollary 3.1).

- One can similarly define the Lebesgue–Stieltjes measure λ_F for an *increasing left-continuous* function $F : \mathbb{R} \to \mathbb{R}$ on the interval semiring

$$\mathscr{S}_2 := \{[a,b) \mid -\infty < a < b < \infty\} \cup \{\emptyset\},$$

the Lebesgue measure λ on $\mathscr{S}_2$ being associated with $F(x) := x$, $x \in \mathbb{R}$.

3.3 Problems

1. Let μ_1 and μ_2 be measures on a semiring $\mathscr{S}$.
   (a) Prove that, for any $a_1, a_2 \ge 0$, $a_1\mu_1 + a_2\mu_2$ is a measure on $\mathscr{S}$.
   (b) Give an example showing that $\max(\mu_1, \mu_2)$ need not be a measure on $\mathscr{S}$.
2. Let μ be an additive finite function on a ring $\mathscr{R}$. Show that, for any $A_1, A_2, A_3 \in \mathscr{R}$,

$$\begin{aligned}\mu(A_1 \cup A_2 \cup A_3) &= \mu(A_1) + \mu(A_2) + \mu(A_3) \\ &\quad - \mu(A_1 \cap A_2) - \mu(A_1 \cap A_3) - \mu(A_2 \cap A_3) \\ &\quad + \mu(A_1 \cup A_2 \cup A_3).\end{aligned}$$

3. Prove

Proposition 3.3 (Characterization of Measure on a Ring). *A set function μ is a measure on a ring $\mathscr{R}$ iff μ is nonnegative, additive, and σ-subadditive on $\mathscr{R}$.*

Hint. First, show that, for a pairwise disjoint sequence $(A_n)_{n\in\mathbb{N}}$ in $\mathscr{R}$ with $\bigcup_{i=1}^{\infty} A_i \in \mathscr{R}$, by the *monotonicity* and *additivity* of μ, for each $n \in \mathbb{N}$,

$$\mu\left(\bigcup_{i=1}^{\infty} A_i\right) \geq \mu\left(\bigcup_{i=1}^{n} A_i\right) = \sum_{i=1}^{n} \mu(A_i).$$

4. Let μ be a measure on a ring $\mathscr{R}$. Show that, for a sequence $(A_n)_{n\in\mathbb{N}}$ in $\mathscr{R}$ with $\mu(A_n) = 0$, $n \in \mathbb{N}$, and $\bigcup_{i=1}^{\infty} A_i \in \mathscr{R}$,

$$\mu\left(\bigcup_{i=1}^{\infty} A_i\right) = 0.$$

5. Let μ be a measure on an algebra $\mathscr{A}$ with $\mu(X) < \infty$. Show that, for a sequence $(A_n)_{n\in\mathbb{N}}$ in $\mathscr{A}$ such that $\mu(A_n) = \mu(X)$, $n \in \mathbb{N}$, and $\bigcap_{i=1}^{\infty} A_i \in \mathscr{A}$,

$$\mu\left(\bigcap_{i=1}^{\infty} A_i\right) = \mu(X).$$

6. Let μ be a measure on an algebra $\mathscr{A}$ such that $\mu(X) = 1$. Show that, if, for $A_1, \dots, A_n \in \mathscr{A}$ $(n \in \mathbb{N})$,

$$\sum_{i=1}^{n} \mu(A_i) > n - 1,$$

then

$$\mu\left(\bigcap_{i=1}^{n} A_i\right) > 0.$$

Hint. Consider the complements A_i^c, $i = 1, \dots, n$.

7. Let μ be a measure on a σ-algebra Σ with $\mu(X) < \infty$ and a sequence $(A_n)_{n\in\mathbb{N}}$ in Σ be such that

$$\sum_{i=1}^{\infty} \mu(A_i) < \infty.$$

Prove that for

$$A := \left\{x \in X \,\middle|\, x \in A_n \text{ for } \textit{finitely many } n \in \mathbb{N} \text{ or } x \notin \bigcup_{i=1}^{\infty} A_i\right\},$$

$\mu(A) = \mu(X)$.

Hint. Show that

$$A^c := \overline{\lim_{n\to\infty}} A_n := \bigcap_{n=1}^{\infty} \bigcup_{k=n}^{\infty} A_k$$

and use the *monotonicity* and σ-*subadditivity* of μ.

8. Prove

   Proposition 3.4 (Sufficient Conditions of σ-Additivity). *A finite, nonnegative, additive, and continuous from above on $\emptyset$ function μ on a ring $\mathscr{R}$ is a measure on $\mathscr{R}$.*

9. Give an example of a measure μ on a ring $\mathscr{R}$, for which there exist *decreasing* sequences $(A_n)_{n\in\mathbb{N}}$ in $\mathscr{R}$ with $\mu(A_n) = \infty$, $n \in \mathbb{N}$, such that

$$\text{(a) } \mu\left(\bigcap_{i=1}^{\infty} A_i\right) = \infty, \quad \text{(b) } \mu\left(\bigcap_{i=1}^{\infty} A_i\right) = 0, \quad \text{(c) } 0 < \mu\left(\bigcap_{i=1}^{\infty} A_i\right) < \infty.$$

10. Prove

    Proposition 3.5 (Lower/Upper Limit). *Let μ be a measure on a σ-algebra Σ. Then, for an arbitrary sequence $(A_n)_{n\in\mathbb{N}}$ in Σ,*
    (1) $\mu(\underline{\lim}_{n\to\infty} A_n) \le \underline{\lim}_{n\to\infty} \mu(A_n)$;
    (2) *if* $\mu\left(\bigcup_{i=1}^{\infty} A_i\right) < \infty$, $\mu(\overline{\lim}_{n\to\infty} A_n) \ge \overline{\lim}_{n\to\infty} \mu(A_n)$.

    and the following.

    Corollary 3.3. *Let μ be a measure on a σ-algebra Σ. If, for a sequence $(A_n)_{n\in\mathbb{N}}$ in Σ, there exists $\lim_{n\to\infty} A_n$ with $\mu\left(\bigcup_{i=1}^{\infty} A_i\right) < \infty$, then*

$$\mu(\lim_{n\to\infty} A_n) = \lim_{n\to\infty} \mu(A_n).$$

11. Prove

    Proposition 3.6 (Approximation of Jordan Measurable Sets). *For an arbitrary $A \in \mathscr{J}_n$ $(n \in \mathbb{N})$ and any $\varepsilon > 0$, there exist a closed set $F \in \mathscr{J}_n$ and open set $G \in \mathscr{J}_n$ such that*

$$F \subseteq A \subseteq G \text{ and } \mu(G) - \mu(F) < \varepsilon.$$

12. Prove that the *Cantor set* C and all its subsets are Jordan measurable and

$$\mu(C) = 0.$$

4 Extension of Measures

In this chapter, we introduce the concept of *outer measure* and prove a number of fundamental statements, including *Carathéodory's*[1] *Extension Theorem* (Theorem 4.5), on the procedure of measure extension central for measure theory.

4.1 Extension of a Set Function

Definition 4.1 (Extension/Restriction of a Set Function). Let $\mathscr{C}_i$, $i = 1, 2$, be a nonempty collections of subsets of a nonempty set X and $\mu_i : \mathscr{C}_i \to \overline{\mathbb{R}}$, $i = 1, 2$, be a set function. The function μ_2 is called an *extension* of μ_1 (μ_1 is called a *restriction* of μ_2) if

$$\mathscr{C}_1 \subseteq \mathscr{C}_2 \text{ and } \mu_1(A) = \mu_2(A), \ A \in \mathscr{C}_1.$$

Remarks 4.1. As is easily seen,
- the restriction of a measure μ from a semiring $\mathscr{S}_2$ to a semiring $\mathscr{S}_1 \subseteq \mathscr{S}_2$ is a measure on $\mathscr{S}_1$;
- the relation of restriction/extension is a *partial order*.

4.2 Extension From a Semiring

Our first step is to furnish a straightforward general procedure of extending a measure from a semiring to its generated ring, which is described in the proof of the following theorem.

Theorem 4.1 (Measure Extension From a Semiring). *Let μ be a measure on a semiring $\mathscr{S}$. Then there exists a unique extension of μ to a measure $\overline{\mu}$ on the generated ring $r(\mathscr{S})$, the extension $\overline{\mu}$ being finite/σ-finite provided the measure μ is finite/σ-finite, respectively.*

Proof. By the *Generated Ring Theorem* (Theorem 2.3) (see Remarks 2.6), for an arbitrary $A \in r(\mathscr{S})$,

$$A = \bigcup_{i=1}^{n} A_i$$

with some $n \in \mathbb{N}$ and *pairwise disjoint* $A_i \in \mathscr{S}$, $i = 1, \ldots, n$, let us set

$$\overline{\mu}(A) := \sum_{i=1}^{n} \mu(A_i). \tag{4.1}$$

1 Constantin Carathéodory (1873–1950).

https://doi.org/10.1515/9783110600995-004

First, we are to show that the set function $\overline{\mu}$ is *well-defined* on $r(\mathscr{S})$, i. e., is independent of the representation of $A \in r(\mathscr{S})$ as a finite union of pairwise disjoint sets from $\mathscr{S}$. Indeed, suppose that

$$A = \bigcup_{j=1}^{m} B_j$$

with some $m \in \mathbb{N}$ and *pairwise disjoint* $B_j \in \mathscr{S}$, $j = 1, \dots, m$.

Then, for each $i = 1, \dots, n$,

$$A_i = A_i \cap A = A_i \cap \left(\bigcup_{j=1}^{m} B_j \right) = \bigcup_{j=1}^{m} (A_i \cap B_j)$$

and, for each $j = 1, \dots, m$,

$$B_j = A \cap B_j = \left(\bigcup_{i=1}^{n} A_i \right) \cap B_j = \bigcup_{i=1}^{n} (A_i \cap B_j),$$

the sets $A_i \cap B_j$, $i = 1, \dots, n$, $j = 1, \dots, m$, being *pairwise disjoint*.

Hence, by the *additivity* of μ on $\mathscr{S}$,

$$\begin{aligned}\sum_{i=1}^{n} \mu(A_i) &= \sum_{i=1}^{n} \mu\left(\bigcup_{j=1}^{m} (A_i \cap B_j) \right) = \sum_{i=1}^{n} \sum_{j=1}^{m} \mu(A_i \cap B_j) = \sum_{j=1}^{m} \sum_{i=1}^{n} \mu(A_i \cap B_j) \\ &= \sum_{j=1}^{m} \mu\left(\bigcup_{i=1}^{n} (A_i \cap B_j) \right) = \sum_{j=1}^{m} \mu(B_j),\end{aligned}$$

which confirms that $\overline{\mu}$ is well-defined on $r(\mathscr{S})$.

By the definition of $\overline{\mu}$ (see (4.1)), for each $A \in \mathscr{S}$,

$$\overline{\mu}(A) = \mu(A),$$

i. e., $\overline{\mu}$ is an extension of μ from $\mathscr{S}$ to $r(\mathscr{S})$.

As can be easily deduced from the definition (see (4.1)), the set function $\overline{\mu}$ is *nonnegative* and *additive* on $r(\mathscr{S})$.

Exercise 4.1. Verify.

Now, let us show that $\overline{\mu}$ is the *unique additive extension* of μ from $\mathscr{S}$ to $r(\mathscr{S})$. Indeed, let λ be an additive extension of μ from $\mathscr{S}$ to $r(\mathscr{S})$. Then, for each $A \in r(\mathscr{S})$, by the *Generated Ring Theorem* (Theorem 2.3),

$$A = \bigcup_{i=1}^{n} A_i$$

with some $n \in \mathbb{N}$ and *pairwise disjoint* $A_i \in \mathscr{S}$, $i = 1, \ldots, n$ $(n \in \mathbb{N})$. Hence, by the additivity of λ,

$$\lambda(A) = \sum_{i=1}^{n} \lambda(A_i) = \sum_{i=1}^{n} \mu(A_i) = \overline{\mu}(A).$$

It remains to show that $\overline{\mu}$ is a measure on $r(\mathscr{S})$, i. e., is *σ-additive* on $r(\mathscr{S})$.

Let $(A_n)_{n\in\mathbb{N}}$ be a pairwise disjoint sequence in $r(\mathscr{S})$ with $\bigcup_{i=1}^{\infty} A_i \in r(\mathscr{S})$. Then, by the *Generated Ring Theorem* (Theorem 2.3),

$$\bigcup_{i=1}^{\infty} A_i = \bigcup_{j=1}^{m} B_j$$

with some $m \in \mathbb{N}$ and *pairwise disjoint* $B_j \in \mathscr{S}$, $j = 1, \ldots, m$ and, for each $n \in \mathbb{N}$,

$$A_n = \bigcup_{k=1}^{l(n)} C_k^{(n)}$$

with some $l(n) \in \mathbb{N}$ and *pairwise disjoint* $C_k^{(n)} \in \mathscr{S}$, $k = 1, \ldots, l(n)$.

Hence,

$$\overline{\mu}\left(\bigcup_{i=1}^{\infty} A_i\right) = \overline{\mu}\left(\bigcup_{j=1}^{m} B_j\right) \qquad \text{by the } \textit{additivity} \text{ of } \overline{\mu} \text{ on } r(\mathscr{S});$$

$$= \sum_{j=1}^{m} \overline{\mu}(B_j) = \sum_{j=1}^{m} \mu(B_j) = \sum_{j=1}^{m} \mu\left(B_j \cap \bigcup_{i=1}^{\infty} A_i\right) = \sum_{j=1}^{m} \mu\left(B_j \cap \bigcup_{i=1}^{\infty} \bigcup_{k=1}^{l(i)} C_k^{(i)}\right)$$

$$= \sum_{j=1}^{m} \mu\left(\bigcup_{i=1}^{\infty} \bigcup_{k=1}^{l(i)} (B_j \cap C_k^{(i)})\right) \qquad \text{by the } \textit{σ-additivity} \text{ of } \mu \text{ on } \mathscr{S};$$

$$= \sum_{j=1}^{m} \sum_{i=1}^{\infty} \sum_{k=1}^{l(i)} \mu(B_j \cap C_k^{(i)}) = \sum_{i=1}^{\infty} \sum_{j=1}^{m} \sum_{k=1}^{l(i)} \overline{\mu}(B_j \cap C_k^{(i)})$$

by the *additivity* of $\overline{\mu}$ on $r(\mathscr{S})$;

$$= \sum_{i=1}^{\infty} \sum_{j=1}^{m} \overline{\mu}(B_j \cap A_i) = \sum_{i=1}^{\infty} \overline{\mu}(A_i).$$

Thus, $\overline{\mu}$ is a *measure* on $r(\mathscr{S})$.

The part about the *finiteness/σ-finiteness* of $\overline{\mu}$ is trivial.

Exercise 4.2. Explain. □

Example 4.1. The procedure described in the proof of the *Measure Extension from a Semiring Theorem* (Theorem 4.1) naturally applies to obtain the *unique finite* and *σ-finite* extension of the finite and σ-finite *Lebesgue–Stieltjes measure*

$$\lambda_F(\emptyset) := 0, \ \lambda_F((a, b]) := F(b) - F(a), \ -\infty < a < b < \infty,$$

associated with an increasing right-continuous function $F : \mathbb{R} \to \mathbb{R}$, from the interval semiring

$$\mathscr{S}_1 := \{(a, b] \mid -\infty < a < b < \infty\} \cup \{\emptyset\}$$

(see Theorem 3.5) to the generated ring $r(\mathscr{S}_1)$.

In particular, for $F(x) = x$, $x \in \mathbb{R}$, we obtain the unique *finite* and *σ-finite* extension of the *Lebesgue measure* (i. e., the *length measure*)

$$\lambda(\emptyset) := 0, \ \lambda((a, b]) := b - a, \ -\infty < a < b < \infty,$$

from $\mathscr{S}_1$ to $r(\mathscr{S}_1)$.

4.3 Outer Measure

The general concept of *outer measure* is pivotal for the procedure of measure extension.

4.3.1 Definition and Examples

Definition 4.2 (Outer Measure). Let X be a nonempty set. An *outer measure* is a set function

$$\mathscr{P}(X) \ni A \mapsto \mu^*(A) \mapsto \overline{\mathbb{R}}$$

such that:
(1) μ^* is *nonnegative* and $\mu^*(\emptyset) = 0$;
(2) μ^* is *monotone*;
(3) μ^* is *σ-subadditive.*

Remarks 4.2.
- An outer measure is defined on the power set $\mathscr{P}(X)$, i. e., assigns a nonnegative value to any subset of X.
- An outer measure is *subadditive.*

Exercise 4.3. Verify the latter.

Examples 4.2. Let X be an arbitrary nonempty set.
1. By the *Properties of Measure* (Theorem 3.1), any *noninfinite measure* on $\mathscr{P}(X)$ is an outer measure, in particular, the following (see Examples 3.2):
   (a) the *zero measure*

$$\mathscr{P}(X) \ni A \mapsto \mu(A) := 0,$$

(b) the *almost infinite measure*

$$\mathscr{P}(X) \ni A \mapsto \mu(A) := \begin{cases} \infty & \text{if } A \neq \emptyset, \\ 0 & \text{if } A = \emptyset, \end{cases}$$

(c) the *counting measure*

$$\mathscr{P}(X) \ni A \mapsto \mu(A) := \begin{cases} \text{number of elements in } A & \text{if } A \text{ is } \textit{finite}, \\ \infty & \text{if } A \text{ is } \textit{infinite}; \end{cases}$$

(d) the *unit point mass measure* at a point $x \in X$

$$\mathscr{P}(X) \ni A \mapsto \delta_x(A) := \begin{cases} 1 & \text{if } x \in A, \\ 0 & \text{if } x \notin A, \end{cases}$$

(e) a *mass distribution measure* over a finite set $\{x_1, \dots, x_n\} \subseteq X$

$$\mathscr{P}(X) \ni A \mapsto \mu(A) := \sum_{i=1}^{n} a_i \delta_{x_i}(A) = \sum_{i \in \mathbb{N}:\, x_i \in A} a_i,$$

where $a_1, \dots, a_n \in [0, \infty)$ $(n \in \mathbb{N})$,

(f) provided X is *infinite*, a *mass distribution measure* over a countably infinite set $\{x_n\}_{n \in \mathbb{N}} \subseteq X$

$$\mathscr{P}(X) \ni A \mapsto \mu(A) := \sum_{i=1}^{\infty} a_i \delta_{x_i}(A) = \sum_{i \in \mathbb{N}:\, x_i \in A} a_i,$$

where $(a_n)_{n \in \mathbb{N}} \subset [0, \infty)$.

2. The set function

$$\mathscr{P}(X) \ni A \mapsto \mu^*(A) := \begin{cases} a & \text{if } A \neq \emptyset, \\ 0 & \text{if } A = \emptyset, \end{cases}$$

where $a > 0$ is fixed, is an outer measure on $\mathscr{P}(X)$ but not a measure provided X has at least two elements.

Exercise 4.4.
(a) Explain 1.
(b) Verify 2.

4.3.2 Construction of Outer Measures

The following statement describes a general procedure of constructing an outer measure.

Proposition 4.1 (Construction of Outer Measure). *Let $\mathscr{C}$ be a nonempty collection of subsets of a nonempty set X containing ∅ and*

$$\mu : \mathscr{C} \to [0, \infty] \text{ with } \mu(\emptyset) = 0.$$

Then the set function

$$\mathscr{P}(X) \ni A \mapsto \mu^*(A) := \begin{cases} \inf\{\sum_{i=1}^{\infty} \mu(A_i) \mid (A_n)_{n\in\mathbb{N}} \text{ in } \mathscr{C},\ A \subseteq \bigcup_{i=1}^{\infty} A_i\} & \text{if such sequences exist,} \\ \infty & \text{otherwise} \end{cases} \tag{4.2}$$

is an outer measure.

Proof. The fact that the set function μ^* defined by (4.2) satisfies conditions (1) and (2) of the definition of outer measure (Definition 4.2) is easily verified.

Exercise 4.5. Verify.

It remains to show that μ^* is *σ-subadditive*.

For an arbitrary sequence $(A_n)_{n\in\mathbb{N}}$ in $\mathscr{P}(X)$, there are two possibilities:
- $\exists N \in \mathbb{N} : \mu^*(A_N) = \infty$, or
- $\forall n \in \mathbb{N} : \mu^*(A_n) < \infty$.

Suppose that

$$\exists N \in \mathbb{N} : \mu^*(A_N) = \infty.$$

Then, since

$$A_N \subseteq \bigcup_{i=1}^{\infty} A_i,$$

by the *monotonicity* of μ^*,

$$\mu^*\left(\bigcup_{i=1}^{\infty} A_i\right) = \infty = \sum_{n=1}^{\infty} \mu^*(A_n).$$

Now, suppose that

$$\forall n \in \mathbb{N} : \mu^*(A_n) < \infty.$$

Then, by the definition of μ^* (see (4.2)),

$$\forall n \in \mathbb{N}\ \forall \varepsilon > 0\ \exists \left(A_j^{(n)}\right)_{j\in\mathbb{N}} \text{ in } \mathscr{C} \text{ with } A_n \subseteq \bigcup_{j=1}^{\infty} A_j^{(n)} : \sum_{j=1}^{\infty} \mu\left(A_j^{(n)}\right) < \mu^*(A_n) + \frac{\varepsilon}{2^n}.$$

Where, in view of the inclusion

$$\bigcup_{n=1}^{\infty}\bigcup_{j=1}^{\infty} A_j^{(n)} \supseteq \bigcup_{n=1}^{\infty} A_n,$$

by the definition of μ^*, we have

$$\mu^*\left(\bigcup_{n=1}^{\infty} A_n\right) \leq \sum_{n=1}^{\infty}\sum_{j=1}^{\infty} \mu(A_{nj}) < \sum_{n=1}^{\infty}\left[\mu^*(A_n) + \frac{\varepsilon}{2^n}\right] = \sum_{n=1}^{\infty} \mu^*(A_n) + \varepsilon.$$

Since $\varepsilon > 0$ is arbitrary, passing to the limit as $\varepsilon \to 0+$, we arrive at

$$\mu^*\left(\bigcup_{n=1}^{\infty} A_n\right) \leq \sum_{n=1}^{\infty} \mu^*(A_n),$$

which completes the proof for the second possibility, and thus, of the entire statement. □

The *Construction of Outer Measure Proposition* (Proposition 4.1) makes the following natural definition possible.

Definition 4.3 (Generated Outer Measure). Let $\mathscr{C}$ be a nonempty collection of subsets of a nonempty set X containing $\emptyset$ and

$$\mu : \mathscr{C} \to [0, \infty] \text{ with } \mu(\emptyset) = 0.$$

The outer measure μ^* defined by (4.2) is called the *outer measure generated by the set function* μ.

Remarks 4.3.
- Under the conditions of the *Construction of Outer Measure Proposition* (Proposition 4.1), for each $A \in \mathscr{C}$,

$$\mu^*(A) \leq \mu(A). \tag{4.3}$$

- As the subsequent example shows, for some $A \in \mathscr{C}$, inequality (4.3) may be strict, and thus, the generated outer measure μ^* need not be an extension of μ (cf. *Carathéodory's Extension Theorem* (Theorem 4.5)).

Exercise 4.6. Verify inequality (4.3).

Examples 4.3.
1. For $X := \{0, 1, 2\}$, the collection

$$\mathscr{C} := \{\emptyset, \{0, 1\}, \{0, 2\}, \{1, 2\}\},$$

and the set function on $\mathscr{C}$ defined as follows:

$$\mu(\emptyset) := 0,\ \mu(\{0,1\}) := 1,\ \mu(\{0,2\}) := 1,\ \mu(\{1,2\}) := 3$$

and satisfying the conditions of the *Construction of Outer Measure Proposition* (Proposition 4.1), the generated outer measure on $\mathscr{P}(X)$ is

$$\begin{aligned}
&\mu^*(\emptyset) = \mu(\emptyset) = 0,\\
&\mu^*(\{0\}) = \mu^*(\{1\}) = \mu^*(\{2\}) = 1,\\
&\mu^*(\{0,1\}) = \mu(\{0,1\}) = 1,\\
&\mu^*(\{0,2\}) = \mu(\{0,2\}) = 1,\\
&\mu^*(\{1,2\}) = 2 < 3 = \mu(\{1,2\}),\\
&\mu^*(X) = 2.
\end{aligned}$$

2. For $X := \mathbb{R}$, the *semiring*

$$\mathscr{S} := \{(n, n+1] \mid n \in \mathbb{Z}\} \cup \{\emptyset\},$$

the length function on $\mathscr{S}$

$$\lambda(\emptyset) := 0,\ \lambda((n, n+1]) := 1,\ n \in \mathbb{Z},$$

satisfying the conditions of the *Construction of Outer Measure Proposition* (Proposition 4.1), and the generated outer measure λ^* on $\mathscr{P}(X)$, we have:

$$\begin{aligned}
&\lambda^*((n, n+1]) = \lambda((n, n+1]) = 1,\ n \in \mathbb{Z},\\
&\lambda^*((1/2, 1]) = \lambda^*([1/2, 1]) = \lambda^*(\{1/2\}) = 1,\\
&\lambda^*(\{1,2\}) = 2,\\
&\lambda^*(\mathbb{N}) = \lambda^*(X) = \infty.
\end{aligned}$$

Exercise 4.7. Verify (cf. Section 4.8, Problem 2).

4.4 μ^*-Measurable Sets, Carathéodory's Theorem

In this section, we define the important notion of μ^*-measurability of a set relative to an outer measure μ^* and prove the collection Σ^* of all such subsets of a nonempty set X is a σ-algebra on X and the restriction μ of μ^* to Σ^* is a measure (see *Carathéodory's Theorem* (Theorems 4.2 and 4.3)).

4.4.1 μ^*-Measurable Sets

Definition 4.4 (μ^*-Measurable Set). Let X be a nonempty set and μ^* be an outer measure on $\mathscr{P}(X)$. A set $A \in \mathscr{P}(X)$ is called μ^*-measurable if, for each $B \in \mathscr{P}(X)$,

$$\mu^*(B) = \mu^*(B \cap A) + \mu^*(B \setminus A) = \mu^*(B \cap A) + \mu^*(B \cap A^c). \tag{4.4}$$

Remarks 4.4.

- The sets $\emptyset$ and X are μ^*-measurable relative to any outer measure μ^*.

  Exercise 4.8. Verify.

- Due to the symmetry of the definition relative to the operation of complement ($(A^c)^c = A$), a set A and its complement A^c are μ^*-measurable or not simultaneously.

  Exercise 4.9. Explain.

- When checking the μ^*-measurability of a set $A \in \mathscr{P}(X)$, one need not verify (4.4) for sets $B \in \mathscr{P}(X)$ *disjoint from* A ($B \cap A = \emptyset$ and $B \cap A^c = B$) or for sets $B \in \mathscr{P}(X)$ that are *subsets* of A ($B \cap A = B$ and $B \cap A^c = \emptyset$).
- Since, for any $A, B \in \mathscr{P}(X)$,

$$B = (B \cap A) \cup (B \setminus A) = (B \cap A) \cup (B \cap A^c),$$

  by the *subadditivity* of μ^* (see Remarks 4.2), we always have the inequality

$$\mu^*(B) \le \mu^*(B \cap A) + \mu^*(B \setminus A) = \mu^*(B \cap A) + \mu^*(B \cap A^c).$$

  Hence, to prove that a set $A \in \mathscr{P}(X)$ is μ^*-measurable, it suffices to show that, for each $B \in \mathscr{P}(X)$,

$$\mu^*(B) \ge \mu^*(B \cap A) + \mu^*(B \setminus A) = \mu^*(B \cap A) + \mu^*(B \cap A^c).$$

  Thus, when checking the μ^*-measurability of a set $A \in \mathscr{P}(X)$, one need not verify the latter for sets $B \in \mathscr{P}(X)$ with $\mu^*(B) = \infty$. Also, by the *monotonicity* of μ^*, this shows that every set $A \in \mathscr{P}(X)$ with $\mu^*(A) = 0$ (called a μ^*-*null set*) is μ^*-measurable.

  Exercise 4.10. Explain.

Examples 4.4. Let X be an arbitrary nonempty set.

1. If μ is a *noninfinite measure* on $\mathscr{P}(X)$, and hence, μ is an outer measure (see Examples 4.2), all sets in $\mathscr{P}(X)$ are μ-measurable.
2. Provided X has at least two elements, relative to the outer

$$\mathscr{P}(X) \ni A \mapsto \mu^*(A) := \begin{cases} a & \text{if } A \ne \emptyset, \\ 0 & \text{if } A = \emptyset, \end{cases}$$

   where $a > 0$ is fixed, only $\emptyset$ and X are μ^*-measurable.
3. Let $X := \{0, 1, 2\}$. For the collection

$$\mathscr{C} := \{\emptyset, \{0, 1\}, \{0, 2\}, \{1, 2\}\}$$

and the set function

$$\mu(\emptyset) := 0,\ \mu(\{0,1\}) := 1,\ \mu(\{0,2\}) := 1,\ \mu(\{1,2\}) := 3,$$

relative to the generated outer measure μ^* on $\mathscr{P}(X)$ (see Examples 4.3), only $\emptyset$ and X are μ^*-measurable.

4. Let $X := \{0,1,2\}$. For the *semiring*,

$$\mathscr{S} := \{\emptyset, \{0\}, \{1,2\}\}$$

(see Examples 2.1) and the *counting measure* on $\mathscr{S}$,

$$\mu(\emptyset) := 0,\ \mu(\{0\}) := 1,\ \mu(\{1,2\}) := 2,$$

(see Examples 3.2 and Remarks 4.1), the generated outer measure on $\mathscr{P}(X)$ is

$$\begin{aligned}
&\mu^*(\emptyset) = \mu(\emptyset) = 0,\\
&\mu^*(\{0\}) = \mu(\{0\}) = 1,\\
&\mu^*(\{1\}) = \mu^*(\{2\}) = 2\\
&\mu^*(\{0,1\}) = \mu^*(\{0,2\}) = 3,\\
&\mu^*(\{1,2\}) = \mu(\{1,2\}) = 2,\\
&\mu^*(X) = 3,
\end{aligned}$$

the μ^*-measurable sets constitute the *algebra*, and hence, also σ-algebra (see Remarks 2.3), generated by $\mathscr{S}$:

$$a(\mathscr{S}) = \{\emptyset, \{0\}, \{1,2\}, X\} = \sigma a(\mathscr{S}).$$

Examples 4.5.
1. Explain 1.
2. Verify 2–4.

4.4.2 Carathéodory's Theorem

As Examples 4.3 and 4.4 demonstrate, an outer measure μ^* need not be a measure on the power set $\mathscr{P}(X)$, but its restriction to the collection of all μ^*-measurable sets, which appears to be a σ-algebra on X, is. The following central theorem explains why this fact is not coincidental.

Theorem 4.2 (Carathéodory's Theorem). *Let X be a nonempty set and μ^* be an outer measure on its power set $\mathscr{P}(X)$. Then the collection Σ^* of all μ^*-measurable subsets of X is a σ-algebra on X and the restriction μ of μ^* to Σ^* is a measure.*

Proof. We prove the statement in three steps.

Step 1. Let us show first that Σ^* is an *algebra* on X.

Observe that we know already that $\emptyset, X \in \Sigma^*$ (see Remarks 4.4), and thus, according to the alternative definition of algebra (Definition 2.3), need to show that Σ^* is closed under *finite unions* and *complements*.

The closedness of Σ^* under complements immediately follows from the symmetry of the definition of μ^*-measurable sets (see (4.4)) relative to the operation of complement (see Remarks 4.4).

Let $A_1, A_2 \in \Sigma^*$ be arbitrary. Then, for any $B \in \mathscr{P}(X)$,

$$\begin{aligned}
&\mu^*(B) && \text{by the } \mu^*\text{-measurability of } A_1;\\
&\quad = \mu^*(B \cap A_1) + \mu^*(B \cap A_1^c) && \text{by the } \mu^*\text{-measurability of } A_2;\\
&\quad = \mu^*(B \cap A_1 \cap A_2) + \mu^*(B \cap A_1 \cap A_2^c) + \mu^*(B \cap A_1^c \cap A_2)\\
&\qquad + \mu^*(B \cap A_1^c \cap A_2^c). && (4.5)
\end{aligned}$$

Since

$$A_1 \cup A_2 = (A_1 \cap A_2) \cup (A_1 \cap A_2^c) \cup (A_1^c \cap A_2),$$

by the *subadditivity* of μ^*,

$$\mu^*(B \cap A_1 \cap A_2) + \mu^*(B \cap A_1 \cap A_2^c) + \mu^*(B \cap A_1^c \cap A_2) \ge \mu^*(B \cap (A_1 \cup A_2)).$$

Hence, by (4.5),

$$\begin{aligned}
\mu^*(B) &\ge \mu^*(B \cap (A_1 \cup A_2)) + \mu^*(B \cap A_1^c \cap A_2^c) && \text{since } A_1^c \cap A_2^c = (A_1 \cup A_2)^c;\\
&= \mu^*(B \cap (A_1 \cup A_2)) + \mu^*(B \cap (A_1 \cup A_2)^c),
\end{aligned}$$

which proves the μ^*-measurability of $A_1 \cup A_2$, and hence, the closedness of Σ^* under finite unions.

Thus, Σ^* is an *algebra* on X.

Step 2. Now, let us show that Σ^* is a σ-algebra on X.

Let $(A_n)_{n\in\mathbb{N}}$ be an arbitrary *pairwise disjoint* sequence in Σ^*.

Let us show inductively that, for any $B \in \mathscr{P}(X)$ and $n \in \mathbb{N}$,

$$\mu^*\left(B \cap \bigcup_{i=1}^{n} A_i\right) = \sum_{i=1}^{n} \mu^*(B \cap A_i). \tag{4.6}$$

Indeed, this is trivially true for $n = 1$. Assume that (4.6) is true for an $n \in \mathbb{N}$. Then

$$\begin{aligned}
&\mu^*\left(B \cap \bigcup_{i=1}^{n+1} A_i\right) && \text{by the } \mu^*\text{-measurability of } A_{n+1};\\
&\quad = \mu^*\left(\left(B \cap \bigcup_{i=1}^{n+1} A_i\right) \cap A_{n+1}\right) + \mu^*\left(\left(B \cap \bigcup_{i=1}^{n+1} A_i\right) \cap A_{n+1}^c\right)\\
&&& \text{in view of the pairwise disjointness of } (A_n)_{n\in\mathbb{N}};
\end{aligned}$$

$$= \mu^* (B \cap A_{n+1}) + \mu^* \left(B \cap \bigcup_{i=1}^{n} A_i \right) \qquad \text{by the inductive assumption;}$$

$$= \mu^* (B \cap A_{n+1}) + \sum_{i=1}^{n} \mu^* (B \cap A_i) = \sum_{i=1}^{n+1} \mu^* (B \cap A_i),$$

which completes the proof of (4.6) by induction.

In view of the fact Σ^* is an *algebra* on X, for each $n \in \mathbb{N}$,

$$\bigcup_{i=1}^{n} A_i \in \Sigma^*,$$

and hence, for any $B \in \mathscr{P}(X)$ and $n \in \mathbb{N}$,

$$\mu^* (B) = \mu^* \left(B \cap \bigcup_{i=1}^{n} A_i \right) + \mu^* \left(B \cap \left(\bigcup_{i=1}^{n} A_i \right)^c \right) \qquad \text{by (4.6);}$$

$$= \sum_{i=1}^{n} \mu^* (B \cap A_i) + \mu^* \left(B \cap \left(\bigcup_{i=1}^{n} A_i \right)^c \right)$$

since $\bigcup_{i=1}^{n} A_i \subseteq \bigcup_{i=1}^{\infty} A_i$, $n \in \mathbb{N}$, by the *monotonicity* of μ^*;

$$\geq \sum_{i=1}^{n} \mu^* (B \cap A_i) + \mu^* \left(B \cap \left(\bigcup_{i=1}^{\infty} A_i \right)^c \right).$$

Where, passing to the limit as $n \to \infty$, we have:

$$\mu^* (B) \geq \sum_{i=1}^{\infty} \mu^* (B \cap A_i) + \mu^* \left(B \cap \left(\bigcup_{i=1}^{\infty} A_i \right)^c \right) \qquad \text{by the } \sigma\text{-subadditivity of } \mu^*;$$

$$\geq \mu^* \left(\bigcup_{i=1}^{\infty} B \cap A_i \right) + \mu^* \left(B \cap \left(\bigcup_{i=1}^{\infty} A_i \right)^c \right)$$

$$= \mu^* \left(B \cap \left(\bigcup_{i=1}^{\infty} A_i \right) \right) + \mu^* \left(B \cap \left(\bigcup_{i=1}^{\infty} A_i \right)^c \right) \qquad \text{by the subadditivity of } \mu^*;$$

$$\geq \mu^* (B), \tag{4.7}$$

which implies that

$$\mu^* (B) = \mu^* \left(B \cap \left(\bigcup_{i=1}^{\infty} A_i \right) \right) + \mu^* \left(B \cap \left(\bigcup_{i=1}^{\infty} A_i \right)^c \right),$$

and hence, shows that

$$\bigcup_{i=1}^{\infty} A_i \in \Sigma^*.$$

Suppose now that $(A_n)_{n\in\mathbb{N}}$ is an arbitrary sequence in Σ^*, which need not be pairwise disjoint.

Since Σ^* is an *algebra* on X, the new sequence

$$B_n := A_n \setminus \bigcup_{i=1}^{n-1} A_i \in \Sigma^*, \ n \in \mathbb{N},$$

with $A_0 := \emptyset$, is *pairwise disjoint*, and hence, as we have shown,

$$\bigcup_{i=1}^{\infty} A_i = \bigcup_{i=1}^{\infty} B_i \in \Sigma^*,$$

which proves that Σ^* is a *σ-algebra* on X.

Step 3. It remains to show that the restriction μ of μ^* to Σ^* is a *measure*.

Inequalities (4.7) imply, in particular, that, for any pairwise disjoint sequence $(A_n)_{n\in\mathbb{N}}$ in Σ^* and each $B \in \mathscr{P}(X)$,

$$\mu^*(B) = \sum_{i=1}^{\infty} \mu^*(B \cap A_i) + \mu^*\left(B \cap \left(\bigcup_{i=1}^{\infty} A_i\right)^c\right),$$

which for $B := \bigcup_{i=1}^{\infty} A_i$, turns into

$$\mu^*\left(\bigcup_{i=1}^{\infty} A_i\right) = \sum_{i=1}^{\infty} \mu^*(A_i)$$

proving that μ^* is *σ-additive* on Σ^*, i. e., that the restriction μ of μ^* to Σ^* is a *measure* and completing the proof. □

4.5 Completeness

In this section, we introduce the useful notions of a *null set* and *completeness* of a measure on a σ-algebra, which, in particular, allows to obtain a more precise version of *Carathéodory's Theorem* (Theorem 4.3). We also provide a general procedure of extending a measure on a σ-algebra to a *complete measure*.

4.5.1 Null Sets, Completeness

Definition 4.5 (Null Set). Let $\mathscr{C}$ be a nonempty collection of subsets of a nonempty set X and $\mu : \mathscr{C} \to \overline{\mathbb{R}}$. A set $A \in \mathscr{C}$ is called a *null set relative to* μ (or a *μ-null set*) if $\mu(A) = 0$.

Examples 4.6.

1. For a nonempty collection $\mathscr{C}$ of subsets of a nonempty set X containing $\emptyset$, $\emptyset$ is a null set relative to any *noninfinite* and *additive* or *σ-additive* set function (see Exercise 3.1 (b)) on $\mathscr{C}$.
2. A *finite set* is a null set relative to the n-dimensional Jordan measure ($n \in \mathbb{N}$) (see Examples 3.4).
3. A *countably infinite* set $\{x_n\}_{n\in\mathbb{N}}$ in $\mathbb{R}^n$ ($n \in \mathbb{N}$) such that there exists
$$\lim_{n\to\infty} x_n = x \in \mathbb{R}^n$$
is a null set relative to the n-dimensional Jordan measure ($n \in \mathbb{N}$) (see Examples 3.4).
4. The *Cantor set* and all its subsets are null sets relative to the one-dimensional Jordan measure (see Section 3.3, Problem 12).
5. The segment $\{0\} \times [0, 1]$ in $\mathbb{R}^2$ is a null set relative to the two-dimensional Jordan measure, whereas the segment $[0, 1]$ in $\mathbb{R}$ is *not* a null set relative to the one-dimensional Jordan measure.

Exercise 4.11. Verify the latter.

Remark 4.5. Observe that, if $A \in \mathscr{R}$ is a null set relative to a measure μ on a ring $\mathscr{R}$ and $A \supseteq C \in \mathscr{R}$, then, by the *monotonicity* of μ (see the *Properties of Measure* (Theorem 3.1)),
$$\mu(C) = 0,$$
i. e., any set in a ring of that is a subset of a null set relative to a measure on the ring is also a null set.

Definition 4.6 (Complete Measure). A measure μ on a σ-algebra Σ is called *complete* if
$$\forall A \in \Sigma \text{ with } \mu(A) = 0 \;\forall C \subseteq A : \; C \in \Sigma,$$
i. e., all subsets of the μ-null sets are Σ-measurable.

Example 4.7. The n-dimensional Jordan measure ($n \in \mathbb{N}$) is complete (cf. Section 3.3, Problem 12).

Exercise 4.12. Explain.

4.5.2 Addendum to Carathéodory's Theorem

Now, we can obtain a more precise version of *Carathéodory's Theorem* (Theorem 4.2).

Proposition 4.2 (Addendum to Carathéodory's Theorem). *Under the conditions of Carathéodory's Theorem (Theorem 4.2), the measure μ on the σ-algebra Σ^* of all μ^*-measurable sets is complete.*

Proof. Let $A \in \Sigma^*$ with $\mu(A) = 0$. Then, for arbitrary $C \subseteq A$, by the *nonnegativity* and *monotonicity* of μ^*,

$$0 \le \mu^*(C) \le \mu^*(A) = \mu(A) = 0,$$

and hence,

$$\mu^*(C) = 0,$$

which implies that C, being a μ^*-*null set*, is μ^*-measurable (see Remarks 4.4), i. e., $C \in \Sigma^*$.

Thus, μ is a *complete measure* on Σ^*. □

Adding the statement of the prior proposition to *Carathéodory's Theorem* (Theorem 4.2), we obtain the following more "complete" version.

Theorem 4.3 (Carathéodory's Theorem). *Let X be a nonempty set and μ^* be an outer measure on its power set $\mathscr{P}(X)$. Then the collection Σ^* of all μ^*-measurable subsets of X is a σ-algebra on X and the restriction μ of μ^* to Σ^* is a complete measure.*

4.5.3 Completion

The following statement describes a general procedure of extending a measure on a σ-algebra to a complete measure.

Theorem 4.4 (Completion Theorem). *If μ is a measure on a σ-algebra Σ, then the collection*

$$\overline{\Sigma} := \{A \triangle N \mid A \in \Sigma,\ N \subseteq B \in \Sigma \text{ with } \mu(B) = 0\}$$

is a σ-algebra and the set function

$$\overline{\Sigma} \ni A \triangle N \mapsto \overline{\mu}(A \triangle N) := \mu(A) \tag{4.8}$$

is a complete measure on $\overline{\Sigma}$ extending μ and called the completion of μ.

If μ' is a complete measure extending μ from Σ to a σ-algebra Σ', then μ' is also an extension of $\overline{\mu}$, i. e., the completion of μ is the smallest complete measure extension of μ.

Proof. Obviously,

$$\Sigma \subseteq \overline{\Sigma}.$$

Exercise 4.13. Verify.

This, in particular implies, that $\emptyset, X \in \overline{\Sigma}$.

Let $A_i \in \Sigma$ and $N_i \subseteq B_i \in \Sigma$ with $\mu(B_i) = 0$, $i = 1, 2$. Then, by the *associativity* and *commutativity* of symmetric set difference,

$$(A_1 \triangle N_1) \triangle (A_2 \triangle N_2) = A_1 \triangle (N_1 \triangle (A_2 \triangle N_2)) = A_1 \triangle ((A_2 \triangle N_2) \triangle N_1)$$
$$= A_1 \triangle (A_2 \triangle (N_2 \triangle N_1)) = (A_1 \triangle A_2) \triangle (N_1 \triangle N_2), \tag{4.9}$$

which implies that $\overline{\Sigma}$ is closed under *symmetric set differences.*

For $A \in \Sigma$ and $N \subseteq B \in \Sigma$ with $\mu(B) = 0$,

$$A \cup N = (A \setminus B) \triangle (B \cap (A \cup N)) \text{ and } A \triangle N = (A \setminus B) \cup (B \cap (A \triangle N)).$$

Exercise 4.14. Verify.

Hence,

$$\overline{\Sigma} = \{A \cup N \mid A \in \Sigma,\ N \subseteq B \in \Sigma \text{ with } \mu(B) = 0\}.$$

Via this union description of $\overline{\Sigma}$, it is easily verified that $\overline{\Sigma}$ is closed under *countable unions*.

Exercise 4.15. Verify.

Being closed under symmetric set differences and countable unions, $\overline{\Sigma}$ is also closed under *set differences*.

Exercise 4.16. Verify.

Hence, $\overline{\Sigma}$ is a *σ-algebra*.

Now, let us show that the set function $\overline{\mu}$ is *well-defined* on $\overline{\Sigma}$ by (4.8), i. e., is independent of the representation $A \triangle N$, where $A \in \Sigma$ and $N \subseteq B \in \Sigma$ with $\mu(B) = 0$.

If

$$A_1 \triangle N_1 = A_2 \triangle N_2,$$

where $A_i \in \Sigma$ and $N_i \subseteq B_i \in \Sigma$ with $\mu(B_i) = 0$, $i = 1, 2$, then, as immediately follows from (4.9),

$$A_1 \triangle A_2 = N_1 \triangle N_2 \subseteq B_1 \cup B_2.$$

Exercise 4.17. Explain.

By the *monotonicity* of the measure μ (see the *Properties of Measure* (Theorem 3.1)),

$$\mu(A_1 \triangle A_2) = 0$$

and further

$$\mu(A_1 \setminus A_2) = \mu(A_2 \setminus A_1) = 0.$$

Hence, by the *additivity* of μ (see the *Properties of Measure* (Theorem 3.1)),

$$\begin{aligned}\overline{\mu}(A_1 \triangle N_1) &:= \mu(A_1) = \mu(A_1 \setminus A_2) + \mu(A_1 \cap A_2) = \mu(A_1 \cap A_2) = \mu(A_2 \cap A_1) + \mu(A_2 \setminus A_1)\\ &= \mu(A_2) =: \overline{\mu}(A_2 \triangle N_2),\end{aligned}$$

which confirms that the set function $\overline{\mu}$ is well-defined on $\overline{\Sigma}$ by (4.8).

In particular, for each $A \in \Sigma$,

$$\overline{\mu}(A) = \overline{\mu}(A \cup \emptyset) := \mu(A),$$

i. e., $\overline{\mu}$ is an extension of μ.

Using the union description of $\overline{\Sigma}$, it is easily verified that $\overline{\mu}$ is a *measure* on $\overline{\Sigma}$.

Exercise 4.18. Verify.

The *completeness* of $\overline{\mu}$ follows from the fact that $\overline{\Sigma}$ contains all subsets of the μ-null sets.

Exercise 4.19. Explain.

To prove that $\overline{\mu}$ on $\overline{\Sigma}$ is the *smallest complete measure extension* of μ, we need to show that, if μ' is a complete measure extending μ from Σ to a σ-algebra Σ', then

$$\overline{\Sigma} \subseteq \Sigma' \text{ and } \overline{\mu}(A) = \mu'(A),\ A \in \overline{\Sigma}.$$

Exercise 4.20. Show. □

Remarks 4.6.

- In particular, the completion of a complete measure μ coincides with μ.
- Thus,

  $$A \in \overline{\Sigma} \Leftrightarrow A = B \cup N,$$

  where $B \in \Sigma$ and $N \subseteq C \in \Sigma$ with $\mu(C) = 0$. The representation $A = B \cup N$ can, obviously, be regarded to be *disjoint*, and hence,

  $$\overline{\mu}(A) = \overline{\mu}(B \cup N) = \overline{\mu}(B) + \overline{\mu}(N) = \overline{\mu}(B) = \mu(B).$$

4.6 Measure Extension From a Ring

If μ^* is an outer measure on the power set $\mathscr{P}(X)$ of a nonempty set X, according to *Carathéodory's Theorem* (Theorem 4.3), the class Σ^* of all μ^*-measurable sets is a σ-algebra on X. However, such a σ-algebra may be quite meager and even consist of $\emptyset$ and X alone (see Examples 4.3 and 4.4). Here, we consider the setup when an outer measure is generated by a measure on a ring as described in the *Construction of Outer Measure Proposition* (Proposition 4.1).

4.6.1 Carathéodory's Extension Theorem

The following formidable statement is fundamental for the measure extension theory and, in particular, underlies the construct of the Lebesgue–Stieltjes measures.

Theorem 4.5 (Carathéodory's Extension Theorem). *Let μ be a measure on a ring $\mathscr{R}$ on a nonempty set X, μ^* be the outer measure on its power set $\mathscr{P}(X)$ generated by μ, and Σ^* be the collection of all μ^*-measurable subsets of X. Then:*

(1) *Σ^* is a σ-algebra;*

(2) *the restriction $\overline{\mu}$ of μ^* to Σ^* is a complete measure;*

(3) *μ^* is an extension of μ from $\mathscr{R}$ to $\mathscr{P}(X)$, i. e., for each $A \in \mathscr{R}$,*

$$\mu^*(A) = \mu(A);$$

(4) *$\mathscr{R} \subseteq \Sigma^*$;*

(5) *the restriction of the measure $\overline{\mu}$ from Σ^* to the generated σ-algebra $\sigma a(\mathscr{R})$ is a measure extending the measure μ and such an extension of μ from $\mathscr{R}$ to $\sigma a(\mathscr{R})$ is unique provided the measure μ is σ-finite;*

(6) *provided the measure μ is σ-finite, the measure $\overline{\mu}$ on Σ^* is the completion of the measure $\overline{\mu}$ on $\sigma a(\mathscr{R})$.*

Proof. Parts (1) and (2) follow directly from *Carathéodory's Theorem* (Theorem 4.3).

To prove (3), observe that, as follows from the *Construction of Outer Measure Proposition* (Proposition 4.1) (see Remarks 4.3), for each $A \in \mathscr{R}$, we have the inequality

$$\mu^*(A) \le \mu(A), \tag{4.10}$$

which, obviously, turns into equality whenever $\mu^*(A) = \infty$.

Now, let $A \in \mathscr{R}$ with $\mu^*(A) < \infty$ be arbitrary. Then, for any sequence $(A_n)_{n\in\mathbb{N}}$ in $\mathscr{R}$ such that

$$A \subseteq \bigcup_{i=1}^{\infty} A_i,$$

we have

$$A = A \cap \bigcup_{i=1}^{\infty} A_i = \bigcup_{i=1}^{\infty} A \cap A_i.$$

Since $\mathscr{R}$ is a *ring*,

$$A \cap A_n \in \mathscr{R},\ n \in \mathbb{N},$$

and hence, by the *σ-subadditivity* and *monotonicity* of μ (see the *Properties of Measure* (Theorem 3.1)),

$$\mu(A) \le \sum_{i=1}^{\infty} \mu(A \cap A_i) \le \sum_{i=1}^{\infty} \mu(A_i),$$

which, by the *Construction of Outer Measure Proposition* (Proposition 4.1), implies that

$$\mu(A) \le \mu^*(A). \tag{4.11}$$

Inequalities (4.10) and (4.11) jointly imply that, for each $A \in \mathscr{R}$ with $\mu^*(A) < \infty$,

$$\mu(A) = \mu^*(A),$$

which completes the proof of (3).

To prove (4), let $A \in \mathscr{R}$ and $\varepsilon > 0$ be arbitrary. Then, for any $B \in \mathscr{P}(X)$ with $\mu^*(B) < \infty$, by the *Construction of Outer Measure Proposition* (Proposition 4.1), there is a sequence $(B_n)_{n\in\mathbb{N}}$ in $\mathscr{R}$ such that

$$B \subseteq \bigcup_{i=1}^{\infty} B_i \quad \text{and} \quad \mu^*(B) + \varepsilon > \sum_{i=1}^{\infty} \mu(B_i).$$

Since $\mathscr{R}$ is a *ring*,

$$B_n \cap A \in \mathscr{R} \text{ and } B_n \setminus A = B_n \cap A^c \in \mathscr{R},\ n \in \mathbb{N}.$$

Further,

$$B \cap A \subseteq \bigcup_{i=1}^{\infty} (B_i \cap A) \quad \text{and} \quad B \cap A^c \subseteq \bigcup_{i=1}^{\infty} (B_i \cap A^c).$$

Therefore, by the *additivity* of μ (see the *Properties of Measure* (Theorem 3.1)) and the *Construction of Outer Measure Proposition* (Proposition 4.1),

$$\begin{aligned}\mu^*(B) + \varepsilon &> \sum_{i=1}^{\infty} \mu(B_i) = \sum_{i=1}^{\infty} [\mu(B_i \cap A) + \mu(B_i \cap A^c)] \\ &= \sum_{i=1}^{\infty} \mu(B_i \cap A) + \sum_{i=1}^{\infty} \mu(B_i \cap A^c) \ge \mu^*(B \cap A) + \mu^*(B \cap A^c).\end{aligned}$$

Since $\varepsilon > 0$ is arbitrary, passing to the limit as $\varepsilon \to 0+$, we infer that, for each $B \in \mathscr{P}(X)$ with $\mu^*(B) < \infty$,

$$\mu^*(B) \ge \mu^*(B \cap A) + \mu^*(B \cap A^c).$$

The latter, obviously, being true also for each $B \in \mathscr{P}(X)$ with $\mu^*(B) = \infty$ (see Remarks 4.4), we conclude that A is μ^*-measurable, i. e., $A \in \Sigma^*$, and hence, we have the inclusion

$$\mathscr{R} \subseteq \Sigma^*.$$

To prove (5), observe that the prior inclusion immediately implies that

$$\sigma a(\mathscr{R}) \subseteq \Sigma^*,$$

and hence, by (3), the restriction of the measure $\bar{\mu}$ from Σ^* to $\sigma a(\mathscr{R})$ is a measure extending μ from $\mathscr{R}$ to $\sigma a(\mathscr{R})$ (see Remarks 4.1).

Now, let us suppose the measure μ to be *σ-finite* on $\mathscr{R}$ and prove the uniqueness of such an extension of μ from $\mathscr{R}$ to $\sigma a(\mathscr{R})$. The σ-finiteness of μ on $\mathscr{R}$ (see Definition 3.1), in particular, implies that

$$X \in \sigma r(\mathscr{R}).$$

Exercise 4.21. Explain.

Hence,

$$\sigma a(\mathscr{R}) = \sigma r(\mathscr{R}).$$

The σ-finiteness of μ^* on $\mathscr{R}$ also immediately implies that the outer measure μ^* on $\mathscr{P}(X)$ and the measure $\bar{\mu}$ on $\sigma a(\mathscr{R})$ are *σ-finite*.

Exercise 4.22. Explain.

Let λ be a *measure* extending the measure μ from $\mathscr{R}$ to $\sigma a(\mathscr{R})$ and

$$\mathscr{E} := \{A \in \sigma a(\mathscr{R}) \mid \lambda(A) = \bar{\mu}(A)\}.$$

Obviously,

$$\mathscr{R} \subseteq \mathscr{E} \subseteq \sigma a(\mathscr{R}) = \sigma r(\mathscr{R}). \tag{4.12}$$

Suppose that $\bar{\mu}$ or λ is *finite* on $\sigma a(\mathscr{R})$ and let us show that then $\mathscr{E}$ is a *monotone class*. Indeed, for an *increasing* sequence $(A_n)_{n\in\mathbb{N}}$ in $\mathscr{E}$, by the *continuity of measure from below* (see the *Continuity of Measure* (Theorem 3.2)),

$$\lambda\left(\bigcup_{i=1}^{\infty} A_i\right) = \lim_{n\to\infty} \lambda(A_n) = \lim_{n\to\infty} \bar{\mu}(A_n) = \bar{\mu}\left(\bigcup_{i=1}^{\infty} A_i\right),$$

which implies that

$$\lim_{n\to\infty} A_n = \bigcup_{i=1}^{\infty} A_i \in \mathscr{E}.$$

In view of the finiteness of $\bar{\mu}$ or λ on $\sigma a(\mathscr{R})$, the case of a *decreasing* sequence $(A_n)_{n\in\mathbb{N}}$ in $\mathscr{E}$ is considered in absolutely the same fashion via the *continuity of measure from above* (see the *Continuity of Measure* (Theorem 3.2)).

Exercise 4.23. Consider.

Thus, under the assumption of the *finiteness* for $\bar{\mu}$ or λ on $\sigma a(\mathscr{R})$, $\mathscr{E}$ is a *monotone class*, and hence, inclusions (4.12), imply

$$\mathscr{R} \subseteq m(\mathscr{R}) \subseteq \mathscr{E} \subseteq \sigma a(\mathscr{R}) = \sigma r(\mathscr{R}).$$

Where, since, by the *Monotone Class Theorem* (Theorem 2.4),

$$m(\mathscr{R}) = \sigma r(\mathscr{R}),$$

we infer that

$$\mathscr{E} = \sigma a(\mathscr{R}) = \sigma r(\mathscr{R}),$$

e. i., $\lambda = \overline{\mu}$ on $\sigma a(\mathscr{R})$.

Now, let us drop the assumption of the finiteness of $\overline{\mu}$ or λ on $\sigma a(\mathscr{R})$. By the *σ-finiteness* of μ on $\mathscr{R}$ (see Definition 3.1), there exists a sequence $(A_n)_{n\in\mathbb{N}}$ in $\mathscr{R}$ with

$$\overline{\mu}(A_n) = \mu(A_n) < \infty, \; n \in \mathbb{N},$$

such that

$$\bigcup_{i=1}^{\infty} A_i = X.$$

Since $\mathscr{R}$ is a *ring*, without loss of generality, the sequence $(A_n)_{n\in\mathbb{N}}$ can be regarded to be *pairwise disjoint*.

Exercise 4.24. Explain.

Also, for each $n \in \mathbb{N}$, $\mathscr{R} \cap A_n$ is a *ring*.

Exercise 4.25. Explain (cf. Section 2.6, Problem 12).

By the *monotonicity* of $\overline{\mu}$, for each $n \in \mathbb{N}$, the measures $\overline{\mu}$ and λ are *finite* on the ring $\mathscr{R} \cap A_n$.

Exercise 4.26. Verify.

By the proven above, for each $n \in \mathbb{N}$, the measures $\overline{\mu}$ and λ coincide on

$$\sigma r(\mathscr{R} \cap A_n) = \sigma r(\mathscr{R}) \cap A_n$$

(see Section 2.6, Problem 12 (b)), which implies that, for any $A \in \sigma a(\mathscr{R}) = \sigma r(\mathscr{R})$ and $n \in \mathbb{N}$,

$$\overline{\mu}(A \cap A_n) = \lambda(A \cap A_n).$$

Hence, by the *σ-additivity* of $\overline{\mu}$ and λ, for each $A \in \sigma a(\mathscr{R}) = \sigma r(\mathscr{R})$,

$$\begin{aligned}\overline{\mu}(A) &= \overline{\mu}\left(A \cap \bigcup_{i=1}^{\infty} A_i\right) = \overline{\mu}\left(\bigcup_{i=1}^{\infty} A \cap A_i\right) = \sum_{i=1}^{\infty} \overline{\mu}(A \cap A_i) = \sum_{i=1}^{\infty} \lambda(A \cap A_i) \\ &= \lambda\left(\bigcup_{i=1}^{\infty} A \cap A_i\right) = \lambda\left(A \cap \bigcup_{i=1}^{\infty} A_i\right) = \lambda(A).\end{aligned}$$

which completes the proof of (5).

Now, continuing to suppose the measure μ to be *σ-finite* on $\mathscr{R}$, we are to prove remaining part (6). Let us designate by $\overline{\Sigma}$ the *domain* (σ-algebra) of the *completion* (measure) of $\overline{\mu}$ on $\sigma a(\mathscr{R})$.

Since, by (2), $\overline{\mu}$ is a *complete* measure on Σ^* extending the measure $\overline{\mu}$ on $\sigma a(\mathscr{R})$, by the *Completion Theorem* (Theorem 4.4), it also extends the completion of μ, i. e.,

$$\overline{\Sigma} \subseteq \Sigma^* \tag{4.13}$$

and the completion coincides with $\overline{\mu} = \mu^*$ on $\overline{\Sigma}$.

Thus, it only remains to prove that

$$\overline{\Sigma} = \Sigma^*.$$

Let $A \in \Sigma^*$ with $\mu^*(A) = \overline{\mu}(A) < \infty$ and $\varepsilon > 0$ be arbitrary. Then, by the *Construction of Outer Measure Proposition* (Proposition 4.1), there exists a sequence $(A_n^{(\varepsilon)})_{n\in\mathbb{N}}$ in $\mathscr{R}$ such that

$$A \subseteq \bigcup_{i=1}^{\infty} A_i^{(\varepsilon)} =: B_\varepsilon \in \sigma a(\mathscr{R}) \subseteq \Sigma^* \text{ and } \mu^*(A) \le \sum_{i=1}^{\infty} \mu\left(A_i^{(\varepsilon)}\right) < \mu^*(A) + \varepsilon.$$

Hence, by the *Construction of Outer Measure Proposition* (Proposition 4.1),

$$\mu^*(A) \le \mu^*(B_\varepsilon) < \mu^*(A) + \varepsilon.$$

Setting

$$B := \bigcap_{n=1}^{\infty} B_{1/n} \in \sigma a(\mathscr{R}),$$

since

$$A \subseteq B \subseteq B_{1/n},\ n \in \mathbb{N},$$

by the *monotonicity* of μ^*, we have

$$\mu^*(A) \le \mu^*(B) \le \mu^*(B_{1/n}) < \mu^*(A) + 1/n.$$

Passing to the limit as $n \to \infty$, we arrive at

$$\mu^*(A) \le \mu^*(B) \le \mu^*(A),$$

which implies that

$$\overline{\mu}(B) = \mu^*(B) = \mu^*(A) = \overline{\mu}(A).$$

Since $A, B \in \Sigma^*$ and $\overline{\mu}$ is a measure on Σ^*, in view of $\overline{\mu}(A) < \infty$, by the *Properties of Measure* (Theorem 3.1),

$$B \setminus A \in \Sigma^* \text{ and } \mu^*(B \setminus A) = \overline{\mu}(B \setminus A) = \overline{\mu}(B) - \overline{\mu}(A) = 0.$$

Hence, similarly, there exists a $C \in \sigma a(\mathscr{R}) \subseteq \Sigma^*$ such that

$$B \setminus A \subseteq C \text{ and } \overline{\mu}(C) = \overline{\mu}(B \setminus A) = 0.$$

Since, in view of the inclusion $A \subseteq B$,

$$A = (B \setminus C) \cup (A \cap C),$$

where

$$B \setminus C \in \sigma a(\mathscr{R}) \text{ and } A \cap C \subseteq C \in \sigma a(\mathscr{R}) \text{ with } \overline{\mu}(C) = 0,$$

by the *Completion Theorem* (Theorem 4.4),

$$A \in \overline{\Sigma},$$

and hence

$$\Sigma^* \subseteq \overline{\Sigma}. \tag{4.14}$$

Inclusions (4.13) and (4.14) jointly imply that

$$\overline{\Sigma} = \Sigma^*,$$

which concludes the proof for the case of $A \in \Sigma^*$ with $\mu^*(A) = \overline{\mu}(A) < \infty$.

Now, let $A \in \Sigma^*$ be arbitrary. By the *σ-finiteness* of μ on $\mathscr{R}$, there exists a sequence $(A_n)_{n\in\mathbb{N}}$ in $\mathscr{R}$ with

$$\overline{\mu}(A_n) = \mu(A_n) < \infty, \; n \in \mathbb{N},$$

such that

$$\bigcup_{i=1}^{\infty} A_i = X.$$

For each $n \in \mathbb{N}$,

$$A \cap A_n \in \Sigma^*$$

and, by the *monotonicity* of $\overline{\mu}$,

$$\overline{\mu}(A \cap A_n) \le \overline{\mu}(A_n) < \infty.$$

Hence, by the proven above, for each $n \in \mathbb{N}$,

$$A \cap A_n \in \overline{\Sigma},$$

which implies that

$$A = A \cap \bigcup_{i=1}^{\infty} A_n = \bigcup_{i=1}^{\infty} A \cap A_n \in \overline{\Sigma},$$

and hence,

$$\Sigma^* \subseteq \overline{\Sigma}. \tag{4.15}$$

Inclusions (4.13) and (4.15) jointly imply that

$$\overline{\Sigma} = \Sigma^*,$$

which concludes the proof of (6), and thus, of the entire statement. □

Remarks 4.7.

– When proving part (6) of *Carathéodory's Extension Theorem* (Theorem 4.5), we have proved the following profound fact:

$$\forall A \in \mathscr{P}(X) \text{ with } \mu^*(A) < \infty \ \exists C \in \sigma a(\mathscr{R}) \subseteq \Sigma^* : \quad A \subseteq C \quad \text{and} \quad \mu^*(A) = \mu^*(C) = \overline{\mu}(C).$$

In particular, for an arbitrary μ^**-null set* $A \in \mathscr{P}(X)$ (i. e., $\mu^*(A) = 0$), which is automatically μ^*-measurable, i. e., $A \in \Sigma^*$ (see Remarks 4.4), there exists a $\overline{\mu}$*-null set* $C \in \sigma a(\mathscr{R}) \subseteq \Sigma^*$ (i. e., $\overline{\mu}(C) = 0$) such that

$$A \subseteq C.$$

– Under the conditions of *Carathéodory's Extension Theorem* (Theorem 4.5), a set $A \subseteq X$ is μ^*-measurable, i. e., $A \in \Sigma^*$, *iff*

$$A = B \cup N, \tag{4.16}$$

where $B \in \sigma a(\mathscr{R})$ and N is a subset of a $\overline{\mu}$-null set $C \in \sigma a(\mathscr{R})$ *disjoint* from B (see Remark 4.6). Since, as is noted in the prior remark, any μ^*-null set disjoint from B is such a subset, the set N in representation (4.16) can be regarded to be a μ^**-null set* (i. e., $\mu^*(N) = 0$) *disjoint from* B.

– As the subsequent examples show, the condition of *σ-finiteness* in parts (5) and (6) of *Carathéodory's Extension Theorem* (Theorem 4.5) is essential and cannot be dropped.

Examples 4.8.

1. Let $X := \{0, 1, 2\}$. For the *ring*,

$$\mathscr{R} := \{\emptyset, \{0\}, \{1\}, \{0, 1\}\}$$

on X, the *counting measure* μ on $\mathscr{R}$

$$\mu(\emptyset) := 0,\ \mu(\{0\}) := 1 =: \mu(\{1\}),\ \mu(\{0,1\}) := 2,$$

(see Examples 3.2 and Remarks 4.1) is *not σ-finite*; the generated outer measure on $\mathscr{P}(X)$ is

$$\begin{aligned}
&\mu^*(\emptyset) = \mu(\emptyset) = 0,\\
&\mu^*(\{0\}) = \mu(\{0\}) = 1,\\
&\mu^*(\{1\}) = \mu(\{1\}) = 1,\\
&\mu^*(\{2\}) = \infty,\\
&\mu^*(\{0,1\}) = \mu(\{0,1\}) = 2,\\
&\mu^*(\{0,2\}) = \mu^*(\{1,2\}) = \infty,\\
&\mu^*(X) = \infty.
\end{aligned}$$

By *Carathéodory's Extension Theorem* (Theorem 4.5) and in view of the *finiteness* of X,

$$\mathscr{P}(X) = a(\mathscr{R}) = \sigma a(\mathscr{R}) \subseteq \Sigma^*.$$

Hence,

$$\Sigma^* = \mathscr{P}(X) = \sigma a(\mathscr{R}).$$

By *Carathéodory's Extension Theorem* (Theorem 4.5), the outer μ^* is a measure on $\sigma a(\mathscr{R})$ extending the measure μ on $\mathscr{R}$. However, the *counting measure* on $\sigma a(\mathscr{R})$ (see Examples 3.2 and Remarks 4.1) is another extension of the measure μ on $\mathscr{R}$ that is distinct from μ^*.

2. Let X be an *uncountable* set. On the σ-algebra

$$\Sigma := \{A \subseteq X \mid A \text{ or } A^c \text{ is } \textit{countable}\}$$

(see Examples 2.3), the *counting measure* μ (see Examples 3.2 and Remarks 4.1) is not *σ-finite* since X is uncountable, but it is trivially *complete* since the only μ-null set is $\emptyset$. As is easily verified, the generated outer measure μ^* on $\mathscr{P}(X)$ is the *counting measure* and the σ-algebra of μ^*-measurable sets is

$$\Sigma^* = \mathscr{P}(X).$$

In view of the *uncountability* of X, the inclusion

$$\Sigma \subset \Sigma^* = \mathscr{P}(X)$$

is strict.

By *Carathéodory's Extension Theorem* (Theorem 4.5), μ^* is a complete measure on $\Sigma^* = \mathscr{P}(X)$ that is not the completion of the already complete measure μ on $\sigma a(\Sigma) = \Sigma$ (see Remarks 4.6).

3. Let $X := \{0, 1, 2\}$. On the *algebra,*

$$\mathscr{A} := \{\emptyset, \{0\}, \{1, 2\}, X\}$$

on X (see Examples 2.2), the *counting measure* (see Examples 3.2 and Remarks 4.1) μ is *σ-finite*; the generated outer measure on $\mathscr{P}(X)$ is

$$\begin{aligned}
&\mu^*(\emptyset) = \mu(\emptyset) = 0,\\
&\mu^*(\{0\}) = \mu(\{0\}) = 1,\\
&\mu^*(\{1\}) = \mu^*(\{2\}) = 2\\
&\mu^*(\{0, 1\}) = \mu^*(\{0, 2\}) = 3,\\
&\mu^*(\{1, 2\}) = \mu(\{1, 2\}) = 2,\\
&\mu^*(X) = \mu(X) = 3.
\end{aligned}$$

In view of the *finiteness* of X,

$$\Sigma^* = \mathscr{A} = a(\mathscr{A}) = \sigma a(\mathscr{A}).$$

In this case, all parts (1)–(6) of *Carathéodory's Extension Theorem* (Theorem 4.5) are nicely in place.

Exercise 4.27. Verify and explain.

4.6.2 Approximation

The following statement shows that, for a σ-finite measure μ on a ring $\mathscr{R}$ on a nonempty set X, certain sets from the generated σ-algebra $\sigma a(\mathscr{R})$ can be approximated, in a sense, by the sets from $\mathscr{R}$. Similar approach allows us to characterize μ^*-measurable sets (see the *Characterization of μ^*-Measurability* (Theorem 4.4), Section 4.8, Problem 6).

Theorem 4.6 (Approximation Theorem). *Let μ be a σ-finite measure on a ring $\mathscr{R}$ on a nonempty set X and the measure $\bar{\mu}$ be its unique extension to the generated σ-algebra $\sigma a(\mathscr{R})$. Then*

$$\forall A \in \sigma a(\mathscr{R}) \text{ with } \bar{\mu}(A) < \infty \ \forall \varepsilon > 0 \ \exists B \in \mathscr{R} : \bar{\mu}(A \triangle B) = \bar{\mu}(A \setminus B) + \bar{\mu}(B \setminus A) < \varepsilon.$$

Proof. Let μ^* be the generated outer measure on $\mathscr{P}(X)$. Then, by *Carathéodory's Extension Theorem* (Theorem 4.5), the measure $\bar{\mu}$ is the restriction of the outer measure μ^* from the power set $\mathscr{P}(X)$ to $\sigma a(\mathscr{R})$.

Let $A \in \sigma a(\mathscr{R})$ with $\mu^*(A) = \overline{\mu}(A) < \infty$ and $\varepsilon > 0$ be arbitrary, then by the *Construction of Outer Measure Proposition* (Proposition 4.1), there is a sequence $(A_n)_{n\in\mathbb{N}}$ in $\mathscr{R}$ such that

$$A \subseteq \bigcup_{i=1}^{\infty} A_i \in \sigma a(\mathscr{R})$$

and

$$\overline{\mu}(A) + \varepsilon/2 = \mu^*(A) + \varepsilon/2 > \sum_{i=1}^{\infty} \mu(A_i) = \sum_{i=1}^{\infty} \overline{\mu}(A_i).$$

Where, by the *σ-subadditivity* and *monotonicity* of $\overline{\mu}$, for each $n \in \mathbb{N}$,

$$\overline{\mu}(A) + \varepsilon/2 > \sum_{i=1}^{\infty} \overline{\mu}(A_i) \ge \overline{\mu}\left(\bigcup_{i=1}^{\infty} A_i\right) \ge \overline{\mu}\left(\bigcup_{i=1}^{n} A_i\right). \tag{4.17}$$

By the *continuity of measure from below* (see the *Continuity of Measure* (Theorem 3.2)),

$$\overline{\mu}\left(\bigcup_{i=1}^{\infty} A_i\right) = \lim_{n\to\infty} \overline{\mu}\left(\bigcup_{i=1}^{n} A_i\right).$$

Exercise 4.28. Explain.

Hence, there exists an $N \in \mathbb{N}$ such that

$$\overline{\mu}\left(\bigcup_{i=1}^{\infty} A_i\right) - \overline{\mu}\left(\bigcup_{i=1}^{N} A_i\right) < \varepsilon/2. \tag{4.18}$$

Setting

$$B := \bigcup_{i=1}^{N} A_i \in \mathscr{R},$$

from inequalities (4.18) and (4.17), by the *Properties of Measure* (Theorem 3.1), we infer that

$$\overline{\mu}(A \setminus B) \le \overline{\mu}\left(\left(\bigcup_{i=1}^{\infty} A_i\right) \setminus B\right) = \overline{\mu}\left(\bigcup_{i=1}^{\infty} A_i\right) - \overline{\mu}(B) < \varepsilon/2,$$

$$\overline{\mu}(B \setminus A) \le \overline{\mu}\left(\left(\bigcup_{i=1}^{\infty} A_i\right) \setminus A\right) = \overline{\mu}\left(\bigcup_{i=1}^{\infty} A_i\right) - \overline{\mu}(A) < \varepsilon/2.$$

Hence, by the *additivity* of $\overline{\mu}$,

$$\overline{\mu}(A \triangle B) = \overline{\mu}(A \setminus B) + \overline{\mu}(B \setminus A) < \varepsilon,$$

which completes the proof. □

Remarks 4.8.

- As follows from the proof, without the requirement of σ-finiteness for the measure μ on $\mathscr{R}$, the prior theorem remains in place for the restriction $\overline{\mu}$ of the generated outer measure μ^* from $\mathscr{P}(X)$ to $\sigma a(\mathscr{R})$, which need not be the unique measure on $\sigma a(\mathscr{R})$ extending the measure μ on $\mathscr{R}$ (see Examples 4.8).
- Furthermore, without the requirement of σ-finiteness for the measure μ on $\mathscr{R}$, the proof of the prior theorem applies to a set $A \in \Sigma^*$ with $\mu^*(A) < \infty$ (cf. the *Characterization of μ^*-Measurability* (Theorem 4.4), Section 4.8, Problem 6).

4.7 Lebesgue–Stieltjes Measures

Here, we apply the power of *Carathéodory's Extension Theorem* (Theorem 4.5) to extend the *σ-finite* Lebesgue–Stieltjes measures further from the generated ring $r(\mathscr{S}_1)$ (see Example 4.1) to a larger σ-algebra of subsets of $\mathbb{R}$ containing the Borel σ-algebra $\mathscr{B}(\mathbb{R})$, which would allow us to meaningfully assign such measures to sets like $\mathbb{N}$, $\mathbb{Q}$, $\mathbb{R}$, or the *Cantor set*.

4.7.1 The Construct

Let

$$\lambda_F(\emptyset) := 0, \ \lambda_F((a,b]) := F(b) - F(a), \ -\infty < a < b < \infty,$$

be the *Lebesgue–Stieltjes measure* associated with an increasing right-continuous function $F : \mathbb{R} \to \mathbb{R}$ on the interval semiring

$$\mathscr{S}_1 := \{(a,b] \mid -\infty < a < b < \infty\} \cup \{\emptyset\}$$

(see the *Lebesgue–Stieltjes Measures on Interval Semiring Theorem* (Theorem 3.5)).

Exercise 4.29. Find $\lambda_F((-1,0])$ and $\lambda_F((0,1])$ for λ_F generated by:

(a) $F(x) = c$, $x \in \mathbb{R}$, where $c \in \mathbb{R}$ is fixed;

(b) $F(x) = x$, $x \in \mathbb{R}$;

(c) $F(x) = \begin{cases} 0 & \text{if } x < 0, \\ 1 & \text{if } x \geq 0; \end{cases}$

(d) $F(x) = \begin{cases} x - 1 & \text{if } x < 0, \\ x & \text{if } x \geq 0. \end{cases}$

By the *Measure Extension from a Semiring Theorem* (Theorem 4.1), λ_F on the semiring $\mathscr{S}_1$ is uniquely extended to a measure, also denoted by λ_F, on the generated ring $r(\mathscr{S}_1)$ (see Example 4.1), which is *σ-finite* on $r(\mathscr{S}_1)$.

Exercise 4.30. Explain the latter.

Let λ_F^* be the outer measure on the power set $\mathscr{P}(\mathbb{R})$ generated by the measure λ_F on the ring $r(\mathscr{S}_1)$ and Σ_F^* be the collection of all λ_F^*-measurable subsets of $\mathbb{R}$.

By *Carathéodory's Extension Theorem* (Theorem 4.5), Σ_F^* is a *σ-algebra* on $\mathbb{R}$ and λ_F^* is a *complete measure* on Σ_F^*, which is the *completion* of the unique measure extension of λ_F from the ring $r(\mathscr{S}_1)$ to the generated σ-algebra

$$\sigma a(\mathscr{S}_1) = \sigma a(r(\mathscr{S}_1)) = \mathscr{B}(\mathbb{R}),$$

which coincides with the *Borel σ-algebra* $\mathscr{B}(\mathbb{R})$ on $\mathbb{R}$ (see Section 2.5.3).

Exercise 4.31. Verify the latter (see Exercise 2.17).

The following inclusions hold:

$$\mathscr{S}_1 \subset r(\mathscr{S}_1) \subset \sigma a(\mathscr{S}_1) = \mathscr{B}(\mathbb{R}) \subset \Sigma_F^*.$$

The sets from Σ_F^* are called *Lebesgue–Stieltjes measurable* and the *σ-finite* measure λ_F^* on Σ_F^*, henceforth designated by λ_F, is called the *Lebesgue–Stieltjes measure F*.

In particular, for

$$F(x) := x,\ x \in \mathbb{R},$$

$\lambda^* := \lambda_F^*$ and $\lambda := \lambda_F$ is the *σ-finite Lebesgue measure* on the σ-algebra $\Sigma^* := \Sigma_F^*$ of all *Lebesgue measurable* subsets of $\mathbb{R}$, which is the *completion* of the unique measure extension of λ from the ring $r(\mathscr{S}_1)$ to the *Borel σ-algebra* $\mathscr{B}(\mathbb{R})$ on $\mathbb{R}$.

Remarks 4.9.

- For an arbitrary increasing right-continuous function $F : \mathbb{R} \to \mathbb{R}$, a set $A \subseteq \mathbb{R}$ is Lebesgue–Stieltjes measurable, i. e., $A \in \Sigma_F^*$, *iff*

$$A = B \cup N, \tag{4.19}$$

  where B is a *Borel set* (i. e., $B \in \mathscr{B}(\mathbb{R})$) and N is a *λ_F^*-null set* (i. e., $\lambda_F(N) = 0$) *disjoint from B* (see Remarks 4.7).
- Conversely, any measure μ on the Borel σ-algebra $\mathscr{B}(\mathbb{R})$ generates the *increasing right-continuous* function

$$F(x) := \mu((-\infty, x]),\ x \in \mathbb{R},$$

  such that μ *coincides* with the Lebesgue–Stieltjes measure λ_F associated with F on $\mathscr{B}(\mathbb{R})$, i. e.,

$$\forall A \in \mathscr{B}(\mathbb{R}) : \ \mu(A) = \lambda_F(A).$$

  Thus, any measure on the Borel σ-algebra $\mathscr{B}(\mathbb{R})$ is a Lebesgue–Stieltjes measure.

Exercise 4.32. Apply *Carathéodory's Extension Theorem* (Theorem 4.5) to verify the latter.

Examples 4.9.

1. For the Lebesgue–Stieltjes measure λ_F associated with an increasing right-continuous function $F : \mathbb{R} \to \mathbb{R}$,
   (a) for each $x \in \mathbb{R}$, $\{x\} \in \Sigma_F^*$ and
   $$\lambda_F(\{x\}) = F(x) - F(x-);$$
   (b) for each *countable* (in particular, *finite*) set $C := \{x_n\}_{n\in I}$, $C \in \Sigma_F^*$ and
   $$\lambda_F(C) = \sum_{i\in I}[F(x_i) - F(x_i-)];$$
   (c) for any $-\infty < a < b < \infty$, $[a,b] \in \Sigma_F^*$ and
   $$\lambda_F([a,b]) = F(b) - F(a-);$$
   (d) for any $-\infty < a < b < \infty$, $(a,b) \in \Sigma_F^*$ and
   $$\lambda_F((a,b)) = F(b-) - F(a);$$
   (e) for any $a, b \in \mathbb{R}$, $(a,\infty), (-\infty,b], \mathbb{R} \in \Sigma_F^*$ and
   $$\begin{aligned}\lambda_F((a,\infty)) &= F(\infty) - F(a),\\ \lambda_F((-\infty,b]) &= F(b) - F(-\infty),\\ \lambda_F(\mathbb{R}) &= F(\infty) - F(-\infty),\end{aligned}$$
   where $F(\pm\infty) := \lim_{x\to\pm\infty} F(x)$.
   Thus, the Lebesgue–Stieltjes measure λ_F is *finite* on Σ_F^* *iff*
   $$\lambda_F(\mathbb{R}) = F(\infty) - F(-\infty) < \infty.$$

2. For the Lebesgue measure λ associated with the function $F(x) := x$, $x \in \mathbb{R}$,
   (a) for each $x \in \mathbb{R}$, $\{x\} \in \Sigma^*$ and
   $$\lambda(\{x\}) = 0;$$
   (b) for each *countable* (in particular, *finite*) set $C := \{x_n\}_{n\in I}$, $C \in \Sigma^*$ and
   $$\lambda(C) = 0,$$
   in particular,
   $$\lambda(\mathbb{N}) = \lambda(\mathbb{Z}) = \lambda(\mathbb{Q}) = 0;$$

(c) for any $-\infty < a < b < \infty$, $[a,b], [a,b), (a,b), (a,b] \in \Sigma^*$ and

$$\lambda([a,b]) = \lambda([a,b)) = \lambda((a,b)) = \lambda((a,b]) := b - a;$$

(d) $(a,\infty), [a,\infty), (-\infty,b), (-\infty,b], \mathbb{R} \in \Sigma^*$ $(a, b \in \mathbb{R})$ and

$$\lambda((a,\infty)) = \lambda([a,\infty)) = \lambda((-\infty,b)) = \lambda((-\infty,b]) = \lambda(\mathbb{R}) = \infty.$$

Remark 4.10. Thus, we conclude that the Lebesgue measure of an arbitrary interval $I \subseteq \mathbb{R}$ is its conventional *length* and the Lebesgue measure, being *σ-finite*, is *not* finite on Σ^* and $\mathscr{B}(\mathbb{R})$.

3. For an open set $G \subseteq \mathbb{R}$, there is a countable pairwise disjoint collection

$$\{(a_i,b_i) \mid -\infty < a_i < b_i < \infty\}_{i\in I}$$

such that

$$G = \bigcup_{i\in I}(a_i,b_i)$$

(see the *Open Sets of the Real Line Proposition* (Proposition 1.5)), and hence,

$$\lambda_F(G) = \sum_{i\in I}(F(b_i-) - F(a_i)),$$

in particular,

$$\lambda(G) = \sum_{i\in I}(b_i - a_i).$$

Remark 4.11. Using the latter as a definition, Émile Borel built the extension of length to $\mathscr{B}(\mathbb{R})$ in 1898 (see, e. g., [3, 7]).

Exercise 4.33.
(a) Verify.
(b) Find the values for $\lambda_F([a,b))$, $\lambda_F((-\infty,b))$, and $\lambda_F([a,\infty))$.
(c) Find $\lambda_F([0,1))$, $\lambda_F((-\infty,0))$, $\lambda_F([0,\infty))$, and $\lambda_F(\mathbb{R})$ for λ_F associated with the function

$$F(x) = \begin{cases} 0 & \text{if } x < 0, \\ 1 & \text{if } x \geq 0. \end{cases}$$

4.7.2 Relationships Between Various Extensions of Length

The following diagram (see Figure 4.1) demonstrates the relationships between various extensions of length and provides corresponding chronology (cf. [3]).

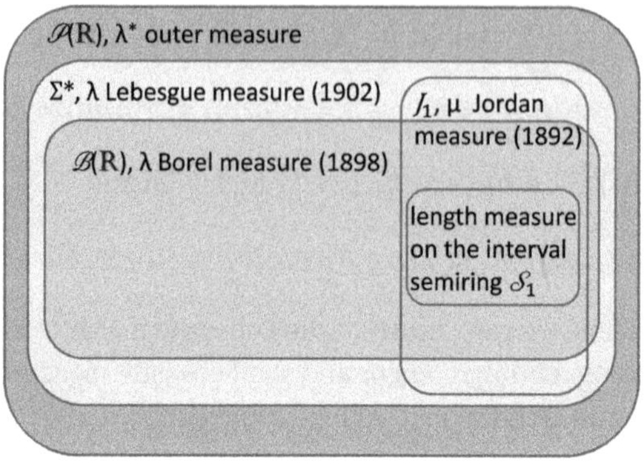

Figure 4.1: Relationships between various extensions of length.

The summary is as follows:
- The Lebesgue measure is an extension of the Jordan measure (see Section 4.8, Problem 8).
- There are Jordan measurable sets, which are not Borel (follows from the subsequent proposition).
- There are Lebesgue measurable sets, which are not Borel (follows from the prior statements).
- There are sets in $\mathbb{R}$, which are not Lebesgue measurable (follows from the subsequent theorem).

Proposition 4.3 (Cardinality of the Jordan Ring). *The cardinality of the ring $\mathscr{J}_1$ of all Jordan measurable subsets of $\mathbb{R}$ is the same as that of $\mathscr{P}(\mathbb{R})$, i. e., $|\mathscr{J}_1| = |\mathscr{P}(\mathbb{R})| > \mathfrak{c}$.*

Exercise 4.34. Prove.

Hint. Use *Cantor's Theorem* (Theorem 1.2) via considering the power set of the *Cantor set C*, for which $|C| = \mathfrak{c}$, and the fact that $|\mathscr{B}(\mathbb{R})| = \mathfrak{c}$ (see Remark 2.7).

4.7.3 Existence of a Non-Lebesgue Measurable Set in $\mathbb{R}$

As follows from the prior statement, the cardinality of the σ-algebra Σ^* of all Lebesgue measurable subsets of $\mathbb{R}$ is the same as that of $\mathscr{P}(\mathbb{R})$, i. e., $|\Sigma| = |\mathscr{P}(\mathbb{R})| > \mathfrak{c}$. However, as the following celebrated theorem shows not all subsets of $\mathbb{R}$ are Lebesgue measurable.

Theorem 4.7 (Vitali Theorem). *There exists a set in $\mathbb{R}$, which is not Lebesgue measurable.*[2]

2 Giuseppe Vitali (1875–1932).

Proof. Let us define an *equivalence relation* (*reflexive*, *symmetric*, and *transitive*) on the unit interval $[0,1]$ as follows: for $x, y \in [0,1]$, we say that x is *similar* to y and write $x \sim y$ if $y - x \in \mathbb{Q}$.

As any equivalence relation, $\sim$ partitions $[0,1]$ into pairwise disjoint *equivalence classes*

$$[x] := \{y \in [0,1] \mid x \sim y\}, \ x \in [0,1].$$

Exercise 4.35.
(a) Verify that $\sim$ is an equivalence relation on $[0,1]$.
(b) Show that, for each $x \in [0,1]$, $[x]$ is *countably infinite*.
(c) Describe the equivalence class $[0]$.
(d) Explain why $[0] = [1/2] = [1]$ and $[0] \cap [\sqrt{2}/2] = \emptyset$.

This partition has the following properties:
(i) for each $x \in [0,1]$, $[x] \neq \emptyset$;
(ii) for any $x, y \in [0,1]$, either $[x] = [y]$ or $[x] \cap [y] = \emptyset$;
(iii) each $x \in [0,1]$ belongs to precisely one equivalence class.

Exercise 4.36.
(a) Verify.
(b) Show that the collection of all distinct equivalence classes is *uncountable*, more precisely, it has the cardinality $\mathfrak{c}$ of the continuum (see Examples 1.3).

By the *Axiom of Choice* (see Appendix A), we can choose one representative from each of the distinct equivalence classes and form a *choice set* V out of them, which is an *uncountable proper subset* of $[0,1]$.

Exercise 4.37.
(a) Verify the latter.
(b) Verify that, for any distinct $x, y \in V$, $y - x \notin \mathbb{Q}$?

The rational translations of V:

$$V + r := \{y + r \mid y \in V\}, \ r \in \mathbb{Q},$$

form a *partition* of $\mathbb{R}$ since
(i) for any $r \in \mathbb{Q}$, $V + r \neq \emptyset$;
(ii) for distinct $r_1, r_2 \in \mathbb{Q}$,

$$(V + r_1) \cap (V + r_2) = \emptyset;$$

(iii)

$$\mathbb{R} = \bigcup_{r \in \mathbb{Q}} (V + r).$$

Since, for any $x \in \mathbb{R}$, there is an $r_1 \in \mathbb{Q}$ such that $x - r_1 \in [0,1]$, and hence, there is a $y \in V$, which is the *representative* of the *equivalence class* $[x - r_1]$, i. e., $(x - r_1) - y = r_2 \in \mathbb{Q}$ and $x = y + (r_1 + r_2)$, which implies that

$$x \in V + (r_1 + r_2).$$

Exercise 4.38. Verify (ii).

Thus, $\mathbb{R}$ is partitioned into *countably many* disjoint subsets.

Assume that V is *Lebesgue measurable*, i. e., $V \in \Sigma^*$, then, by the *Translation Invariance of the Lebesgue Measure Proposition* (Proposition 4.5) (see Section 4.8, Problem 7 (a)), for each $r \in \mathbb{Q}$,

$$V + r \in \Sigma^* \text{ and } \lambda(V + r) = \lambda(V).$$

By the *countable additivity* of the Lebesgue measure,

$$\infty = \lambda(\mathbb{R}) = \sum_{r \in \mathbb{Q}} \lambda(V + r).$$

Where, we infer that

$$\lambda(V) > 0,$$

which, by the *monotonicity* of the Lebesgue measure, yields the *contradiction*:

$$2 = \lambda([0,2]) \geq \lambda\left(\bigcup_{r \in \mathbb{Q},\ 0 \leq r \leq 1} \lambda(V + r) \right) = \sum_{r \in \mathbb{Q},\ 0 \leq r \leq 1} \lambda(V + r) = \infty.$$

The source of the obtained contradiction being the assumption of the Lebesgue measurability of the set V, we conclude that the set V is not Lebesgue measurable, which completes the proof. □

4.7.4 Multidimensional Lebesgue Measure

Let $n \in \mathbb{N}$ and λ_n be the *Lebesgue measure* on the product semiring

$$\mathscr{S}_n := \left\{ \prod_{i=1}^{n} (a_i, b_i] \,\middle|\, -\infty < a_i < b_i < \infty,\ i = 1, \dots, n \right\} \cup \{\emptyset\}$$

on $\mathbb{R}^n$ (see the *Product Semiring Proposition* (Proposition 2.1), Section 2.6, Problem 2):

$$\lambda_n(\emptyset) := 0,\ \lambda_n\left(\prod_{i=1}^{n} (a_i, b_i] \right) := \prod_{i=1}^{n} (b_i - a_i).$$

For $n = 1$, λ_1 is the length on the intervals of the form $(a, b]$, considered in the *Lebesgue–Stieltjes Measures on Interval Semiring Theorem* (Theorem 3.5). For $n = 2$, λ_2 is the *area* on the rectangles of the form $(a_1, b_1] \times (a_2, b_2]$. For $n = 3$, λ_3 is the *volume* on the beams of the form $(a_1, b_1] \times (a_2, b_2] \times (a_3, b_3]$.

By the *Measure Extension from a Semiring Theorem* (Theorem 4.1), λ_n on the semiring $\mathscr{S}_n$ is uniquely extended to the measure, also denoted by λ_n, on the generated ring $r(\mathscr{S}_n)$, which is *σ-finite* on $r(\mathscr{S}_n)$.

Exercise 4.39. Explain the latter.

Let λ_n^* be the outer measure on the power set $\mathscr{P}(\mathbb{R}^n)$ generated by the measure λ_n on the ring $r(\mathscr{S}_n)$ and Σ_n^* be the collection of all λ_n^*-measurable subsets of $\mathbb{R}^n$.

By *Carathéodory's Extension Theorem* (Theorem 4.5), Σ_n^* is a *σ-algebra* on $\mathbb{R}^n$ and λ_n^* is a *complete measure* on Σ_n^*, which is the completion of the unique measure extension of λ_n to

$$\sigma a(r(\mathscr{S}_n)) = \sigma a(\mathscr{S}_n) = \mathscr{B}(\mathbb{R}^n).$$

The sets from Σ_n^* are called *Lebesgue measurable* and the measure λ_n^* on Σ_n^*, henceforth designated by λ_n, is called the *(n-dimensional) Lebesgue measure.*

We have the inclusions

$$\mathscr{S}_n \subset r(\mathscr{S}_n) \subset \sigma a(\mathscr{S}_n) = \mathscr{B}(\mathbb{R}^n) \subset \Sigma_n^*.$$

Remark 4.12. The n-dimensional Lebesgue–Stieltjes measures ($n \in \mathbb{N}$) can be defined in the same fashion.

Examples 4.10. For the n-dimensional Lebesgue measure λ_n ($n \in \mathbb{N}$),

1. for each $x := (x_1, \dots, x_n) \in \mathbb{R}^n$, $\{x\} \in \Sigma_n^*$ and
$$\lambda_n(\{x\}) = 0;$$
2. for each countable set $C := \{x_n\}_{n \in I}$, $C \in \Sigma_n^*$ and
$$\lambda_n(C) = 0,$$
in particular,
$$\lambda_n(\mathbb{N}^n) = \lambda_n(\mathbb{Z}^n) = \lambda_n(\mathbb{Q}^n) = 0;$$
3. for any $-\infty < a < b < \infty$ and $c \in \mathbb{R}$, $[a, b] \times \{c\} \in \Sigma_2^*$ and
$$\lambda_2([a, b] \times \{c\}) = 0;$$
4. for a straight line l in $\mathbb{R}^2$, $l \in \Sigma_2^*$ and
$$\lambda_2(l) = 0;$$

5. for a function $f \in C(\mathbb{R})$, the graph

$$G_f := \{(x, f(x)) \mid x \in \mathbb{R}\} \in \Sigma_2^*$$

and

$$\lambda_2(G_f) = 0;$$

6. for a nonnegative function $f \in C[a, b]$ $(-\infty < a < b < \infty)$,

$$R := \{(x, y) \mid x \in [a, b],\ 0 \le y \le f(x)\} \in \Sigma_2^*$$

and

$$\lambda_2(R) = \int_a^b f(x)\, dx.$$

Exercise 4.40. Verify.

4.8 Problems

1. For the interval semiring,

$$\mathscr{S}_1 := \{(a, b] \mid -\infty < a < b < \infty\} \cup \{\emptyset\}$$

and the Lebesgue measure

$$\lambda(\emptyset) := 0,\ \lambda((a, b]) := b - a,\ (a, b] \in \mathscr{S}_1,$$

on $\mathscr{S}_1$.
(a) define the outer measure λ^* generated by λ;
(b) find $\lambda^*(\{x\})$ for any $x \in \mathbb{R}$;
(c) find $\lambda^*(\mathbb{R})$, $\lambda^*(\mathbb{Q})$, and $\lambda^*(\mathbb{R} \setminus \mathbb{Q})$;
(d) find $\lambda^*([a, b])$, $-\infty < a < b < \infty$.

2. Let

$$\mathscr{S} := \{(n, n + 1] \mid n \in \mathbb{Z}\} \cup \{\emptyset\}$$

and

$$\lambda(\emptyset) := 0,\ \lambda((n, n + 1]) := 1,\ n \in \mathbb{Z},$$

(see Examples 4.3).
(a) verify that $\mathscr{S}$ is a *semiring*;
(b) explain why λ is a measure on $\mathscr{S}$;

(c) define the outer measure λ^* generated by μ;
(d) find $\lambda^*(\{x\})$ for $x \in \mathbb{Z}$ and $x \notin \mathbb{Z}$;
(e) find $\lambda^*(\mathbb{N})$;
(f) find $\lambda^*((1/2, 3/2])$.

3. Show that the class of all λ^*-measurable sets relative to the outer measure λ^* generated by the measure λ from the prior problem is the collection of all countable unions of sets from $\mathscr{S}$. Infer that the set $(0, \infty)$ is λ^*-measurable and the set $(1/2, 3/2]$ is not.
4. Let μ be a measure on a σ-algebra Σ on a nonempty set X. Show that
   (a) the set function
   $$\mathscr{P}(X) \ni A \mapsto \mu^*(A) := \inf\{\mu(B) \mid B \in \Sigma,\ A \subseteq B\}$$
   is an outer measure on $\mathscr{P}(X)$;
   (b) each set $A \in \Sigma$ is μ^*-measurable and
   $$\mu^*(A) = \mu(A).$$
5. Let μ be a measure on a ring $\mathscr{R}$ on a nonempty set X, μ^* be the outer measure on $\mathscr{P}(X)$ generated by μ, Σ^* be the σ-algebra of all μ^*-measurable sets, and
   $$\mathscr{P}(X) \ni A \mapsto \mu^{**}(A) := \inf\left\{\sum_{i=1}^{\infty} \mu^*(A_i) \,\middle|\, (A_n)_{n\in\mathbb{N}} \subseteq \Sigma^*,\ A \subseteq \bigcup_{i=1}^{\infty} A_i\right\}.$$
   Show that $\mu^{**} = \mu^*$.
6. Prove

   Proposition 4.4 (Characterization of μ^*-Measurability). *Let μ be a measure on a ring $\mathscr{R}$ on a nonempty set X, μ^* be the outer measure generated by μ, and Σ^* be the σ-algebra of all μ^*-measurable sets on X. Then, for a set $A \in \mathscr{P}(X)$ to be μ^*-measurable ($A \in \Sigma^*$), it is sufficient and, provided $\mu^*(A) < \infty$, necessary that*
   $$\forall \varepsilon > 0\ \exists B \in \mathscr{R} :\ \mu^*(A \triangle B) < \varepsilon.$$
   See the *Approximation Theorem* (Theorem 4.6) and Remarks 4.8.
7. (a) Prove

   Proposition 4.5 (Translation Invariance of the Lebesgue Measure). *Let λ be the Lebesgue measure on $\mathbb{R}$. Then, for any $a \in \mathbb{R}$ and $A \in \Sigma^*$,*
   $$A + a := \{x + a \mid x \in A\} \in \Sigma^* \text{ and } \lambda(A + a) = \lambda(A).$$
   (b) Can the prior statement be generalized to an arbitrary Lebesgue–Stieltjes measure on $\mathbb{R}$?
8. Show that the Lebesgue measure on $\mathbb{R}$ is an extension of the Jordan measure.

Hint. Use the *Approximation of Jordan Measurable Sets Proposition* (Proposition 3.6).

9. Show that, for each bounded Lebesgue measurable set A in $\mathbb{R}$, $\lambda(A) < \infty$.
10. Let $A \in \Sigma^*$ with $\lambda(A) < \infty$. Show that the function
$$F(x) := \lambda(A \cap (-\infty, x)), \ x \in \mathbb{R},$$
is *increasing* and *continuous*.
11. Let $A \in \Sigma^*$ be bounded with $\lambda(A) > 0$. Show that
$$\forall\, 0 < m < \lambda(A)\ \exists\, B \in \Sigma^* : \ B \subseteq A \text{ and } \lambda(B) = m.$$
12. * Give an example of a closed set $A \in \Sigma^*$ with $\lambda(A) > 0$, which is *nowhere dense* in $\mathbb{R}$, i. e., contains no open interval (a, b) $(-\infty < a < b < \infty)$.
13. Prove the following.

Proposition 4.6 (Approximation of Borel Sets). *Let A be a nonempty bounded Borel measurable subset of $\mathbb{R}$. Then, for any $\varepsilon > 0$,*
(1) *there exists a nonempty open set $G \subset \mathbb{R}$ such that*
$$A \subseteq G \quad \textit{and} \quad \lambda(G \setminus A) < \varepsilon;$$
(2) *there exists a nonempty closed set $F \subset \mathbb{R}$ such that*
$$F \subseteq A \quad \textit{and} \quad \lambda(A \setminus F) < \varepsilon;$$
(3) *there exists a set $I \subset \mathbb{R}$, which is a countable intersection of open sets, such that*
$$A \subseteq I \quad \textit{and} \quad \lambda(I \setminus A) = 0;$$
(4) *there exists a set $U \subset \mathbb{R}$, which is a countable union of closed sets, such that*
$$U \subseteq A \quad \textit{and} \quad \lambda(A \setminus U) = 0.$$

Hint. Use an approach similar to proving the *Approximation of Jordan Measurable Sets Proposition* (Proposition 3.6) and the fact that the boundedness of A implies that $A \subseteq [a, b]$ with some $-\infty < a < b < \infty$.

Remark 4.13. The boundedness of the set A is essential for parts (2) and (4) only. For parts (1) and (3), the condition of boundedness for the set A can be tempered to the condition of *finiteness* for its Lebesgue measure: $\lambda(A) < \infty$.

14. In view of Remark 4.13, prove the following corollary of the prior proposition

Corollary 4.1 (Approximation of Sets with Finite Outer Measure). *For an arbitrary set $A \subset \mathbb{R}$ with $\lambda^*(A) < \infty$ and any $\varepsilon > 0$, there exists an open set $G \subset \mathbb{R}$ such that*
$$A \subseteq G \quad \textit{and} \quad \lambda(G) < \lambda^*(A) + \varepsilon.$$

Hint. Using the fact noted in Remarks 4.7, first show that

$$\exists B \in \mathscr{B}(\mathbb{R}),\ A \subseteq B :\ \lambda^*(A) = \lambda(B)$$

and then apply the prior proposition, part (a).

5 Measurable Functions

Here, we introduce and study the concept of *measurability* for functions, relate it to the familiar notion of continuity over an interval $[a, b]$ $(-\infty < a < b < \infty)$ (see *Luzin's Theorem* (Theorem 5.6)), and discuss in detail two new types of convergence of function sequences related to the idea of measure: *convergence almost everywhere* and *convergence in measure*.

5.1 Measurable Space and Measure Space

To proceed, we need to define the notions of *measurable space* and *measure space*.

Definition 5.1 (Measurable Space, Measurable Sets). Let Σ be a σ-algebra on a nonempty set X. The pair (X, Σ) is called a *measurable space*.

A set $A \subseteq X$ is said to be *measurable*, more precisely Σ-*measurable*, if $A \in \Sigma$.

Remark 5.1. In particular, if (X, ρ) is a metric space and $\mathscr{B}(X)$ is the Borel σ-algebra on X (see Definition 2.8), we call $\mathscr{B}(X)$-measurable sets *Borel measurable*.

Definition 5.2 (Measure Space). Let (X, Σ) be measurable space with a measure μ on the σ-algebra Σ. The triple (X, Σ, μ) is called a *measure space*.

Remark 5.2. Each measure space is, of course, a measurable space.

Examples 5.1.
1. For an arbitrary nonempty set X with the power set σ-algebra $\mathscr{P}(X)$ and the *counting measure* μ on $\mathscr{P}(X)$ (see Examples 3.2), $(X, \mathscr{P}(X), \mu)$ is a measure space.
2. Of particular importance are the *Lebesgue measure space* $(\mathbb{R}, \Sigma^*, \lambda)$ and the *Borel measure space* $(\mathbb{R}, \mathscr{B}(\mathbb{R}), \lambda)$, where Σ^* and $\mathscr{B}(\mathbb{R})$ are the σ-algebras of Lebesgue and Borel measurable subsets of $\mathbb{R}$, respectively, and λ is the Lebesgue measure. Observe that the former space is the *completion* of the latter (see Section 4.7.1).

5.2 Definition and Examples

The following central definition is reminiscent of the *Characterization of Continuity* (Theorem 1.19) being also based on preimage operation.

Definition 5.3 (Measurable Function). Let (X, Σ) and (X', Σ') be measurable spaces. A function

$$f : X \to X'$$

is called Σ-Σ'-*measurable* if

$$\forall A' \in \Sigma' : f^{-1}(A') \in \Sigma,$$

https://doi.org/10.1515/9783110600995-005

i. e., the preimage of every Σ'-measurable set is Σ-measurable, or equivalently,

$$f^{-1}(\Sigma') := \left\{f^{-1}(A') \,\middle|\, A \in \Sigma'\right\} \subseteq \Sigma.$$

Provided $X' := \mathbb{R}$ and $\Sigma' := \mathscr{B}(\mathbb{R})$ a Σ-Σ'-measurable function

$$f : X \to \mathbb{R},$$

i. e., such that

$$\forall A' \in \mathscr{B}(\mathbb{R}) : f^{-1}(A') \in \Sigma,$$

is called Σ-*measurable*.

Remarks 5.3.

- For a Σ-measurable function $f : X \to \mathbb{R}$,

$$\forall a \in \mathbb{R} : f^{-1}(\{a\}) \in \Sigma.$$

- A complex-valued function

$$f(x) = \operatorname{Re} f(x) + i \operatorname{Im} f(x),\ x \in X,$$

  is naturally defined to be Σ-measurable if the real-valued functions $\operatorname{Re} f$ and $\operatorname{Im} f$ are both Σ-measurable.

Examples 5.2. For arbitrary nonempty sets X and X',

1. if $\Sigma := \mathscr{P}(X)$ or $\Sigma' := \{\emptyset, X'\}$, any function $f : X \to X'$ is Σ-Σ'-measurable;
2. *constant* functions $f : X \to X'$ are Σ-Σ'-measurable regardless of the nature of the underlying σ-algebras Σ and Σ';
3. if $\Sigma := \{\emptyset, X\}$, the set X' contains at least two elements, and Σ' contains singletons (e. g., $X' := \mathbb{R}$ and $\Sigma' := \mathscr{B}(\mathbb{R})$), only *constant functions* $f : X \to X'$ are Σ-measurable;
4. a function $f : X \to X'$ with a finite range $f(X) = \{a_1, \ldots, a_n\} \subseteq X'$ ($n \in \mathbb{N}$) and such that

$$f^{-1}(\{a_i\}) \in \Sigma,\ i = 1, \ldots, n,$$

   is Σ-Σ'-measurable.

Exercise 5.1. Verify.

Remarks 5.4.

- For arbitrary nonempty sets X and X' and σ-algebra Σ' on X', any function $f : X \to X'$ is Σ_f-Σ'-measurable provided $\Sigma_f := f^{-1}(\Sigma')$ is the *preimage σ-algebra* of Σ' relative to the function f, which is the *smallest* of all σ-algebras Σ on X such that the function f is Σ-Σ'-measurable (see the *Preimage σ-Algebra Proposition* (Proposition 5.3), Section 5.12, Problem 1).

– The restriction f_A of a Σ-Σ'-measurable function $f : X \to X'$ to any nonempty Σ-measurable set $A \in \Sigma$ is $\Sigma \cap A$-Σ'-measurable, where

$$\Sigma \cap A := \{B \cap A \mid B \in \Sigma\}$$

is the *trace σ-algebra* on A (see Section 2.6, Problem 12).

5.3 A Characterization of Σ-Σ′-Measurability

Theorem 5.1 (Characterization of Σ-Σ′-Measurability). *Let (X, Σ) and (X', Σ') be measurable spaces and*

$$\Sigma' = \sigma a(\mathscr{C})$$

with some $\mathscr{C} \subseteq \mathscr{P}(X')$. A function $f : X \to X'$ is Σ-Σ'-measurable iff

$$\forall A' \in \mathscr{C} : f^{-1}(A') \in \Sigma.$$

Proof. *"Only if"* part immediately follows from the definition.

"If" part. Suppose that

$$f^{-1}(\mathscr{C}) := \left\{f^{-1}(A') \,\middle|\, A' \in \mathscr{C}\right\} \subseteq \Sigma,$$

and consider

$$\mathscr{C}' := \left\{A' \in \Sigma' \,\middle|\, f^{-1}(A') \in \Sigma\right\}.$$

Clearly, we have the inclusions

$$\mathscr{C} \subseteq \mathscr{C}' \subseteq \Sigma' = \sigma a(\mathscr{C}).$$

By the *Properties of Inverse Image* (Theorem 1.4), it can be easily shown that $\mathscr{C}'$ is a *σ-algebra* on X'.

Exercise 5.2. Show.

Therefore,

$$\sigma a(\mathscr{C}) \subseteq \mathscr{C}' \subseteq \sigma a(\mathscr{C}),$$

which shows that

$$\mathscr{C}' = \sigma a(\mathscr{C}) = \Sigma'$$

completing the proof. □

Corollary 5.1 (Characterizations of Σ-Measurability). *Let (X, Σ) be a measurable space and $f : X \to \mathbb{R}$. The following statements are equivalent:*
(1) *The function f is Σ-measurable.*
(2) *For each $a \in \mathbb{R}$, $f^{-1}((a, \infty)) \in \Sigma$.*
(3) *For each $a \in \mathbb{R}$, $f^{-1}([a, \infty)) \in \Sigma$.*
(4) *For each $b \in \mathbb{R}$, $f^{-1}((-\infty, b)) \in \Sigma$.*
(5) *For each $b \in \mathbb{R}$, $f^{-1}((-\infty, b]) \in \Sigma$.*
(6) *For any $-\infty < a < b < \infty$, $f^{-1}((a, b)) \in \Sigma$.*
(7) *For any $-\infty < a < b < \infty$, $f^{-1}([a, b)) \in \Sigma$.*
(8) *For any $-\infty < a < b < \infty$, $f^{-1}((a, b]) \in \Sigma$.*
(9) *For any $-\infty < a < b < \infty$, $f^{-1}([a, b]) \in \Sigma$.*

Exercise 5.3. Prove (see Exercise 2.17).

It is convenient to consider the following natural extension of the concept of the Borel σ-algebra on $\mathbb{R}$ (see Definition 2.8) to the extended real line $\overline{\mathbb{R}} := [-\infty, \infty]$, the *extended Borel σ-algebra on* $\overline{\mathbb{R}}$:

$$\mathscr{B}(\overline{\mathbb{R}}) := \{A, A \cup \{-\infty\}, A \cup \{\infty\}, A \cup \{-\infty, \infty\} \mid A \in \mathscr{B}(\mathbb{R})\}$$

generated by any of the following set collections:

$$\mathscr{C}_1 := \{[a, \infty] \mid a \in \overline{\mathbb{R}}\} \subseteq \mathscr{P}(\overline{\mathbb{R}}),$$
$$\mathscr{C}_2 := \{[-\infty, b] \mid b \in \overline{\mathbb{R}}\} \subseteq \mathscr{P}(\overline{\mathbb{R}}),$$
$$\mathscr{C}_3 := \{[a, b] \mid -\infty \le a < b \le \infty\} \subseteq \mathscr{P}(\overline{\mathbb{R}}),$$

and extended real-valued functions

$$f : X \to \overline{\mathbb{R}}.$$

To characterize the Σ-measurability (i. e., Σ-$\mathscr{B}(\overline{\mathbb{R}})$-measurability) of such functions, the prior corollary is modified as follows.

Corollary 5.2 (Characterizations of Σ-Measurability of Extended Real Valued Functions). *Let (X, Σ) be a measurable space and $f : X \to \overline{\mathbb{R}}$. The following statements are equivalent:*
(1) *The function f is Σ-measurable.*
(2) *For each $a \in \overline{\mathbb{R}}$, $f^{-1}([a, \infty]) \in \Sigma$.*
(3) *For each $b \in \overline{\mathbb{R}}$, $f^{-1}([-\infty, b]) \in \Sigma$.*
(4) *For any $-\infty \le a < b \le \infty$, $f^{-1}([a, b]) \in \Sigma$.*

5.4 Borel and Lebesgue Measurable Functions

Two important particular cases of measurability of functions introduced here deserve our special attention.

Definition 5.4 (Borel Measurable Function). Let (X,ρ) be a metric space and $\mathscr{B}(X)$ be the Borel σ-algebra on X. A $\mathscr{B}(X)$-measurable function

$$f : X \to \mathbb{R}$$

is called *Borel measurable*:

$$\forall A' \in \mathscr{B}(\mathbb{R}) : f^{-1}(A') \in \mathscr{B}(X).$$

Examples 5.3.

1. A *continuous* function $f : X \to \mathbb{R}$ ($f \in C(X)$) is Borel measurable.
2. If $X \in \mathscr{B}(\mathbb{R})$, a *monotone* (i. e., *increasing* or *decreasing*) (see Definition 8.3) function $f : X \to \mathbb{R}$ is Borel measurable, i. e., $\mathscr{B}(X)$-measurable, where $\mathscr{B}(X) = \mathscr{B}(\mathbb{R}) \cap X$ (see Remarks 5.4) is the σ-algebra of Borel subsets of X.

Exercise 5.4. Verify.

Definition 5.5 (Lebesgue Measurable Function). Let a set $X \subseteq \mathbb{R}^n$ ($n \in \mathbb{N}$) be Lebesgue measurable ($X \in \Sigma_n^*$) and

$$\Sigma_n^*(X) = \Sigma_n^* \cap X$$

be the σ-algebra of Lebesgue measurable subsets of X. A $\Sigma_n^*(X)$-measurable function

$$f : X \to \mathbb{R},$$

i. e., such that

$$\forall A' \in \mathscr{B}(\mathbb{R}) : f^{-1}(A') \in \Sigma_n^*,$$

is called *Lebesgue measurable*.

Remark 5.5. If $X \in \mathscr{B}(\mathbb{R}^n)$ ($n \in \mathbb{N}$), each Borel measurable function $f : X \to \mathbb{R}$ is Lebesgue measurable, but not vice versa.

Exercise 5.5. Verify. Give a corresponding counterexample.

Example 5.4. Each function $f : [a,b] \to \mathbb{R}$ ($-\infty < a < b < \infty$), whose set of points of discontinuity is a *null-set* relative to the Lebesgue measure λ (i. e., its Lebesgue measure is 0), is Lebesgue measurable.

Exercise 5.6. Verify (cf. the *A. E. Continuous is Lebesgue Measurable Proposition* (Proposition 5.10), Section 5.12, Problem 13).

Remarks 5.6.

- The definitions of Borel and Lebesgue measurability naturally stretch to extended real-valued functions as $\mathscr{B}(X)$-$\mathscr{B}(\overline{\mathbb{R}})$ and $\Sigma_n^*(X)$-$\mathscr{B}(\overline{\mathbb{R}})$ measurabilities, respectively.

- A complex-valued function f is naturally defined to be Borel/Lebesgue measurable if the real-valued functions $\operatorname{Re} f$ and $\operatorname{Im} f$ are both Borel/Lebesgue measurable (see Remarks 5.3).

5.5 Properties of Measurable Functions

5.5.1 Compositions of Measurable Functions

Theorem 5.2 (Composition of Measurable Functions). *Let (X,Σ), (X',Σ'), and (X'',Σ'') be measurable spaces, $f : X \to X'$ be Σ-Σ'-measurable and $g : X' \to X''$ be Σ'-Σ''-measurable. Then their composition*

$$(g \circ f)(x) := g(f(x)),\ x \in X,$$

is Σ-Σ''-measurable.

Exercise 5.7. Prove.

Remark 5.7. In particular, if $f, g : \overline{\mathbb{R}} \to \overline{\mathbb{R}}$ are Borel measurable functions, then the composition $g \circ f : \overline{\mathbb{R}} \to \overline{\mathbb{R}}$ is Borel measurable.

Corollary 5.3 (Composition of Measurable Functions). *Let (X,Σ) be a measurable space, functions*

$$f_i : X \to \mathbb{R},\ i = 1,\dots,n\ (n \in \mathbb{N})$$

be Σ-measurable with

$$(f_1(x),\dots,f_n(x)) \in B,\ x \in X,$$

for some $B \in \mathscr{B}(\mathbb{R}^n)$, and

$$F : B \to \overline{\mathbb{R}}$$

be Borel measurable. Then the function

$$X \ni x \mapsto F(f_1(x),\dots,f_n(x)) \in \overline{\mathbb{R}},$$

is Σ-measurable.

Proof. Let us show that the mapping

$$X \ni x \mapsto f(x) := (f_1(x),\dots,f_n(x)) \in B \in \mathscr{B}(\mathbb{R}^n)$$

is Σ-$\mathscr{B}(B)$-measurable, where

$$\mathscr{B}(B) = \mathscr{B}(\mathbb{R}^n) \cap B$$

is the σ-algebra of Borel measurable subsets of B.

Since

$$\mathscr{B}(\mathbb{R}^n) = \sigma a\left(\left\{\prod_{i=1}^{n}(-\infty, b_i)\,\middle|\, b_i \in \mathbb{R},\ i = 1,\dots,n\right\}\right)$$

(cf. Exercises 2.17), by the *Characterization of* Σ-Σ'*-Measurability* (Theorem 5.1), it suffices to show that

$$f^{-1}\left(\prod_{i=1}^{n}(-\infty, b_i)\right),\ b_i \in \mathbb{R},\ i = 1,\dots,n,$$

which is true since, in view of the Σ-measurability of each f_i, $i = 1,\dots,n$, for all $b_i \in \mathbb{R}$, $i = 1,\dots,n$,

$$\begin{aligned} f^{-1}\left(\prod_{i=1}^{n}(-\infty, b_i)\right) &= \left\{x \in X \,\middle|\, (f_1(x),\dots,f_n(x)) \in \prod_{i=1}^{n}(-\infty, b_i)\right\} \\ &= \bigcap_{i=1}^{n}\{x \in X \mid f_i(x) \in (-\infty, b_i)\} = \bigcap_{i=1}^{n} f_i^{-1}((-\infty, b_i)) \in \Sigma. \end{aligned}$$

Hence, by the *Composition of Measurable Functions Theorem* (Theorem 5.2), the composition

$$F(f_1(x),\dots,f_n(x)) = F(f(x)),\ x \in X$$

is Σ-measurable, which completes the proof. □

Remark 5.8. It is known that the composition $g \circ f$ of a Borel measurable (even continuous) function f with a Lebesgue measurable function g need not be Lebesgue measurable ([7, Chapter 4, § 19]). Thus, in the prior corollary the requirement of Borel measurability for F is essential and cannot be dropped.

5.5.2 Combinations of Measurable Functions

Theorem 5.3 (Combinations of Measurable Functions). *Let* (X, Σ) *be a measurable space and*

$$f, g : X \to \mathbb{R}$$

be Σ*-measurable functions. Then* Σ*-measurable are the following functions:*
(1) cf $(c \in \mathbb{R})$;
(2) $f \pm g$;
(3) $f \cdot g$;
(4) $\frac{f}{g}$ *provided* $g(x) \neq 0$, $x \in X$;
(5) $|f|$;
(6) $\min(f, g)$ *and* $\max(f, g)$.

Proof. Parts (1)–(5) follow immediately from Corollary 5.3 when applied to:
(1) f and the *continuous* $F(y) := cy$, $y \in \mathbb{R}$ $(n = 1)$;
(2) f and g and the *continuous* $F(x_1, x_2) = x_1 \pm x_2$, $(x_1, x_2) \in \mathbb{R}^2$ $(n = 2)$;
(3) f and g and the *continuous* $F(x_1, x_2) = x_1 \cdot x_2$, $(x_1, x_2) \in \mathbb{R}^2$ $(n = 2)$;
(4) f and g and the *continuous* $F(x_1, x_2) = \frac{x_1}{x_2}$, $(x_1, x_2) \in \mathbb{R}^2$, $x_2 \neq 0$ $(n = 2)$;
(5) f and the *continuous* $F(y) := |y|$, $y \in \mathbb{R}$ $(n = 1)$;

Part (6) follows from (1), (2), and (5), in view of the representations

$$\min(f, g) = \frac{f + g - |f - g|}{2}, \quad \max(f, g) = \frac{f + g + |f - g|}{2}.$$ □

Remark 5.9. The prior theorem naturally stretches to extended real-valued functions $f, g : X \to \overline{\mathbb{R}}$ whenever indeterminacies

$$\infty - \infty, \ 0 \cdot \infty, \ \frac{\infty}{\infty}$$

do not occur.

Corollary 5.4 (Positive and Antinegative Parts of a Function). *Let (X, Σ) be a measurable space, and $f : X \to \overline{\mathbb{R}}$ be an extended real-valued Σ-measurable function. Then the functions*

$$f_+(x) := \max(f(x), 0) := \begin{cases} f(x), & x \in X : f(x) \geq 0, \\ 0, & x \in X : f(x) < 0, \end{cases}$$

and

$$f_-(x) := -\min(f(x), 0) := \begin{cases} -f(x), & x \in X : f(x) < 0, \\ 0, & x \in X : f(x) \geq 0, \end{cases}$$

we call the positive and antinegative parts of the function f, respectively, are Σ-measurable and

$$f(x) = f_+(x) - f_-(x) \text{ and } |f(x)| = f_+(x) + f_-(x), \ x \in X.$$

Remark 5.10. In particular, the *Combinations of Measurable Functions Theorem* (Theorem 5.3) and the *Positive and Antinegative Parts of a Function Corollary* (Corollary 5.4) hold for Borel and Lebesgue measurable functions.

5.5.3 Sequences of Measurable Functions

Sequences of measurable functions and their limits in various senses are indispensable for our subsequent discourse.

Theorem 5.4 (Sequences of Measurable Functions). *Let (X, Σ) be a measurable space and $(f_n : X \to \overline{\mathbb{R}})_{n\in\mathbb{N}}$ be a sequence of extended real-valued Σ-measurable functions. Then Σ-measurable are the following functions:*

(1) $\inf_{n\in\mathbb{N}} f_n(x)$, $x \in X$, *and* $\sup_{n\in\mathbb{N}} f_n(x)$, $x \in X$;

(2) $\underline{\lim}_{n\to\infty} f_n(x)$, $x \in X$, *and* $\overline{\lim}_{n\to\infty} f_n(x)$, $x \in X$;

(3) $\lim_{n\to\infty} f_n(x)$, $x \in X$, *provided the (finite or infinite) limit exists for each* $x \in X$, *and furthermore the finite limit set*

$$F := \{x \in X \mid (f_n(x))_{n\in\mathbb{N}} \text{ has a finite limit}\} \in \Sigma.$$

Proof.

(1) For each $a \in \overline{\mathbb{R}}$,

$$\left\{x \in X \,\middle|\, \inf_{n\in\mathbb{N}} f_n(x) \geq a\right\} = \bigcap_{n=1}^{\infty} \{x \in X \mid f_n(x) \geq a\} = \bigcap_{n=1}^{\infty} f_n^{-1}([a, \infty]) \in \Sigma.$$

For each $b \in \overline{\mathbb{R}}$,

$$\left\{x \in X \,\middle|\, \sup_{n\in\mathbb{N}} f_n(x) \leq b\right\} = \bigcap_{n=1}^{\infty} \{x \in X \mid f_n(x) \leq b\} = \bigcap_{n=1}^{\infty} f_n^{-1}([-\infty, b]) \in \Sigma.$$

Hence, by the *Characterizations of Σ-Measurability of Extended Real Valued Functions* (Corollary 5.2), the extended real-valued functions

$$\inf_{n\in\mathbb{N}} f_n(x),\ x \in X, \quad \text{and} \quad \sup_{n\in\mathbb{N}} f_n(x),\ x \in X,$$

are Σ-measurable.

(2) Since

$$\underline{\lim}_{n\to\infty} f_n(x) := \sup_{n\in\mathbb{N}} \inf_{k\geq n} f_k(x),\ x \in X, \quad \overline{\lim}_{n\to\infty} f_n(x) := \inf_{n\in\mathbb{N}} \sup_{k\geq n} f_k(x),\ x \in X,$$

by (1),

$$\underline{\lim}_{n\to\infty} f_n(x),\ x \in X, \quad \text{and} \quad \overline{\lim}_{n\to\infty} f_n(x),\ x \in X,$$

are Σ-measurable.

(3) If $\lim_{n\to\infty} f_n(x)$ exists for each $x \in X$. Then, by the *Characterization of Limit Existence* (Proposition 1.3),

$$\lim_{n\to\infty} f_n(x) = \underline{\lim}_{n\to\infty} f_n(x) = \overline{\lim}_{n\to\infty} f_n(x),\ x \in X,$$

is Σ-measurable.

Furthermore, by (2) and the *Equality Set Proposition* (Proposition 5.4, Section 5.12, Problem 5),

$$\begin{aligned} F &:= \{x \in X \mid (f_n(x))_{n\in\mathbb{N}} \text{ has a finite limit}\} \\ &= \left\{x \in X \,\middle|\, \underline{\lim}_{n\to\infty} f_n(x) = \overline{\lim}_{n\to\infty} f_n(x)\right\} \bigcap \left\{x \in X \,\middle|\, \overline{\lim}_{n\to\infty} f_n(x) \in \mathbb{R}\right\} \in \Sigma. \end{aligned}$$

□

Remark 5.11. By the prior theorem,

$$L_{-\infty} := \left\{x \in X \,\middle|\, \lim_{n\to\infty} f_n(x) = -\infty\right\} = \left\{x \in X \,\middle|\, \overline{\lim_{n\to\infty}} f_n(x) = -\infty\right\} \in \Sigma,$$
$$L_{\infty} := \left\{x \in X \,\middle|\, \lim_{n\to\infty} f_n(x) = \infty\right\} = \left\{x \in X \,\middle|\, \underline{\lim_{n\to\infty}} f_n(x) = \infty\right\} \in \Sigma,$$

and hence, the *infinite limit set*

$$I := \{x \in X \mid (f_n(x))_{n\in\mathbb{N}} \text{ has an infinite limit}\} = L_{-\infty} \cup L_\infty \in \Sigma$$

and the *limit set*

$$L := \{x \in X \mid (f_n(x))_{n\in\mathbb{N}} \text{ has a (finite or infinite) limit}\} = F \cup I \in \Sigma.$$

Corollary 5.5 (Series of Measurable Functions). *Let (X, Σ) be a measurable space.*

(1) *If $(f_n : X \to \mathbb{R})_{n\in\mathbb{N}}$ be a sequence of Σ-measurable functions and the series*

$$\sum_{k=1}^{\infty} f_k(x)$$

converges for each $x \in X$, then

$$s(x) := \sum_{k=1}^{\infty} f_k(x), \; x \in X,$$

is a Σ-measurable function.

(2) *If $(f_n : X \to [0, \infty])_{n\in\mathbb{N}}$ be a sequence of nonnegative extended real-valued Σ-measurable functions, then*

$$s(x) := \sum_{k=1}^{\infty} f_k(x), \; x \in X,$$

is a nonnegative extended real-valued Σ-measurable function.

Exercise 5.8. Prove.

Remark 5.12. In particular, the *Sequences of Measurable Functions Theorem* (Theorem 5.4) and the *Series of Measurable Functions Corollary* (Corollary 5.5) hold for Borel and Lebesgue measurable functions.

5.6 Simple Functions

Simple functions are needed for the construct of general Lebesgue integral introduced and discussed in the following chapter.

Definition 5.6 (Characteristic Function). Let A be a subset of a nonempty set X. The function

$$\chi_A(x) := \begin{cases} 1, & x \in A, \\ 0, & x \in A^c \end{cases}$$

is called the *characteristic function* of A.

Definition 5.7 (Simple Function). Let (X, Σ) be a measurable space. A function $s : X \to \mathbb{R}$ is called *simple* if its range $s(X)$ is a finite subset of $\mathbb{R}$:

$$s(X) = \{a_1, \dots, a_m\} \subset \mathbb{R} \; (m \in \mathbb{N}).$$

Remarks 5.13.

- Every simple function $s : X \to \mathbb{R}$ can be represented as follows:

$$s = \sum_{k=1}^{m} a_k \chi_{A_k}, \tag{5.1}$$

  where $m \in \mathbb{N}$, $a_k \in \mathbb{R}$, $k = 1, \dots, m$, the sets $A_k \in \mathscr{P}(X)$, $k = 1, \dots, m$, form a *partition* of X, i. e.,

$$A_i \cap A_j = \emptyset, \; i \neq j, \quad \text{and} \quad \bigcup_{k=1}^{m} A_k = X, \tag{5.2}$$

  and χ_{A_k} is the characteristic function of the set A_k, $k = 1, \dots, m$.
- Conversely, every function $s : X \to \mathbb{R}$ allowing representation (5.1)–(5.2) is simple.
- Representation (5.1)–(5.2) for a simple function $s : X \to \mathbb{R}$ with *distinct* $a_k \in \mathbb{R}$ $k = 1, \dots, m$, is *unique*, in which case

$$A_k := s^{-1}(\{a_k\}), \; k = 1, \dots, m.$$

- A simple function $s : X \to \mathbb{R}$ with $s(X) = \{a_1, \dots, a_m\} \subset \mathbb{R}$ $(m \in \mathbb{N})$ is Σ-measurable *iff*

$$\forall k = 1, \dots, m : \; A_k := s^{-1}(\{a_k\}) \in \Sigma$$

  (cf. Examples 5.2).
- The linear combinations and products of simple functions are also simple functions, which, by the *Combinations of Measurable Functions Theorem* (Theorem 5.3) are measurable provided the initial functions are.
- The notions of a simple function and its measurability naturally extend to the complex-valued case.

Exercise 5.9. Verify.

Examples 5.5.
1. For the Borel measurable space $(\mathbb{R}, \mathscr{B}(\mathbb{R}))$,

   (a) $f(x) := \chi_{[0,2]} - \chi_{[3,4]} = \begin{cases} 1, & x \in [0,2], \\ -1, & x \in [3,4], \\ 0, & \text{otherwise} \end{cases}$

   is a Borel measurable simple function and

   (b) $f(x) := \sum_{n=1}^{\infty} n\chi_{[n,n+1)} = \begin{cases} n, & x \in [n, n+1),\ n \in \mathbb{N}, \\ 0, & x \in (-\infty, 1) \end{cases}$

   is *not* a simple function.
2. For the measurable space $(\mathbb{N}, \mathscr{P}(\mathbb{N}))$, any real/complex-valued function f on $\mathbb{N}$, i. e., any real/complex sequence $(x_n := f(n))_{n\in\mathbb{N}}$ is $\mathscr{P}(\mathbb{N})$-measurable (see Remarks 5.3) and simple functions are only *eventually constant sequences*, i. e., such that

$$\exists N \in \mathbb{N}\, \forall n \geq N : \ x_n = x_N.$$

Exercise 5.10. Explain.

As the following statements reveal, simple functions are instrumental for the characterization of measurable ones.

Theorem 5.5 (Characterization of Measurability (Extended Nonnegative Case)). *Let (X, Σ) be a measurable space.*

(1) *A function*

$$f : X \to [0, \infty]$$

is Σ-measurable iff there exists an increasing sequence $(s_n)_{n\in\mathbb{N}}$:

$$s_n(x) \leq s_{n+1}(x),\ n \in \mathbb{N}, x \in X,$$

of nonnegative Σ-measurable simple functions on X converging to f pointwise on X:

$$\forall x \in X : \ s_n(x) \to f(x),\ n \to \infty.$$

(2) *Furthermore, if a Σ-measurable function $f : X \to [0, \infty]$ is bounded on X (i. e., $\sup_{x\in X} |f(x)| < \infty$), the sequence $(s_n)_{n\in\mathbb{N}}$ converges to f uniformly on X:*

$$\sup_{x\in X} |s_n(x) - f(x)| \to 0,\ n \to \infty.$$

Proof.
(1) *"If"* part follows directly from the *Sequences of Measurable Functions Theorem* (Theorem 5.4).

"Only if" part. Suppose that $f : X \to [0, \infty]$ is Σ-measurable.
For each $n \in \mathbb{N}$ and $k = 0, 1, \dots, n2^n - 1$, let

$$A_n^k := \left\{ x \in X \,\middle|\, \frac{k}{2^n} \le f(x) < \frac{k+1}{2^n} \right\} = f^{-1}\left(\left[\frac{k}{2^n}, \frac{k+1}{2^n} \right) \right) \in \Sigma$$

and

$$B_n := \{ x \in X \mid f(x) \ge n \} = f^{-1}([n, \infty)) \in \Sigma.$$

By the *Properties of Inverse Image* (Theorem 1.4), for each $n \in \mathbb{N}$, the sets A_n^k, $k = 0, 1, \dots, n2^n - 1$, and B_n are *pairwise disjoint* and

$$X = \bigcup_{k=0}^{n2^n - 1} A_n^k \cup B_n. \tag{5.3}$$

For each $n \in \mathbb{N}$, the *simple function*

$$s_n(x) := \sum_{k=0}^{n2^n-1} \frac{k}{2^n} \chi_{A_n^k}(x) + n\chi_{B_n}(x), \; x \in X,$$

is *nonnegative* and Σ-*measurable* (see Remarks 5.13).
Let $n \in \mathbb{N}$ and $x \in X$ be arbitrary.
If $x \in A_n^k$ for some $k = 0, 1, \dots, n2^n - 1$, then

$$x \in A_n^k = A_{n+1}^{2k} \cup A_{n+1}^{2k+1}.$$

Exercise 5.11. Verify.

Hence, by the definition,

$$s_n(x) = \frac{k}{2^n} \quad \text{and} \quad s_{n+1}(x) = \begin{cases} \frac{2k}{2^{n+1}}, & x \in A_{n+1}^{2k}, \\ \frac{2k+1}{2^{n+1}}, & x \in A_{n+1}^{2k+1}, \end{cases}$$

which implies that

$$s_n(x) \le s_{n+1}(x). \tag{5.4}$$

If $x \in B_n$, then

$$x \in B_n = \bigcup_{k=n2^{n+1}}^{(n+1)2^{n+1}-1} A_{n+1}^k \cup B_{n+1}.$$

Exercise 5.12. Verify.

Hence, by the definition,

$$s_n(x) = n$$

and

$$s_{n+1}(x) = \begin{cases} \frac{k}{2^{n+1}}, & x \in A_{n+1}^k,\ k = n2^{n+1}, \dots, (n+1)2^{n+1} - 1, \\ n+1, & x \in B_{n+1}, \end{cases}$$

which implies that

$$s_n(x) \le s_{n+1}(x). \tag{5.5}$$

In view of (5.3), inequalities (5.4) and (5.5) jointly imply that

$$s_n(x) \le s_{n+1}(x),\ n \in \mathbb{N}, x \in X.$$

Thus, $(s_n)_{n\in\mathbb{N}}$ is an *increasing* sequence.
It remains to show that $(s_n)_{n\in\mathbb{N}}$ converges to f *pointwise* on X.
If, for $x \in X, f(x) = \infty$, then

$$\forall\, n \in \mathbb{N} : f(x) \ge n,$$

i. e.,

$$\forall\, n \in \mathbb{N} : x \in B_n,$$

which implies that

$$s_n(x) = n \to \infty = f(x),\ n \to \infty.$$

If, for $x \in X, f(x) < \infty$, then

$$\exists\, N \in \mathbb{N} : f(x) < N,$$

and hence,

$$\forall\, n \ge N\ \exists\, 0 \le k \le n2^n - 1 : x \in A_n^k,$$

which implies that

$$\forall\, n \ge N : |s_n(x) - f(x)| < \frac{k+1}{2^n} - \frac{k}{2^n} = \frac{1}{2^n}. \tag{5.6}$$

Where, by the *Squeeze Theorem*, we infer that

$$s_n(x) \to f(x),\ n \to \infty.$$

(2) a Σ-measurable function $f : X \to [0,\infty]$ is *bounded* on X:

$$\sup_{x\in X} |f(x)| < \infty,$$

then

$$\exists N \in \mathbb{N}\ \forall x \in X : f(x) < N,$$

and hence,

$$\forall n \geq N : X = \bigcup_{k=0}^{n2^n-1} A_n^k,$$

which, as in (5.6), implies that

$$\forall n \geq N\ \forall x \in X : |s_n(x) - f(x)| < \frac{1}{2^n}.$$

Where, by the *Squeeze Theorem*, we infer that

$$\sup_{x \in X} |s_n(x) - f(x)| \to 0,\ n \to \infty,$$

i. e., the sequence $(s_n)_{n\in\mathbb{N}}$ converges to f uniformly on X. □

Corollary 5.6 (Characterization of Measurability (Extended Real Case)). *Let (X, Σ) be a measurable space.*

(1) *A function*

$$f : X \to \overline{\mathbb{R}}$$

is Σ-measurable iff there exists a sequence $(s_n)_{n\in\mathbb{N}}$ of real-valued Σ-measurable simple functions converging to f pointwise on X.

(2) *Furthermore, if a Σ-measurable function $f : X \to \overline{\mathbb{R}}$ is bounded on X, the sequence $(s_n)_{n\in\mathbb{N}}$ converges to f uniformly on X.*

Exercise 5.13. Prove.

Hint. Apply the prior theorem and the *Positive and Negative Parts of a Function Corollary* (Corollary 5.4).

Corollary 5.7 (Characterization of Measurability (Complex Case)). *Let (X, Σ) be a measurable space.*

(1) *A function*

$$f : X \to \mathbb{C}$$

is Σ-measurable iff there exists a sequence $(s_n)_{n\in\mathbb{N}}$ of complex-valued Σ-measurable simple functions converging to f pointwise on X.

(2) *Furthermore, if a Σ-measurable function $f : X \to \mathbb{C}$ is bounded on X, the sequence $(s_n)_{n\in\mathbb{N}}$ converges to f uniformly on X.*

Exercise 5.14. Prove.

Hint. Use the definition (see Remarks 5.3) and the prior corollary.

5.7 Luzin's Theorem

The following celebrated theorem due to Nikolay Luzin[1] establishes that a Borel measurable function $f : [a,b] \to \mathbb{R}$ $(-\infty < a < b < \infty)$ is, in a manner of speaking, *continuous on almost entire interval* $[a,b]$ or *almost continuous* on $[a,b]$.

Theorem 5.6 (Luzin's Theorem). *Let*

$$f : [a,b] \to \mathbb{R} \ (-\infty < a < b < \infty)$$

be a Borel measurable function. Then, for any $\varepsilon > 0$, there exists a closed set $E \subseteq [a,b]$ such that

$$\lambda([a,b] \setminus E) < \varepsilon,$$

where λ is the Lebesgue measure, and the restriction of f to E is a continuous on E function.

Proof. First, suppose that

$$f(x) = \chi_A(x), \ x \in X,$$

where A is a nonempty Borel subset of $[a,b]$.

Since A is *bounded*, by the *Approximation of Borel Sets Proposition* (Proposition 4.6) (see Section 4.8, Problem 13), for any $\varepsilon > 0$, there exist nonempty open set G and closed set F such that

$$F \subseteq A \subseteq G \quad \text{and} \quad \lambda(G \setminus A) < \varepsilon/2, \ \lambda(A \setminus F) < \varepsilon/2. \tag{5.7}$$

Since for an arbitrary nonempty subset C of $\mathbb{R}$, the distance to C function

$$\operatorname{dist}(x,C) := \inf_{y \in C} |x-y|, \ x \in \mathbb{R},$$

is *continuous* on $\mathbb{R}$ and the *closed* sets F and G^c are *disjoint*, the function

$$g(x) := \frac{\operatorname{dist}(x,G^c)}{\operatorname{dist}(x,F) + \operatorname{dist}(x,G^c)}, \ x \in \mathbb{R},$$

is
(i) *continuous* on $\mathbb{R}$,
(ii) takes values in $[0,1]$,
(iii) is equal to 0 on G^c and to 1 on F.

1 Nikolay Luzin (1883–1950).

The set

$$E := (F \cup G^c) \cap [a,b] \subseteq [a,b]$$

is *closed* and by *De Morgan's laws*

$$[a,b] \setminus E = (G \setminus F) \cap [a,b].$$

Exercise 5.15. Verify.

By the *monotonicity* and *subadditivity* of the Lebesgue measure λ and in view of (5.7),

$$\lambda([a,b] \setminus E) \le \lambda_F(G \setminus F) = \lambda((G \setminus A) \cup (A \setminus F)) \le \lambda(G \setminus A) + \lambda(A \setminus F) < \varepsilon/2 + \varepsilon/2 = \varepsilon.$$

Furthermore, $f = g$ on E.

Exercise 5.16. Verify.

Hence, the restriction of $f = \chi_A$ to E is *continuous* on E.

Now, suppose that $f : X \to \mathbb{R}$ is a simple Borel measurable function, i. e.,

$$f(x) = \sum_{k=1}^{n} a_k \chi_{A_k}(x), \; x \in [a,b],$$

with some $n \in \mathbb{N}$, *distinct* $a_k \in \mathbb{R}$, $k = 1, \dots, n$, and pairwise disjoint Borel subsets A_k of $[a,b]$, $k = 1, \dots, n$, and let $\varepsilon > 0$ be arbitrary.

We can choose closed sets $E_k \subseteq [a,b]$, $k = 1, \dots, n$, such that

$$\lambda([a,b] \setminus E_k) < \varepsilon/n, \; k = 1, \dots, n$$

and the restriction of χ_{A_k} to E_k is continuous on E_k, $k = 1, \dots, n$.

The closed set

$$E := \bigcap_{k=1}^{n} E_k \subseteq [a,b]$$

is such that, by *De Morgan's laws* and the *subadditivity* of λ,

$$\lambda\left([a,b] \setminus \bigcap_{k=1}^{n} E_k\right) = \lambda\left(\bigcup_{k=1}^{n} [a,b] \setminus E_k\right) \le \sum_{k=1}^{n} \lambda([a,b] \setminus E_k) < n(\varepsilon/n) = \varepsilon$$

and the restriction of f to E is *continuous* on E.

Exercise 5.17. Explain the latter.

Now, suppose that $f : X \to \mathbb{R}$ is a *bounded* Borel measurable function and let $\varepsilon > 0$ be arbitrary.

By the *Characterization of Measurability* (Corollary 5.6), there is a sequence $(s_n)_{n\in\mathbb{N}}$ of real-valued Borel measurable simple functions, which converges to f *uniformly* on $[a,b]$.

For each $n \in \mathbb{N}$, there is a closed set

$$E_n \subseteq [a,b]$$

such that

$$\lambda([a,b] \setminus E_n) < \varepsilon/2^n$$

and the restriction of s_n to E_n is *continuous* on E_n.

The closed set

$$E := \bigcap_{n=1}^{\infty} E_n \subseteq [a,b]$$

is such that, by *De Morgan's laws* and the *σ-subadditivity* of λ,

$$\lambda\left([a,b] \setminus \bigcap_{n=1}^{\infty} E_n \right) = \lambda\left(\bigcup_{n=1}^{\infty} [a,b] \setminus E_n \right) \leq \sum_{n=1}^{\infty} \lambda([a,b] \setminus E_n) < \sum_{n=1}^{\infty} \frac{\varepsilon}{2^n} = \varepsilon$$

and the restrictions of all s_n, $n \in \mathbb{N}$, to E are *continuous* on E.

Hence, the restriction of f to E is *continuous* on E as the *uniform limit* on E of the *continuous* on E restrictions of s_n, $n \in \mathbb{N}$.

Now, suppose that $f : X \to \mathbb{R}$ is a Borel measurable function and let $\varepsilon > 0$ be arbitrary.

Then, for each $n \in \mathbb{N}$,

$$A_n := \{x \in [a,b] \mid |f(x)| \leq n\} = |f|^{-1}([0,n]) \in \mathscr{B}(\mathbb{R}).$$

Exercise 5.18. Explain.

Since f does not take infinite values,

$$A_n \uparrow \bigcup_{k=1}^{\infty} A_k = [a,b].$$

Exercise 5.19. Explain.

Hence, by the *continuity of measure from below* (see the *Continuity of Measure* (Theorem 3.2)),

$$\exists N \in \mathbb{N} : \lambda(A_N) > b - a - \varepsilon/3,$$

and hence, by the *Properties of Measure* (Theorem 3.1),

$$\lambda([a,b] \setminus A_N) = \lambda([a,b]) - \lambda(A_N) < \varepsilon/3.$$

By the *Approximation of Borel Sets Proposition* (Proposition 4.6) (see Section 4.8, Problem 13), there exist a closed set F_N such that

$$F_N \subseteq A_N \quad \text{and} \quad \lambda(A_N \setminus F_N) < \varepsilon/3.$$

The function $f \cdot \chi_{A_N}$ (this is where it is essential again that f does not take infinite values) is *bounded* and Borel measurable on $[a, b]$.

Exercise 5.20. Explain.

Hence, by the proven above, there is a closed set $E_N \subseteq [a, b]$ such that

$$\lambda([a, b] \setminus E_N) < \varepsilon/3$$

and the restriction of $f \cdot \chi_{A_N}$ to E_N is continuous on E_N.

Then the closed set

$$E := F_N \cap E_N \subseteq [a, b]$$

is such that, by De Morgan's laws and the *monotonicity* and *subadditivity* of λ,

$$\begin{aligned}\lambda([a, b] \setminus E) &= \lambda\left(([a, b] \setminus F_N) \cup ([a, b] \setminus E_N)\right) \leq \lambda([a, b] \setminus F_N) + \lambda([a, b] \setminus E_N) \\ &= \lambda\left(([a, b] \setminus A_N) \cup (A_N \setminus F_N)\right) + \lambda([a, b] \setminus E_N) \leq \lambda([a, b] \setminus A_N) + \lambda(A_N \setminus F_N) \\ &\quad + \lambda([a, b] \setminus E_N) < \varepsilon/3 + \varepsilon/3 + \varepsilon/3 = \varepsilon\end{aligned}$$

and the restriction of f to E is continuous on E, which completes the proof. □

Remarks 5.14.

- *Luzin's Theorem* naturally extends to a Σ-measurable complex-valued function

  $$f(x) = \operatorname{Re} f(x) + i \operatorname{Im} f(x), \; x \in [a, b],$$

  via separately applying the proven real-valued version to the real-valued functions $\operatorname{Re} f$ and $\operatorname{Im} f$.
- There are more general forms of *Luzin's Theorem*, e. g., for a Borel measurable function $f : X \to \mathbb{C}$ on a metric space X with $\mu(X) < \infty$.
- By *Luzin's Theorem*, every Borel measurable function $f : [a, b] \to \mathbb{C}$ is, in a manner of speaking, *almost continuous* on $[a, b]$. However, as the subsequent example shows, a Borel measurable a function $f : [a, b] \to \mathbb{C}$, even being *almost continuous* on $[a, b]$, can be discontinuous everywhere on $[a, b]$.

Example 5.6. The characteristic function of the set of all irrationals of the interval $[0, 1]$:

$$f(x) := \chi_A(x), \; x \in [0, 1],$$

where $A := [0, 1] \setminus \mathbb{Q}$, is Borel measurable but is discontinuous at every point $x \in [0, 1]$.

Exercise 5.21. Explain.

Let $\{r_n\}_{n\in\mathbb{N}}$ be an enumeration of the countable set $[0,1]\cap\mathbb{Q}$, $\varepsilon > 0$ be arbitrary, and let

$$I_n := (r_n - \varepsilon/2^{n+1}, r_n + \varepsilon/2^{n+1}),\ n \in \mathbb{N}.$$

Then, for the *closed* set,

$$E := [0,1] \setminus \bigcup_{n=1}^{\infty} I_n \subset [0,1] \setminus \mathbb{Q} = A \subset [0,1]$$

by the *monotonicity* and *σ-subadditivity* of λ,

$$\lambda([0,1] \setminus E) \le \lambda\left(\bigcup_{n=1}^{\infty} I_n\right) \le \sum_{n=1}^{\infty} \lambda(I_n) < \sum_{n=1}^{\infty} \frac{\varepsilon}{2^n} = \varepsilon,$$

and the restriction of f to E, being identically 1, is *continuous* on E, which is consistent with *Luzin's Theorem* (Theorem 5.6).

5.8 Notion of Almost Everywhere

To proceed, we need the following useful notion of *almost everywhere* relative to a measure.

5.8.1 Definition and Examples

Definition 5.8 (Notion of Almost Everywhere). Let (X, Σ, μ) be a measure space. A condition $C(x)$, which every element $x \in X$ either satisfies or not, is said to be satisfied *almost everywhere (a. e.) relative to measure μ on X* if

$$N := \{x \in X \mid C(x) \text{ is not satisfied}\} \in \Sigma \quad \text{and} \quad \mu(N) = 0,$$

i. e., the condition $C(x)$ is satisfied everywhere on X except on a μ-null set.

Notation. $C(x)$ a. e. μ on X or $C(x)$ $(\text{mod } \mu)$.

Remarks 5.15.

- Certainly, if a condition $C(x)$ is satisfied *everywhere* on X, i. e.,

$$N := \{x \in X \mid C(x) \text{ is not satisfied}\} = \emptyset,$$

it is satisfied *a. e.* relative to any measure μ, except the *infinite* one (see Examples 3.2), on the power set $\mathscr{P}(X)$.

- For that matter, no condition is satisfied a. e. relative the infinite measure and any condition is satisfied a. e. relative the *zero measure* (see Examples 3.2).
- For a measure space (X, Σ, μ) such that $\emptyset$ is the only μ-null set, in particular for $(X, \mathscr{P}(X), \mu)$ with the *counting measure* μ (see Examples 3.2), a. e. μ on X is merely everywhere on X.

Examples 5.7. For $(\mathbb{R}, \Sigma^*, \lambda)$ (see Examples 5.1),

(a) the condition $C(x)$ of $x \in \mathbb{R}$ being an irrational number is satisfied a. e. λ on $\mathbb{R}$ and can be equivalently restated as follows:

$$\chi_{\mathbb{R}\setminus\mathbb{Q}}(x) = 1 \pmod{\lambda} \quad \text{or} \quad \chi_{\mathbb{Q}}(x) = 0 \pmod{\lambda};$$

(b) the function

$$f(x) := \begin{cases} |\csc x|, & x \neq n\pi,\ n \in \mathbb{Z}, \\ \infty, & x = n\pi,\ n \in \mathbb{Z}, \end{cases}$$

is *finite* and *continuous* a. e. λ on $\mathbb{R}$.

Exercise 5.22. Verify.

5.8.2 Equivalence of Functions

Definition 5.9 (Equivalence of Functions). Let (X, Σ, μ) be a measure space and X' be a nonempty set. Functions

$$f, g : X \to X'$$

are called *equivalent relative to measure* μ *on* X if they are equal a. e. μ:

$$f(x) = g(x) \text{ a. e. } \mu \text{ on } X \quad \text{or} \quad f = g \pmod{\mu}.$$

Examples 5.8. For the Lebesgue measure space $(\mathbb{R}, \Sigma^*, \lambda)$,

1. the functions $f(x) := \chi_{\mathbb{R}\setminus\mathbb{Q}}(x)$, $x \in \mathbb{R}$, and $g(x) := 1$, $x \in \mathbb{R}$, are equivalent relative to the Lebesgue measure λ on $\mathbb{R}$;
2. the functions $f(x) := \chi_{\mathbb{Q}}(x)$, $x \in \mathbb{R}$, and $g(x) := 0$, $x \in \mathbb{R}$, are equivalent relative to the Lebesgue measure λ on $\mathbb{R}$;
3. the functions $f(x) := \chi_{[0,\infty)}(x)$, $x \in \mathbb{R}$, and $g(x) := 0$, $x \in \mathbb{R}$, are not equivalent relative to the Lebesgue measure λ on $\mathbb{R}$.

Remarks 5.16.

- The equivalence of functions so defined is indeed an *equivalence relation* (reflexive, symmetric, and transitive).

- For $f, g : X \to X'$, not being equivalent relative to μ on X is *not* the same as
$$f(x) \neq g(x) \text{ a. e. } \mu \text{ on } X$$
- For a measure space (X, Σ, μ) such that $\emptyset$ is the only μ-null set, in particular for $(X, \mathscr{P}(X), \mu)$ with the *counting measure* μ (see Examples 3.2), the equivalence of functions relative to measure μ on X, i. e., their equality a. e. μ on X, is merely equality everywhere on X.
- Although $\chi_{\mathbb{R}\setminus\mathbb{Q}}(x) = 1$ a. e. λ on $\mathbb{R}$, $\chi_{\mathbb{R}\setminus\mathbb{Q}}$ is discontinuous for each $x \in \mathbb{R}$ (cf. Example 5.6).
  Thus, provided (X, ρ) is a metric space, a function f equal a. e. on X to a continuous on X function g need not be continuous on X and may even be discontinuous at every point of X.
- A continuous a. e. on X function f need not be equal a. e. on X to a continuous on X function g.

Exercise 5.23. Verify. Give corresponding examples.

5.8.3 A. E. Characterization of Measurability

Theorem 5.7 (A. E. Characterization of Measurability). *Let (X, Σ, μ) be a complete measure space and (X', Σ') be a measurable space. If functions*
$$f, g : X \to X'$$
are equivalent relative to μ on X and one of them is Σ-Σ'-measurable, then so is the other one.

Proof. Without loss of generality, suppose that f is Σ-Σ'-measurable. Then, for each $A' \in \Sigma'$,
$$\begin{aligned} g^{-1}(A') = \{x \in X \,|\, g(x) \in A'\} &= (\{x \in X \,|\, g(x) \in A'\} \cap \{x \in X \,|\, g(x) = f(x)\}) \\ &\quad \cup (\{x \in X \,|\, g(x) \in A'\} \cap \{x \in X \,|\, g(x) \neq f(x)\}) \\ &= (\{x \in X \,|\, f(x) \in A'\} \cap \{x \in X \,|\, g(x) = f(x)\}) \\ &\quad \cup (\{x \in X \,|\, g(x) \in A'\} \cap \{x \in X \,|\, g(x) \neq f(x)\}) \in \Sigma. \end{aligned}$$

Indeed, by the Σ-Σ'-measurability of f
$$\{x \in X \,|\, f(x) \in A'\} = f^{-1}(A') \in \Sigma,$$
by the equivalence of f and g relative to μ on X,

$$\{x \in X \mid g(x) \neq f(x)\} \in \Sigma \quad \text{and} \quad \mu(\{x \in X \mid g(x) \neq f(x)\}) = 0,$$

and hence,

$$\{x \in X \mid g(x) = f(x)\} = \{x \in X \mid g(x) \neq f(x)\}^c \in \Sigma$$

and, by the *completeness* of μ,

$$\{x \in X \mid g(x) \in A'\} \cap \{x \in X \mid g(x) \neq f(x)\} \in \Sigma,$$

which concludes the proof. □

In view of the *completeness* of the Lebesgue measure, we immediately obtain the following.

Corollary 5.8 (A. E. Characterization of Lebesgue Measurability). *Let $X \subseteq \mathbb{R}^n$ ($n \in \mathbb{N}$) be Lebesgue measurable ($X \in \Sigma_n^*$). If functions*

$$f, g : X \to \overline{\mathbb{R}}$$

are equivalent relative to the Lebesgue measure λ_n on X and one of them is Lebesgue measurable, then so is the other one.

Remark 5.17. As the subsequent example shows, the *completeness* condition for measure in the prior theorem is essential and cannot be dropped.

Example 5.9. For the *incomplete* Borel measure space $(\mathbb{R}, \mathscr{B}(\mathbb{R}), \lambda)$ (see Examples 5.1 and Section 4.7.1) and a Lebesgue measurable subset A of the Cantor set C, which is not Borel measurable:

$$\mathscr{B}(\mathbb{R}) \ni C \supseteq A \in \Sigma^* \setminus \mathscr{B}(\mathbb{R})$$

(see Section 4.7.2 and Exercise 4.34), the functions

$$f(x) := \chi_C(x),\ x \in \mathbb{R}, \quad \text{and} \quad g(x) := -\chi_A(x),\ x \in \mathbb{R},$$

are equivalent relative to the Lebesgue measure λ on $\mathbb{R}$ since

$$\{x \in \mathbb{R} \mid f(x) \neq g(x)\} = C \in \mathscr{B}(\mathbb{R}) = \Sigma \quad \text{and} \quad \lambda(C) = 0.$$

The function f is Borel measurable and g is not (see Section 4.7.1), being, however, *Lebesgue measurable* and even *continuous a. e. λ on* $\mathbb{R}$ ("*a. e. λ on* $\mathbb{R}$" is understood relative to the *complete* Lebesgue measure λ on the Lebesgue σ-algebra Σ^*).

Exercise 5.24. Verify, explain.

5.9 Convergence Almost Everywhere

5.9.1 Definition, Examples, and Properties

Combining the notion of pointwise convergence of a function sequence with the notion of almost everywhere, we arrive at a new type of convergence: *convergence almost everywhere*.

Definition 5.10 (Convergence Almost Everywhere). Let (X,Σ,μ) be a measure space and

$$f, f_n : X \to \overline{\mathbb{R}},\ n \in \mathbb{N}.$$

The function sequence $(f_n)_{n\in\mathbb{N}}$ is said to *converge to the function f almost everywhere (a. e.) relative to measure μ on X* if

$$N := \left\{x \in X \,\middle|\, \lim_{n\to\infty} f_n(x) \neq f(x)\right\} \in \Sigma \quad \text{and} \quad \mu(N) = 0,$$

i. e., $(f_n)_{n\in\mathbb{N}}$ converges to f everywhere on X except on a μ-null set.

Notation. $f_n \to f$ a. e. μ on X or $f_n \to f \pmod{\mu}$.

Remarks 5.18.

- If functions $f, f_n : X \to \overline{\mathbb{R}}$, $n \in \mathbb{N}$, are Σ-measurable, the condition

$$N := \left\{x \in X \,\middle|\, \lim_{n\to\infty} f_n(x) \neq f(x)\right\} \in \Sigma$$

  is satisfied automatically.
  Indeed, by the *Sequences of Measurable Functions Theorem* (Theorem 5.4) (see also Remarks 5.11),

$$F := \{x \in X \mid (f_n(x))_{n\in\mathbb{N}} \text{ has a finite limit}\} \in \Sigma,$$
$$L_{-\infty} := \left\{x \in X \,\middle|\, \lim_{n\to\infty} f_n(x) = -\infty\right\} = \left\{x \in X \,\middle|\, \overline{\lim_{n\to\infty}} f_n(x) = -\infty\right\} \in \Sigma,$$

  and

$$L_{\infty} := \left\{x \in X \,\middle|\, \lim_{n\to\infty} f_n(x) = \infty\right\} = \left\{x \in X \,\middle|\, \underline{\lim_{n\to\infty}} f_n(x) = \infty\right\} \in \Sigma,$$

  and hence,

$$L := \{x \in X \mid (f_n(x))_{n\in\mathbb{N}} \text{ has a (finite or infinite) limit}\} = F \cup L_{-\infty} \cup L_{\infty} \in \Sigma.$$

  By the Σ-measurability of f,

$$F_{-\infty} := \{x \in X \mid f(x) = -\infty\} = f^{-1}(\{-\infty\}) \in \Sigma$$

and

$$F_\infty := \{x \in X \mid f(x) = \infty\} = f^{-1}(\{\infty\}) \in \Sigma.$$

Therefore,

$$N = L^c \cup (L_{-\infty} \cap F_\infty) \cup (L_\infty \cap F_{-\infty}) \cup ((L_{-\infty} \cup L_\infty) \cap F^c_{-\infty} \cap F^c_\infty)$$
$$\cup \left\{x \in F \,\middle|\, \lim_{n\to\infty} f_n(x) \neq f(x)\right\} \in \Sigma.$$

The latter set is Σ-measurable, by the *Equality Set Proposition* (Proposition 5.4, Section 5.12, Problem 5), since the functions

$$\lim_{n\to\infty} f_n(x),\ x \in F, \quad \text{and} \quad f(x),\ x \in F,$$

are $\Sigma \cap F$*-measurable*, being the restrictions of Σ-measurable functions

$$\overline{\lim_{n\to\infty}} f_n(x),\ x \in X,$$

and f to the set $F \in \Sigma$, respectively (see Remarks 5.4).

Exercise 5.25. Explain.

– The analogues of the definition of convergence almost everywhere and of the prior remark are also in place for Σ-measurable functions $f, f_n : X \to \mathbb{C}$, $n \in \mathbb{N}$.

Exercise 5.26. Explain.

– Provided measure μ is not *infinite* (see Examples 3.2), a function sequence $(f_n)_{n\in\mathbb{N}}$ converging to f *pointwise* on X (everywhere) also converges a. e. μ on X but not vice versa.

Examples 5.10.

1. For $(\mathbb{R}, \mathscr{B}(\mathbb{R}), \lambda)$, the function sequences

$$f_n(x) := \chi_{[n,n+1]}(x),\ n \in \mathbb{N}, x \in \mathbb{R},$$

and

$$g_n(x) := \chi_{[n,\infty)}(x),\ n \in \mathbb{N}, x \in \mathbb{R},$$

converge to the function

$$f(x) := 0,\ x \in \mathbb{R},$$

pointwise on $\mathbb{R}$, and hence, a. e. λ on $\mathbb{R}$.

2. For $(\mathbb{R}, \mathscr{B}(\mathbb{R}), \lambda)$, the function sequence

$$f_n(x) := n\chi_{[-1/n,1/n]}(x), \; n \in \mathbb{N}, x \in \mathbb{R},$$

converges to the function

$$f(x) := 0, \; x \in \mathbb{R},$$

for all $x \in \mathbb{R} \setminus \{0\}$, and hence, a. e. λ on $\mathbb{R}$ and converges to

$$g(x) := \begin{cases} 0, & x \neq 0, \\ \infty, & x = 0, \end{cases}$$

pointwise on $\mathbb{R}$.
3. For $([0,1], \mathscr{B}([0,1]), \lambda)$, the function sequence

$$f_n(x) := x^n, \; n \in \mathbb{N}, x \in [0,1],$$

converges to the function

$$f(x) := 0, \; x \in [0,1],$$

for all $x \in [0,1)$, and hence, a. e. λ on $[0,1]$ and converges to

$$g(x) := \begin{cases} 0, & x \in [0,1), \\ 1, & x = 1, \end{cases}$$

pointwise on $[0,1]$.
4. For $(\mathbb{R}, \mathscr{B}(\mathbb{R}), \lambda)$, the function sequence

$$f_n(x) := \sin^n x, \; n \in \mathbb{N}, x \in \mathbb{R},$$

converges to the function

$$f(x) := 0, \; x \in \mathbb{R},$$

for all $x \in \mathbb{R} \setminus \{\pi/2 + n\pi \mid n \in \mathbb{Z}\}$, and hence, a. e. λ on $\mathbb{R}$, but does not converge pointwise on $\mathbb{R}$ to any extended real-valued function.
5. For the *incomplete* Borel measure space $(\mathbb{R}, \mathscr{B}(\mathbb{R}), \lambda)$ and a Lebesgue measurable subset A of the Cantor set C, which is not Borel measurable:

$$\mathscr{B}(\mathbb{R}) \ni C \supseteq A \in \Sigma^* \setminus \mathscr{B}(\mathbb{R})$$

(see Example 5.9), the function sequence

$$f_n(x) := \chi_A(x), \; n \in \mathbb{N}, x \in \mathbb{R},$$

each f_n, $n \in \mathbb{N}$, being Lebesgue but not Borel measurable (see Example 5.9), does not converge a. e. λ on $\mathbb{R}$ to the Borel measurable function

$$f(x) := 0,\ x \in \mathbb{R},$$

since

$$\left\{x \in X \,\middle|\, \lim_{n\to\infty} f_n(x) \neq f(x)\right\} = A \notin \mathscr{B}(\mathbb{R}).$$

However, in the *complete* Lebesgue measure space $(\mathbb{R}, \Sigma^*, \lambda)$, which is the *completion* of the Borel measure space $(\mathbb{R}, \mathscr{B}(\mathbb{R}), \lambda)$ (see Section 4.7.1),

$$f_n \to f \quad (\text{mod } \lambda).$$

Exercise 5.27. Verify.

Remark 5.19. Examples 5.10 1–4 remain in place if the underlying Borel σ-algebra $\mathscr{B}(\mathbb{R})$ or $\mathscr{B}([0,1])$ is replaced with its *completion* Σ^* or $\Sigma^*([0,1])$, respectively, relative to the Lebesgue measure λ (see Section 4.7.1).

The natural step now is to establish analogues of the properties of pointwise convergence for convergence almost everywhere.

Proposition 5.1 (Properties of Convergence Almost Everywhere). *Let* (X, Σ, μ) *be a measure space and*

$$f, f_n, g, g_n : X \to \mathbb{R},\ n \in \mathbb{N}.$$

If

$$f_n \to f \quad (\text{mod } \mu) \quad \textit{and} \quad g_n \to g \quad (\text{mod } \mu).$$

Then:

1. *for any* $c \in \mathbb{R}$, $cf_n \to cf$ $(\text{mod } \mu)$;
2. $f_n \pm g_n \to f \pm g$ $(\text{mod } \mu)$;
3. $f_n g_n \to fg$ $(\text{mod } \mu)$;
4. $|f_n| \to |f|$ $(\text{mod } \mu)$.
5. $(f_n)_\pm \to f_\pm$ $(\text{mod } \mu)$.

Exercise 5.28. Prove.

Remark 5.20. Properties (1) (with $c \in \mathbb{C}$)–(2) remain in place for

$$f, f_n, g, g_n : X \to \mathbb{C},\ n \in \mathbb{N}.$$

5.9.2 Uniqueness A. E. of Limit A. E.

With the additional condition of the *completeness* of the underlying measure, the uniqueness of the limit is also in place in the almost-everywhere sense.

Theorem 5.8 (Uniqueness A. E. of Limit A. E.). *Let* (X, Σ, μ) *be a complete measure space and*

$$f, g, f_n : X \to \overline{\mathbb{R}},\ n \in \mathbb{N}.$$

If

$$f_n \to f \pmod{\mu} \quad \textit{and} \quad f_n \to g \pmod{\mu},$$

then $f = g \pmod{\mu}$.

Proof. Since $f_n \to f \pmod{\mu}$ and $f_n \to g \pmod{\mu}$,

$$N_f := \left\{x \in X \;\middle|\; \lim_{n\to\infty} f_n(x) \neq f(x)\right\} \in \Sigma \quad \text{and} \quad \mu(N_f) = 0$$

and

$$N_g := \left\{x \in X \;\middle|\; \lim_{n\to\infty} f_n(x) \neq g(x)\right\} \in \Sigma \quad \text{and} \quad \mu(N_g) = 0.$$

Hence, for $N := N_f \cup N_g$, by the *subadditivity* of measure μ,

$$\mu(N) \leq \mu(N_f) + \mu(N_g) = 0$$

and, for each $x \in N^c$,

$$f_n(x) \to f(x) \text{ and } f_n(x) \to g(x),\ n \to \infty.$$

Where, by the uniqueness of the limit of an extended real-valued sequence, we infer that

$$\{x \in X \,|\, f(x) \neq g(x)\} \subseteq N,$$

which, by the *completeness* of measure μ, implies that

$$\{x \in X \,|\, f(x) \neq g(x)\} \in \Sigma \quad \text{and} \quad \mu\left(\{x \in X \,|\, f(x) \neq g(x)\}\right) = 0,$$

i. e., $f = g \pmod{\mu}$. □

Remarks 5.21.

– The analogue of the prior theorem remains in place for

$$f, f_n, g, g_n : X \to \mathbb{C},\ n \in \mathbb{N}.$$

– The subsequent example shows that the *completeness* condition for measure in the prior theorem is essential and cannot be dropped.

Example 5.11. For the *incomplete* Borel measure space $(\mathbb{R}, \mathscr{B}(\mathbb{R}), \lambda)$ and a Lebesgue measurable subset A of the Cantor set C, which is not Borel measurable:

$$\mathscr{B}(\mathbb{R}) \ni C \supseteq A \in \Sigma^* \setminus \mathscr{B}(\mathbb{R})$$

(see Example 5.9), the sequence of Borel measurable functions

$$f_n(x) := \chi_C(x), \; n \in \mathbb{N}, x \in \mathbb{R},$$

and the Lebesgue measurable but not Borel measurable function

$$g(x) := -\chi_A(x), \; x \in \mathbb{R},$$

we have

$$f_n \to 0 \pmod{\lambda} \quad \text{and} \quad f_n \to g \pmod{\lambda},$$

but the limit functions 0 and g are *not* equivalent on $\mathbb{R}$ relative to the *incomplete* Lebesgue measure on $\mathscr{B}(\mathbb{R})$ since

$$\{x \in \mathbb{R} \,|\, g(x) \neq 0\} = A \notin \mathscr{B}(\mathbb{R}).$$

Exercise 5.29. Verify.

5.9.3 Measurability of Limit A. E.

Now, the question is whether the limit function inherits the property of measurability. As we see from the following statement, with the additional condition of the *completeness* of the underlying measure, the answer is affirmative.

Theorem 5.9 (Measurability of Limit A. E.). *Let (X, Σ, μ) be a complete measure space and functions*

$$f_n : X \to \overline{\mathbb{R}}, \; n \in \mathbb{N},$$

be Σ-measurable. If

$$f : X \to \overline{\mathbb{R}}$$

and

$$f_n \to f \pmod{\mu},$$

then f is Σ-measurable.

Proof. Since $f_n \to f \pmod{\mu}$,

$$N := \left\{x \in X \,\middle|\, \lim_{n\to\infty} f_n(x) \neq f(x)\right\} \in \Sigma \quad \text{and} \quad \mu(N) = 0.$$

Since, for each $x \in N^c \in \Sigma$,

$$f_n(x) \to f(x), \; n \to \infty,$$

by the *Sequences of Measurable Functions Theorem* (Theorem 5.4) (see also Remarks 5.4), the restriction f_{N^c} of the limit function f to N^c is $\Sigma \cap N^c$-measurable. The restriction f_N of the limit function f to N is $\Sigma \cap N$-measurable by the *completeness* of μ.

Exercise 5.30. Explain.

Hence, the limit function f is Σ-measurable. Indeed,

$$\forall A' \in \mathscr{B}(\overline{\mathbb{R}}) : f^{-1}(A') = f_{N^c}^{-1}(A') \cup f_N^{-1}(A') \in \Sigma.$$

Exercise 5.31. Explain. □

Remarks 5.22.

- The analogue of the prior theorem remains in place for

$$f, f_n, g, g_n : X \to \mathbb{C}, \; n \in \mathbb{N}.$$

- Example 5.11 shows that the *completeness* condition for measure in the prior theorem is essential and cannot be dropped.

5.9.4 Egorov's Theorem

The following famed theorem due to Dmitriy Egorov[2] establishes that, under the condition of the *finiteness* of the underlying measure, convergence almost everywhere is actually uniform convergence on *almost entire set* or *almost uniform convergence* (cf. *Luzin's Theorem* (Theorem 5.6)).

Theorem 5.10 (Egorov's Theorem). *Let (X, Σ, μ) be a measure space with $\mu(X) < \infty$ and functions*

$$f, f_n : X \to \mathbb{R}, \; n \in \mathbb{N},$$

be Σ-measurable. If

$$f_n \to f \pmod{\mu},$$

then, for any $\varepsilon > 0$, there exists a set $E \in \Sigma$ such that

$$\mu(E^c) < \varepsilon$$

2 Dmitriy Egorov (1869–1931).

and the function sequence $(f_n)_{n\in\mathbb{N}}$ converges to the function f uniformly on E, i. e.,

$$\sup_{x\in E} |f_n(x) - f(x)| \to 0,\ n \to \infty.$$

Proof. Since $f_n \to f \pmod{\mu}$,

$$N := \left\{x \in X \,\middle|\, \lim_{n\to\infty} f_n(x) \neq f(x)\right\} \in \Sigma \quad \text{and} \quad \mu(N) = 0$$

and, for each $x \in N^c$,

$$f_n(x) \to f(x),\ n \to \infty.$$

For any $j, k \in \mathbb{N}$,

$$E_j^{(k)} := \bigcap_{i=j}^{\infty} \{x \in X \mid |f_i(x) - f(x)| < 1/k\} \in \Sigma. \tag{5.8}$$

Exercise 5.32. Explain.

For each $k \in \mathbb{N}$,

$$E_1^{(k)} \subseteq E_2^{(k)} \subseteq \dots \quad \text{and} \quad N^c \subseteq \bigcup_{j=1}^{\infty} E_j^{(k)}.$$

Exercise 5.33. Explain.

Hence, by *De Morgan's laws*,

$$E_1^{(k)^c} \supseteq E_2^{(k)^c} \supseteq \dots \quad \text{and} \quad N \supseteq \bigcap_{j=1}^{\infty} E_j^{(k)^c}.$$

In view of $\mu(X) < \infty$, by the *continuity of measure from above* (see the *Continuity of Measure* (Theorem 3.2)) and the *monotonicity* of μ, for each $k \in \mathbb{N}$,

$$0 \le \lim_{n\to\infty} \mu\left(E_j^{(k)^c}\right) = \mu\left(\bigcap_{j=1}^{\infty} E_j^{(k)^c}\right) \le \mu(N) = 0,$$

and hence, for each $k \in \mathbb{N}$,

$$\lim_{n\to\infty} \mu\left(E_j^{(k)^c}\right) = 0.$$

Let $\varepsilon > 0$ be arbitrary. Then the latter implies that

$$\forall\, k \in \mathbb{N}\ \exists\, j(k,\varepsilon) \in \mathbb{N} :\ \mu\left(E_{j(k,\varepsilon)}^{(k)}{}^{c}\right) < \frac{\varepsilon}{2^k}. \tag{5.9}$$

Consider the set

$$E := \bigcap_{k=1}^{\infty} E_{j(k,\varepsilon)}^{(k)} \in \Sigma.$$

In view of (5.9), by *De Morgan's laws* and the *σ-subadditivity* of μ, we have

$$\mu(E^c) = \mu\left(\bigcup_{k=1}^{\infty} E_{j(k,\varepsilon)}^{(k)\ c}\right) \le \sum_{k=1}^{\infty} \mu\left(E_{j(k,\varepsilon)}^{(k)\ c}\right) < \sum_{k=1}^{\infty} \frac{\varepsilon}{2^k} = \varepsilon.$$

Further, in view of the inclusion

$$E \subseteq E_{j(k,\varepsilon)}^{(k)},\ k \in \mathbb{N},$$

by (5.8), for each $k \in N$ and every $n \ge j(k,\varepsilon)$, we have

$$\sup_{x \in E} |f_n(x) - f(x)| \le \sup_{x \in E_{j(k,\varepsilon)}^{(k)}} |f_n(x) - f(x)| \le 1/k,$$

which implies that

$$\sup_{x \in E} |f_n(x) - f(x)| \to 0,\ n \to \infty,$$

completing the proof. □

Example 5.12. For $([0,1], \Sigma^*([0,1]), \lambda)$, the sequence

$$f_n(x) := x^n,\ n \in \mathbb{N}, x \in [0,1],$$

converges to

$$f(x) := 0,\ x \in [0,1],$$

for all $x \in [0,1)$ and, for any $\varepsilon \in (0,1)$,

$$\sup_{x \in [0,1-\varepsilon]} |f_n(x) - f(x)| = \sup_{x \in [0,1-\varepsilon]} x^n = (1-\varepsilon)^n \to 0,\ n \to \infty,$$

which is consistent with *Egorov's Theorem*.

Egorov's Theorem gives rise to the following notion of *almost uniform convergence*.

Definition 5.11 (Almost Uniform Convergence). Let (X, Σ, μ) be a measure space and

$$f, f_n : X \to \mathbb{R},\ n \in \mathbb{N}.$$

The function sequence $(f_n)_{n \in \mathbb{N}}$ is said to converge to the function f *almost uniformly* on X if, for any $\varepsilon > 0$, there exists a set $E \in \Sigma$ such that

$$\mu(X \setminus E) < \varepsilon$$

and the function sequence $(f_n)_{n \in \mathbb{N}}$ converges to the function f uniformly on E, i. e.,

$$\sup_{x \in E} |f_n(x) - f(x)| \to 0,\ n \to \infty.$$

Remarks 5.23.

- Thus, *Egorov's Theorem* establishes the fact that every sequence of real-valued Σ-measurable functions, convergent to a real-valued Σ-measurable function a. e.

on a set of finite measure, converges to the limit function *almost uniformly* on this set.

- As well as *Luzin's Theorem* (Theorem 5.6) (see Remarks 5.14), *Egorov's Theorem* naturally extends to Σ-measurable functions

$$f, f_n : X \to \mathbb{C}, \ n \in \mathbb{N},$$

via separately applying the proven real-valued version to the real and imaginary parts.

- *Egorov's Theorem* also remains in place for Σ-measurable functions

$$f, f_n : X \to \overline{\mathbb{R}}, \ n \in \mathbb{N}$$

finite a. e. μ on X.

Exercise 5.34. Explain.

- As the subsequent example shows, the *finiteness* condition for measure in *Egorov's Theorem* is essential and cannot be dropped.

Example 5.13. For the Borel measure space $(\mathbb{R}, \mathscr{B}(\mathbb{R}), \lambda)$ with $\lambda(\mathbb{R}) = \infty$, the sequence of Borel measurable functions

$$f_n(x) := \chi_{[n,\infty)}(x), \ n \in \mathbb{N}, x \in \mathbb{R},$$

converges to the Borel measurable function

$$f(x) := 0, \ x \in \mathbb{R},$$

a. e. λ on $\mathbb{R}$ (see Examples 5.10), but the convergence is *not* almost uniform on $\mathbb{R}$.

Exercise 5.35.
(a) Verify.
(b) Give another counterexample.

5.10 Convergence in Measure

Here, we consider another type of convergence of a sequence of measurable functions: *convergence in measure.*

5.10.1 Definition, Examples, and Properties

Definition 5.12 (Convergence in Measure). Let (X, Σ, μ) be a measure space and functions

$$f, f_n : X \to \mathbb{R}$$

be Σ-measurable. The function sequence $(f_n)_{n\in\mathbb{N}}$ is said to *converge to the function f in measure μ on X* if

$$\forall \varepsilon > 0: \ \mu(\{x \in X \mid |f_n(x) - f(x)| \ge \varepsilon\}) \to 0, \ n \to \infty. \tag{5.10}$$

Notation. $f_n \xrightarrow{\mu} f$.

Remarks 5.24. As is seen from the definition,
- *uniform* convergence on X implies convergence in measure;
- to show that $f_n \xrightarrow{\mu} f$, one only needs to verify that (5.10) holds for arbitrary $\varepsilon \in (0,1)$.

Exercise 5.36. Explain.

Examples 5.14.
1. For $(\mathbb{R}, \mathscr{B}(\mathbb{R}), \lambda)$, the function sequences

$$f_n(x) := \chi_{[n,n+1/n]}(x), \ n \in \mathbb{N}, x \in \mathbb{R},$$
$$g_n(x) := n\chi_{[0,1/n]}(x), \ n \in \mathbb{N}, x \in \mathbb{R},$$
$$h_n(x) := \frac{1}{n}\chi_{[n,n+1]}(x), \ n \in \mathbb{N}, x \in \mathbb{R},$$
$$u_n(x) := \frac{1}{n}\chi_{[n,\infty)}(x), \ n \in \mathbb{N}, x \in \mathbb{R},$$

   converge both in measure λ and pointwise on $\mathbb{R}$ to the function

$$f(x) := 0, \ x \in \mathbb{R},$$

   with $(h_n)_{n\in\mathbb{N}}$ and $(u_n)_{n\in\mathbb{N}}$ converging to f *uniformly* on $\mathbb{R}$.
2. For $(\mathbb{R}, \mathscr{B}(\mathbb{R}), \lambda)$, the function sequences

$$f_n(x) := \chi_{[n,n+1]}(x), \ n \in \mathbb{N}, x \in \mathbb{R},$$

   and

$$g_n(x) := \chi_{[n,\infty)}(x), \ n \in \mathbb{N}, x \in \mathbb{R},$$

   do not converge in measure λ to the function

$$f(x) := 0, \ x \in \mathbb{R},$$

   although converge to f pointwise on $\mathbb{R}$.
3. For $([0,1], \mathscr{B}([0,1]), \lambda)$, the function sequence

$$f_n(x) := x^n, \ n \in \mathbb{N}, x \in [0,1],$$

converges in measure λ to the functions

$$f(x) := 0,\ x \in [0,1],$$

$$g(x) := \begin{cases} 0, & x \in [0,1), \\ 1, & x = 1, \end{cases}$$

and

$$h(x) := \begin{cases} 1, & x = 0, \\ 0, & x \in (0,1]. \end{cases}$$

Exercise 5.37. Verify.

Remarks 5.25.

- As follows from the prior examples, a function sequence convergent a. e. μ need not converge in measure μ.
- As the subsequent example shows, a function sequence convergent in measure μ need not converge a. e. μ and can even diverge at any point.

Example 5.15 (Dancing Steps Example). Let the underlying measure space be $([0,1], \Sigma^*([0,1]), \lambda)$.

For each $k \in \mathbb{Z}_+$, let

$$f_n(x) = \chi_{[(n-2^k)2^{-k},(n+1-2^k)2^{-k}]}(x),\ n = 2^k, \dots, 2^{k+1} - 1, x \in [0,1],$$

and

$$f(x) := 0,\ x \in [0,1].$$

Let $\varepsilon \in (0,1)$ be arbitrary.

For each $n \in \mathbb{N}$, there is an $k(n) \in \mathbb{Z}_+$ such that

$$2^{k(n)} \le n \le 2^{k(n)+1} - 1,$$

and hence, considering that $k(n) \to \infty$ as $n \to \infty$, we have

$$\begin{aligned} &\lambda(\{x \in [0,1] \mid |f_n(x) - f(x)| \ge \varepsilon\}) = \lambda(\{x \in [0,1] \mid f_n(x) \ge \varepsilon\}) \\ &= \lambda\left(\left[\left(n - 2^{k(n)}\right)2^{-k(n)}, \left(n + 1 - 2^{k(n)}\right)2^{-k(n)}\right]\right) = 2^{-k(n)} \to 0,\ n \to \infty. \end{aligned}$$

Thus,

$$f_n \xrightarrow{\lambda} f.$$

However, by the construct of the function sequence $(f_n)_{n\in\mathbb{N}}$, for each $x \in [0,1]$, the numeric sequence $(f_n(x))_{n\in\mathbb{N}}$, assuming each of the values 0 and 1 for *infinitely many* $n \in \mathbb{N}$, diverges.

Exercise 5.38. Explain the latter.

Proposition 5.2 (Properties of Convergence in Measure). *Let (X, Σ, μ) be a measure space and*

$$f, f_n, g, g_n : X \to \mathbb{R},\ n \in \mathbb{N},$$

be Σ-measurable and

$$f_n \xrightarrow{\mu} f \text{ and } g_n \xrightarrow{\mu} g.$$

Then:
(1) *for any $c \in \mathbb{R}$, $cf_n \xrightarrow{\mu} cf$;*
(2) $f_n \pm g_n \xrightarrow{\mu} f \pm g$;
(3) $|f_n| \xrightarrow{\mu} |f|$;
(4) $(f_n)_\pm \xrightarrow{\mu} f_\pm$.

Exercise 5.39.
(a) Prove.
(b) Give an example showing that, under the conditions of the prior proposition $f_n g$ need not converge in measure μ to fg (cf. the *Properties of Convergence Almost Everywhere* (Theorem 5.1) and Section 5.12, Problem 22).

Remark 5.26. The analogue of the definition of convergence in measure remains in place for complex-valued functions as well as properties (1) (with $c \in \mathbb{C}$)–(3).

5.10.2 Uniqueness A. E. of Limit in Measure

The limit in measure is unique in the almost-everywhere sense subject to no additional conditions (cf. the *Uniqueness A. E. of Limit A. E. Theorem* (Theorem 5.8)).

Theorem 5.11 (Uniqueness A. E. of Limit in Measure). *Let (X, Σ, μ) be a measure space and*

$$f, g, f_n : X \to \mathbb{R},\ n \in \mathbb{N},$$

be Σ-measurable. If

$$f_n \xrightarrow{\mu} f \text{ and } f_n \xrightarrow{\mu} g,$$

then $f = g \pmod{\mu}$.

Proof. For an arbitrary $\varepsilon > 0$ and each $n \in \mathbb{N}$, we have the inclusion

$$\begin{aligned}\{x \in X \mid |f(x) - g(x)| \geq \varepsilon\} &= \{x \in X \mid |f(x) - f_n(x) + f_n(x) - g(x)| \geq \varepsilon\} \\ &\subseteq \{x \in X \mid |f(x) - f_n(x)| \geq \varepsilon/2\} \cup \{x \in X \mid |f_n(x) - g(x)| \geq \varepsilon/2\}.\end{aligned}$$

Exercise 5.40. Explain.

Hence, by the *subadditivity* of μ,

$$\mu(\{x \in X \mid |f(x) - g(x)| \geq \varepsilon\}) \leq \mu(\{x \in X \mid |f(x) - f_n(x)| \geq \varepsilon/2\}) + \mu(\{x \in X \mid |f_n(x) - g(x)| \geq \varepsilon/2\}).$$

Where, passing to the limit as $n \to \infty$, we infer that, for any $\varepsilon > 0$,

$$\mu(\{x \in X \mid |f(x) - g(x)| \geq \varepsilon\}) = 0.$$

We have

$$\{x \in X \mid f(x) \neq g(x)\} = \bigcup_{n=1}^{\infty} \{x \in X \mid |f(x) - g(x)| \geq 1/n\}.$$

Exercise 5.41. Verify.

Hence, by the *σ-subadditivity* of μ,

$$\mu(\{x \in X \mid f(x) \neq g(x)\}) \leq \sum_{n=1}^{\infty} \mu(\{x \in X \mid |f(x) - g(x)| \geq 1/n\}) = 0. \qquad \square$$

Remark 5.27. For the uniqueness a. e. of the limit in measure, unlike for the uniqueness a. e. of the limit a. e. (cf. the *Uniqueness A. E. of the Limit A. E. Theorem* (Theorem 5.8)), the *completeness* condition for the measure is superfluous.

5.10.3 Lebesgue and Riesz Theorems

The following two theorems elucidate connections between the two types of convergence: convergence almost everywhere and convergence in measure.

Theorem 5.12 (Lebesgue Theorem). *Let (X, Σ, μ) be a measure space with $\mu(X) < \infty$ and $f, f_n : X \to \mathbb{R}$, $n \in \mathbb{N}$, be Σ-measurable functions. If*

$$f_n \to f \pmod{\mu},$$

then

$$f_n \xrightarrow{\mu} f.$$

Proof. Since $f_n \to f \pmod{\mu}$,

$$N := \left\{x \in X \,\middle|\, \lim_{n\to\infty} f_n(x) \neq f(x)\right\} \in \Sigma \quad \text{and} \quad \mu(N) = 0.$$

Let $\varepsilon > 0$ be arbitrary, then, for each $n \in \mathbb{N}$,

$$\{x \in X \mid |f_n(x) - f(x)| \geq \varepsilon\} \in \Sigma$$

and we have the following inclusion:

$$\Sigma \ni \varlimsup_{n\to\infty} \{x \in X \mid |f_n(x) - f(x)| \geq \varepsilon\} := \bigcap_{n=1}^{\infty} \bigcup_{k=n}^{\infty} \{x \in X \mid |f_k(x) - f(x)| \geq \varepsilon\} \subseteq N.$$

Exercise 5.42. Explain.

The latter, by the *monotonicity* of μ, implies that, for any $\varepsilon > 0$,

$$\mu\left(\varlimsup_{n\to\infty} \{x \in X \mid |f_n(x) - f(x)| \geq \varepsilon\}\right) = 0.$$

In view of $\mu(X) < \infty$, by the *continuity of measure from above* (see the *Continuity of Measure* (Theorem 3.2)) (see also Remarks 1.1), for any $\varepsilon > 0$,

$$\begin{aligned}\lim_{n\to\infty} \mu&\left(\bigcup_{k=n}^{\infty} \{x \in X \mid |f_k(x) - f(x)| \geq \varepsilon\}\right) \\ &= \mu\left(\varlimsup_{n\to\infty} \{x \in X \mid |f_n(x) - f(x)| \geq \varepsilon\}\right) = 0.\end{aligned}$$

Where, since, for any $\varepsilon > 0$ and each $n \in \mathbb{N}$,

$$\{x \in X \mid |f_n(x) - f(x)| \geq \varepsilon\} \subseteq \bigcup_{k=n}^{\infty} \{x \in X \mid |f_k(x) - f(x)| \geq \varepsilon\},$$

by the *monotonicity* of μ, we infer that, for any $\varepsilon > 0$,

$$\lim_{n\to\infty} \mu(\{x \in X \mid |f_n(x) - f(x)| \geq \varepsilon\}) = 0,$$

which completes the proof. □

Remarks 5.28.

- Under the conditions of the *Lebesgue Theorem*, we have not just proved that

$$\forall \varepsilon > 0 : \mu(\{x \in X \mid |f_n(x) - f(x)| \geq \varepsilon\}) \to 0, \ n \to \infty.$$

In fact, we have proved the following stronger statement:

$$\forall \varepsilon > 0 : \mu\left(\bigcup_{k=n}^{\infty} \{x \in X \mid |f_k(x) - f(x)| \geq \varepsilon\}\right) \to 0, \ n \to \infty,$$

(see the *Characterization of Convergence A. E.* (Proposition 5.12), Section 5.12, Problem 18).

- The *Lebesgue Theorem* naturally extends to Σ-measurable functions

$$f, f_n : X \to \mathbb{C},\ n \in \mathbb{N},$$

via separately applying the proven real-valued version to the real and imaginary parts.
- The definition of convergence in measure can be naturally extended to Σ-measurable functions

$$f, f_n : X \to \overline{\mathbb{R}},\ n \in \mathbb{N}$$

finite a. e. λ on X and the *Lebesgue theorem* remains in place for such functions.
- As the *Dancing Steps Example* (Example 5.15) shows, the converse to the *Lebesgue Theorem* is not true, i. e., a function sequence convergent in finite measure μ need not converge a. e. μ and can even diverge at any point (see Remarks 5.25).
- As Examples 5.14 show, the *finiteness* condition for measure in the *Lebesgue Theorem* is essential and cannot be dropped, i. e., a function sequence convergent a. e. relative to infinite measure μ need not converge in measure μ.

Theorem 5.13 (Riesz Theorem). *Let (X, Σ, μ) be a measure space and $f, f_n : X \to \mathbb{R}$, $n \in \mathbb{N}$, be Σ-measurable functions. If*

$$f_n \xrightarrow{\mu} f,$$

there exists a subsequence $(f_{n(k)})_{k\in\mathbb{N}}$ of the function sequence $(f_n)_{n\in\mathbb{N}}$ such that[3]

$$f_{n(k)} \to f \pmod{\mu}.$$

Proof. Since

$$f_n \xrightarrow{\mu} f,$$

then there exists a sequence $(n(k))_{k\in\mathbb{N}} \subseteq \mathbb{N}$ such that $n(k) < n(k+1)$, $k \in \mathbb{N}$, and

$$\forall\, k \in \mathbb{N} : \mu\left(\left\{x \in X \,\middle|\, |f_{n(k)}(x) - f(x)| \ge 2^{-k}\right\}\right) < 2^{-k}. \tag{5.11}$$

Exercise 5.43. Explain.

By the Σ-measurability of f and $f_{n(k)}$, $k \in \mathbb{N}$,

$$N := \left\{x \in X \,\middle|\, \lim_{k\to\infty} f_{n(k)}(x) \ne f(x)\right\} \in \Sigma$$

(see Remarks 5.18).

3 Frigyes Riesz (1880–1956).

Further, the following inclusion holds:

$$N \subseteq \overline{\lim_{k\to\infty}} \left\{x \in X \,\middle|\, |f_{n(k)}(x) - f(x)| \ge 2^{-k}\right\}$$
$$:= \bigcap_{k=1}^{\infty} \bigcup_{i=k}^{\infty} \left\{x \in X \,\middle|\, |f_{n(i)}(x) - f(x)| \ge 2^{-i}\right\} \in \Sigma.$$

Exercise 5.44. Explain (cf. the proof of the *Lebesgue Theorem* (Theorem 5.12)).

Hence, in view of (5.11), by the *monotonicity* and *σ-subadditivity* of μ, for each $k \in \mathbb{N}$,

$$\mu(N) \le \mu\left(\overline{\lim_{k\to\infty}} \left\{x \in X \,\middle|\, |f_{n(k)}(x) - f(x)| \ge 2^{-k}\right\}\right)$$
$$\le \mu\left(\bigcup_{i=k}^{\infty} \left\{x \in X \,\middle|\, |f_{n(i)}(x) - f(x)| \ge 2^{-i}\right\}\right)$$
$$\le \sum_{i=k}^{\infty} \mu\left(\left\{x \in X \,\middle|\, |f_{n(i)}(x) - f(x)| \ge 2^{-i}\right\}\right) < \sum_{i=k}^{\infty} 2^{-i} = 2^{-k+1}.$$

Passing to the limit as $k \to \infty$, we infer that

$$\mu(N) = 0,$$

which implies that, for the *subsequence* $(f_{n(k)})_{k\in\mathbb{N}}$ of the function sequence $(f_n)_{n\in\mathbb{N}}$,

$$f_{n(k)} \to f \pmod{\mu}$$

completing the proof. □

Remarks 5.29.

- The *Riesz Theorem* naturally extends to Σ-measurable functions

  $$f, f_n : X \to \mathbb{C},\ n \in \mathbb{N},$$

  via separately applying the proven real-valued version to the real and imaginary parts.
- Although the function sequence $(f_n)_{n\in\mathbb{N}}$ from the *Dancing Steps Example* (Example 5.15), being convergent to 0 in measure λ, diverges everywhere on $[0, 1]$, for the subsequence $(f_{2^k})_{k\in\mathbb{Z}_+}$, we have

  $$f_{2^k} \to 0 \pmod{\lambda},$$

  which is consistent with the *Riesz Theorem*.

Exercise 5.45.

(a) Verify.

(b) Find another subsequence of the sequence $(f_n)_{n\in\mathbb{N}}$ from the *Dancing Steps Example* (Example 5.15), which converges to 0 a. e. λ on $[0, 1]$.

5.11 Probabilistic Terminology

The theory of measure and integration underlies contemporary probability theory (see, e. g., [20]). Here, we parallel certain terminology between the two.

Definition 5.13 (Probability Space). A measure space (X, Σ, μ) with $\mu(X) = 1$ is called a *probability space* and μ is called a *probability measure.*

Example 5.16. The measure space $([0,1], \Sigma^*([0,1]), \lambda)$, where

$$\Sigma^*([0,1]) = \Sigma^* \cap [0,1]$$

is the σ-algebra of Lebesgue measurable subsets of $[0,1]$ and λ is the Lebesgue measure, and is a probability space.

Remark 5.30. Any measure space (X, Σ, μ) with $0 < \mu(X) < \infty$ can be turned into a probability space with the probability measure

$$\overline{\mu}(A) := \frac{1}{\mu(X)}\mu(A), \; A \in \Sigma.$$

In probability theory,

- a point $x \in X$ is called an *elementary event*;
- a set $A \in \Sigma$ is called a *random event*;
- for a random event $A \in \Sigma$, $\mu(A)$ is called the *probability* of A;
- a random event $A \in \Sigma$ with $\mu(A) = 0$ is called *almost impossible*, in particular $\emptyset$ is the impossible event;
- a Σ-measurable function $f : X \to \mathbb{R}$ is called a *random variable*;
- a sequence $(f_n)_{n \in \mathbb{N}}$ of random variables converging to a random variable f a. e. μ on X is said to *converge with probability* 1;
- a sequence $(f_n)_{n \in \mathbb{N}}$ of random variables converging to a random variable f in measure μ on X is said to *converge in probability*.

5.12 Problems

1. Prove

   Proposition 5.3 (Preimage σ-Algebra). *Let X be a nonempty set, (X', Σ') be a measurable space, and $f : X \to X'$ be a function, then*
   (1)
   $$\Sigma_f := f^{-1}(\Sigma') := \left\{ f^{-1}(A) \,\middle|\, A \in \Sigma' \right\}$$
   is a σ-algebra on X, the preimage σ-algebra of Σ' relative to the function f;
   (2) *Σ_f is the smallest of all σ-algebras Σ on X such that the function f is Σ-Σ'-measurable.*

   Hint. Use the *Properties of Inverse Image* (Theorem 1.4).

2. Let (X,Σ) be a measurable space, $(A_n)_{n\in\mathbb{N}}$ be a sequence in Σ such that

$$\bigcup_{n=1}^{\infty} A_n = X,$$

and $f : X \to \mathbb{R}$ be a function such that, for each $n \in \mathbb{N}$, the function $f \cdot \chi_{A_n}$, where χ_{A_n} is the characteristic function of A_n, $n \in \mathbb{N}$, is Σ-measurable. Prove that f is Σ-measurable.

3. Let (X,Σ) be a measurable space. Prove that $f : X \to \mathbb{R}$ is Σ-measurable *iff*,

$$\forall\, r \in \mathbb{Q} : f^{-1}((r,\infty)) \in \Sigma.$$

4. Let (X,Σ) be a measurable space, and $f : X \to \overline{\mathbb{R}}$ be an extended real-valued Σ-measurable function. Then, for each $-\infty < a < b < \infty$, the function

$$f_{a,b}(x) := \begin{cases} b, & x \in X : f(x) > b, \\ f(x), & x \in X : a \le f(x) \le b, \\ a, & x \in X : f(x) < a \end{cases}$$

is Σ-measurable.

5. Prove

Proposition 5.4 (Equality Set). *Let (X,Σ) be a measurable space. Then, for Σ-measurable functions $f,g : X \to \overline{\mathbb{R}}$,*

$$\{x \in X \,|\, f(x) = g(x)\} \in \Sigma.$$

6. Let (X,Σ) be a measurable space and $f,g : X \to \mathbb{R}$ be Σ-measurable functions. Prove that

$$\left\{x \in X \,\middle|\, e^{f(x)} \ge g(x) + 1\right\} \in \Sigma.$$

7. Prove

Proposition 5.5 (Finite Limit Set Structure). *Let (X,Σ) be a measurable space and $(f_n : X \to \mathbb{R})_{n\in\mathbb{N}}$ be a sequence of Σ-measurable functions. Then, for an arbitrary sequence $(\varepsilon_n)_{n\in\mathbb{N}} \subset (0,\infty)$ such that $\varepsilon_n \to 0$, $n \to \infty$,*

$$\begin{aligned} F := &\left\{x \in X \,\middle|\, (f_n(x))_{n\in\mathbb{N}} \text{ has a finite limit}\right\} \\ = &\bigcap_{n=1}^{\infty} \bigcup_{m=1}^{\infty} \bigcap_{k,l=m}^{\infty} \{x \in X \,|\, |f_k(x) - f_l(x)| < \varepsilon_n\} \in \Sigma. \end{aligned}$$

(Cf. the *Sequences of Measurable Functions Theorem* (Theorem 5.4).)

Hint. Use the *Cauchy Convergence Criterion* for numeric sequences.

8. Prove

Proposition 5.6 (Divergence Set Structure). *Let (X,Σ) be a measurable space and $f, f_n : X \to \mathbb{R}$, $n \in \mathbb{N}$, be Σ-measurable functions. Then, for an arbitrary sequence $(\varepsilon_n)_{n\in\mathbb{N}} \subset (0,\infty)$ such that $\varepsilon_n \to 0$, $n \to \infty$,*

$$N := \left\{x \in X \,\middle|\, \lim_{k\to\infty} f_n(x) \neq f(x)\right\} = \bigcup_{n=1}^{\infty}\bigcap_{m=1}^{\infty}\bigcup_{k=m}^{\infty} \{x \in X \mid |f_k(x) - f(x)| \geq \varepsilon_n\} \in \Sigma.$$

9. Prove

Proposition 5.7 (Borel Measurability of Derivative). *If a function $f : I \to \mathbb{R}$, where $I \subseteq \mathbb{R}$ is an interval, is differentiable on I, then its derivative $f' : I \to \mathbb{R}$ is a Borel measurable function.*

Hint. First, consider the case of an *open* interval I.

10. (a) Prove

Proposition 5.8 (Finiteness A. E. on a Finite-Measure Space). *Let (X,Σ,μ) be a measure space with $\mu(X) < \infty$ and a function*

$$f : X \to \overline{\mathbb{R}},$$

be Σ-measurable and finite a. e. μ on X, i. e.,

$$\mu(\{x \in X \mid |f(x)| = \infty\}) = 0.$$

Then

$$\mu(\{x \in X \mid |f(x)| \geq n\}) \to 0, \ n \to \infty.$$

(b) Give an example showing that the *finiteness* condition for the measure μ in the prior proposition is essential and cannot be dropped.

Hint. First, show that

$$\{x \in X \mid |f(x)| = \infty\} = \bigcap_{n=1}^{\infty} \{x \in X \mid |f(x)| \geq n\}.$$

11. Let (X,Σ,μ) be a measure space and $f_n : X \to \overline{\mathbb{R}}$, $n \in \mathbb{N}$, are such that

$$f_n = 0 \text{ a. e. } \mu \text{ on } X, \ n \in \mathbb{N}.$$

Prove that

(a) $g_1(x) := \sup_{n\in\mathbb{N}} |f_n(x)| = 0$ a. e. μ on X,

(b) $g_2(x) := \inf_{n\in\mathbb{N}} |f_n(x)| = 0$ a. e. μ on X, and
(c) $g_3(x) := \sum_{n=1}^{\infty} |f_n(x)| = 0$ a. e. μ on X.

12. Prove

Proposition 5.9 (Equivalence of Continuous Functions). *If $f, g : \mathbb{R} \to \mathbb{R}$ are continuous on $\mathbb{R}$ and $f(x) = g(x)$ a. e. λ on $\mathbb{R}$, then $f(x) = g(x)$, $x \in \mathbb{R}$.*

13. (a) Prove

Proposition 5.10 (A. E. Continuous is Lebesgue Measurable). *If a function $f : I \to \mathbb{R}$, where $I \subseteq \mathbb{R}$ is an interval, is continuous almost everywhere relative to the Lebesgue measure λ on I, then f is Lebesgue measurable.*

(b) Give an example of a function f continuous almost everywhere relative to the Lebesgue measure λ on an interval $I \subseteq \mathbb{R}$ (*a. e. λ on I* is understood relative to the *complete* Lebesgue measure λ on the Lebesgue σ-algebra $\Sigma^*(I)$) but *not* Borel measurable.

Hint. For (a), use the completeness of on the σ-algebra $\Sigma^*(I)$ Lebesgue measurable subsets of I. For (b), see Example 5.9.

14. Prove

Proposition 5.11 (Lebesgue Measurable vs. Borel Measurable). *For each Lebesgue measurable function $f : I \to \mathbb{R}$, where $I \subseteq \mathbb{R}$ is an interval, there exists a Borel measurable function $g : I \to \mathbb{R}$ such that*

$$f(x) = g(x) \text{ a. e. } \lambda \text{ on } I.$$

Hint. Start with a Lebesgue measurable characteristic function, then extend to a Lebesgue measurable simple function, then extend to an arbitrary Lebesgue measurable function (cf. the proof of *Luzin's Theorem* (Theorem 5.6)).

15. Let (X, Σ, μ) be a measure space and $f : X \to \mathbb{R}$ be a Σ-measurable function. Prove that, if

$$\forall \varepsilon > 0 : \mu(\{x \in X \mid |f(x)| \geq \varepsilon\}) = 0,$$

then

$$f(x) = 0 \text{ a. e. } \mu \text{ on } X.$$

16. For the measure space $(\mathbb{R}, \Sigma^*, \lambda)$, show that the function sequence

$$f_n(x) := e^{-n\sin^2 \pi x}, \quad n \in \mathbb{N}, x \in \mathbb{R},$$

converges a. e. λ on $\mathbb{R}$ and describe all limit functions.

17. ⋆ Apply *Egorov's Theorem* (Theorem 5.10) to provide an alternative proof to the *Lebesgue Theorem* (Theorem 5.12).
18. Prove

    Proposition 5.12 (Characterization of Convergence A. E.). *Let (X, Σ, μ) be a measure space and $f, f_n : X \to \mathbb{R}$, $n \in \mathbb{N}$, be Σ-measurable functions. Then for*

    $$f_n \to f \pmod{\mu},$$

    it is sufficient and, provided $\mu(X) < \infty$, necessary that

    $$\forall \varepsilon > 0 : \mu\left(\bigcup_{k=n}^{\infty} \{x \in X \mid |f_k(x) - f(x)| \geq \varepsilon\} \right) \to 0, \ n \to \infty.$$

    Hint. For *sufficiency*, apply the *Divergence Set Structure Proposition* (Proposition 5.6) (Problem 8). For *necessity*, see Remarks 5.28.

19. Let (X, Σ, μ) be a measure space with $\mu(X) < \infty$ and functions

    $$f, f_n : X \to \mathbb{R}$$

    be Σ-measurable. Prove that, if, for any $\varepsilon > 0$,

    $$\sum_{k=1}^{\infty} \mu\left(\{x \in X \mid |f_k(x) - f(x)| \geq \varepsilon\}\right) < \infty,$$

    then $f_n \to f \pmod{\mu}$.
20. Let (X, Σ, μ) be a measure space and

    $$f, g, f_n : X \to \mathbb{R}, \ n \in \mathbb{N},$$

    be Σ-measurable. Prove that, if

    $$f_n \xrightarrow{\mu} f \quad \text{and} \quad f_n \to g \pmod{\mu},$$

    then $f = g \pmod{\mu}$.
21. Prove

    Proposition 5.13 (Characterization of Convergence in Measure). *Let (X, Σ, μ) be a measure space and*

    $$f, f_n : X \to \mathbb{R}, \ n \in \mathbb{N},$$

    be Σ-measurable. Then for

    $$f_n \xrightarrow{\mu} f,$$

    it is necessary and, provided $\mu(X) < \infty$, sufficient that every subsequence $(f_{n(k)})_{k \in \mathbb{N}}$ of $(f_n)_{n \in \mathbb{N}}$ contain a subsequence $(f_{n(k(j))})_{j \in \mathbb{N}}$ such that

    $$f_{n(k(j))} \to f \pmod{\mu}.$$

Hint. For *necessity*, apply the *Riesz Theorem* (Theorem 5.13). For *sufficiency*, apply the *Lebesgue Theorem* (Theorem 5.12) and the *Characterization of Convergence* (Theorem 1.6).

22. Let (X, Σ, μ) be a measure space with $\mu(X) < \infty$ and

$$f, f_n, g, g_n : X \to \mathbb{R}, \; n \in \mathbb{N},$$

be Σ-measurable. Prove that, if

$$f_n \xrightarrow{\mu} f \text{ and } g_n \xrightarrow{\mu} g,$$

then $f_n g_n \xrightarrow{\mu} fg$.

Hint. Apply the *Characterization of Convergence in Measure Proposition* form Problem 21.

23. Let (X, Σ, μ) be a measure space and

$$f, f_n : X \to \mathbb{R}, \; n \in \mathbb{N}.$$

(a) Prove that, if $f_n \to f \pmod{\mu}$ and $F \in C(\mathbb{R})$, then $F(f_n) \to F(f) \pmod{\mu}$.
(b) Give an example showing that the analogue of the prior statement does not hold for $f_n \xrightarrow{\mu} f$, provided f and f_n, $n \in \mathbb{N}$, are Σ-measurable.

Hint. For (b), choose $F(x) := \sin x$, $x \in \mathbb{R}$.

6 Abstract Lebesgue Integral

In this chapter, we consider the construct of the abstract *Lebesgue integral*, which is based on the concept of *measure*, study the inherent properties of such integration, including the celebrated *limit theorems*, and compare its important particular case relative to the one-dimensional Lebesgue measure to the classical *Riemann*[1] *integral*.

6.1 Definitions and Examples

The construct of the abstract Lebesgue integral is developed in several consecutive stages. First, we define it for nonnegative measurable simple functions (see Definition 5.7 and Remarks 5.13).

Definition 6.1 (Lebesgue Integral of a Nonnegative Measurable Simple Function). Let (X, Σ, μ) be a measure space and $s : X \to [0, \infty)$ be a nonnegative Σ-measurable simple function. Any such a function can be represented in the form

$$s = \sum_{k=1}^{m} a_k \chi_{A_k}, \tag{6.1}$$

where $m \in \mathbb{N}$, $a_k \geq 0$, $k = 1, \ldots, m$, and

$$A_k \in \Sigma,\ k = 1, \ldots, m, \quad A_i \cap A_j = \emptyset,\ i \neq j, \quad \text{and} \quad \bigcup_{k=1}^{m} A_k = X, \tag{6.2}$$

and χ_{A_k} is the characteristic function of the set A_k, $k = 1, \ldots, m$ (see Remarks 5.13).

For a set $A \in \Sigma$, the *Lebesgue integral of s over A relative to μ* is the nonnegative value

$$\int_A s\, d\mu = \int_A s(x)\, d\mu(x) := \sum_{k=1}^{m} a_k \mu(A_k \cap A) \in [0, \infty],$$

with

$$a_k \mu(A_k \cap A) := \begin{cases} 0 & \text{if } a_k = 0 \text{ and } \mu(A_k \cap A) = \infty, \\ \infty & \text{if } a_k > 0 \text{ and } \mu(A_k \cap A) = \infty. \end{cases}$$

In particular, for $A = X$,

$$\int_X s\, d\mu = \int_X s(x)\, d\mu(x) := \sum_{k=1}^{m} a_k \mu(A_k).$$

1 Bernhard Riemann (1826–1866).

https://doi.org/10.1515/9783110600995-006

Notation. $\int_A s\,d\mu$ or $\int_A s(x)\,d\mu(x)$.

Remarks 6.1.

- Generally, representation (6.1)–(6.2) for a nonnegative Σ-measurable simple function $s : X \to [0,\infty)$ is *not unique*, e. g., for the Lebesgue measure space $(\mathbb{R}, \Sigma^*, \lambda)$,

$$\chi_{[0,\infty)} = \chi_{[0,n)} + \chi_{[n,\infty)}$$

with an arbitrary $n \in \mathbb{N}$ and

$$\chi_{[0,1)} + 2\chi_{[1,\infty)} = \chi_{[0,1/2)} + \chi_{[1/2,1)} + 2\chi_{[1,2)} + 2\chi_{[2,\infty)}.$$

Exercise 6.1. Give some more examples of this nature.

Thus, we are to show that the notion of the integral in the prior definition is *well-defined*, i. e., is independent of representation (6.1)–(6.2).

Indeed, let

$$s = \sum_{k=1}^{n} b_k \chi_{B_k},$$

where $n \in \mathbb{N}$, $b_k \ge 0$, $k = 1, \dots, n$, and

$$B_k \in \Sigma,\ k = 1, \dots, n, \quad B_i \cap B_j = \emptyset,\ i \ne j, \quad \text{and} \quad \bigcup_{k=1}^{n} B_k = X.$$

Then, in view of the fact that both the sets A_i, $i = 1, \dots, m$, and the sets B_j, $j = 1, \dots, n$, *partition* X, and hence, for any $i = 1, \dots, m$ and $j = 1, \dots, n$,

$$a_i = b_j \text{ whenever } A_i \cap B_j \ne \emptyset, \tag{6.3}$$

we have:

$$\begin{aligned}
&\sum_{i=1}^{m} a_i \mu(A_i \cap A) && \text{by the } \textit{additivity} \text{ of } \mu;\\
&= \sum_{i=1}^{m} a_i \sum_{j=1}^{n} \mu((A_i \cap A) \cap B_j) = \sum_{i=1}^{m} \sum_{j=1}^{n} a_i \mu((A_i \cap A) \cap B_j)\\
&= \sum_{j=1}^{n} \sum_{i=1}^{m} a_i \mu((A_i \cap A) \cap B_j) && \text{by (6.3);}\\
&= \sum_{j=1}^{n} \sum_{i=1}^{m} b_j \mu((A_i \cap A) \cap B_j)\\
&= \sum_{j=1}^{n} b_j \sum_{i=1}^{m} \mu(A_i \cap (B_j \cap A)) && \text{by the } \textit{additivity} \text{ of } \mu;\\
&= \sum_{j=1}^{n} b_j \mu(B_j \cap A).
\end{aligned}$$

Thus, for any nonnegative Σ-measurable simple function $s : X \to [0, \infty)$ and an arbitrary set $A \in \Sigma$, the Lebesgue integral $\int_A s\, d\mu$ of s over A relative to μ has the same value regardless of representation (6.1)–(6.2).

- Therefore, without loss of generality, for any nonnegative Σ-measurable simple function $s : X \to [0, \infty)$, the numbers $a_k \geq 0$, $k = 1, \dots, m$, in representation (6.1)–(6.2) can be regarded to be *distinct*, in which case

$$A_k := s^{-1}(\{a_k\}) \in \Sigma,\ k = 1, \dots, m,$$

(see Remarks 5.13).

- If $A \in \Sigma$ is a *μ-null set*, i. e., $\mu(A) = 0$, for any nonnegative Σ-measurable simple function $s : X \to [0, \infty)$,

$$\int_A s\, d\mu = 0.$$

Exercise 6.2. Verify.

Examples 6.1.

1. Let (X, Σ, μ) be a measure space and $A, B \in \Sigma$. Then
   (a) $\int_A 0\, d\mu = \int_A 0\chi_X\, d\mu = 0\mu(X \cap A) = 0$;
   (b) $\int_A 1\, d\mu = \int_A \chi_X\, d\mu = \mu(A \cap X) = \mu(A)$;
   (c) $\int_B \chi_A\, d\mu = 1\mu(A \cap B) + 0\mu(A^c \cap B) = \mu(A \cap B)$, and hence, with $A \subseteq B$,

$$\int_B \chi_A\, d\mu = \mu(A \cap B) = \mu(A),$$

   in particular,

$$\int_X \chi_A\, d\mu = \mu(A) = \int_A 1\, d\mu.$$

2. For the Lebesgue measure space $(\mathbb{R}, \Sigma^*, \lambda)$ and $A, B \in \Sigma^*$ with $A \subseteq B$,
   (a) $\int_A 0\, d\lambda = 0$;
   (b) $\int_A 1\, d\lambda = \int_B \chi_A\, d\lambda = \int_{\mathbb{R}} \chi_A\, d\lambda = \lambda(A)$, in particular, for $A = \mathbb{R}$,

$$\int_{\mathbb{R}} 1\, d\lambda = \lambda(\mathbb{R}) = \infty,$$

   for $A = \mathbb{Q}$ and any $\Sigma^* \ni B \supseteq A$,

$$\int_{\mathbb{Q}} 1\, d\lambda = \int_B \chi_{\mathbb{Q}}\, d\lambda = \int_{\mathbb{R}} \chi_{\mathbb{Q}}\, d\lambda = \lambda(\mathbb{Q}) = 0,$$

for $A = \mathbb{Q}^c$ and any $\Sigma^* \ni B \supseteq A$,

$$\int_{\mathbb{Q}^c} 1\,d\lambda = \int_B \chi_{\mathbb{Q}^c}\,d\lambda = \int_{\mathbb{R}} \chi_{\mathbb{Q}^c}\,d\lambda = \lambda(\mathbb{Q}^c) = \infty,$$

for $A := \bigcup_{n=1}^{\infty}[n, n+2^{-n})$ and any $\Sigma^* \ni B \supseteq A$,

$$\int_A 1\,d\lambda = \int_B \chi_A\,d\lambda = \int_{\mathbb{R}} \chi_A\,d\lambda = \lambda(A) = \sum_{n=1}^{\infty} \lambda([n, n+2^{-n})) = \sum_{n=1}^{\infty} 2^{-n} = 1,$$

for $A := \bigcup_{n=1}^{\infty}[n, n+1/n^2)$ and any $\Sigma^* \ni B \supseteq A$,

$$\int_A 1\,d\lambda = \int_B \chi_A\,d\lambda = \int_{\mathbb{R}} \chi_A\,d\lambda = \lambda(A) = \sum_{n=1}^{\infty} \lambda([n, n+1/n^2)) = \sum_{n=1}^{\infty} \frac{1}{n^2} = \frac{\pi^2}{6},$$

and, for $A := \bigcup_{n=1}^{\infty}[n, n+1/n)$ and any $\Sigma^* \ni B \supseteq A$,

$$\int_A 1\,d\lambda = \int_B \chi_A\,d\lambda = \int_{\mathbb{R}} \chi_A\,d\lambda = \lambda(A) = \sum_{n=1}^{\infty} \lambda([n, n+1/n)) = \sum_{n=1}^{\infty} \frac{1}{n} = \infty.$$

3. For the measure space $(\mathbb{N}, \mathscr{P}(\mathbb{N}), \mu)$, where μ is a *mass distribution measure over* $\mathbb{N}$:

$$\mathscr{P}(X) \ni A \mapsto \mu(A) := \sum_{k\in\mathbb{N}:\, k\in A} a_k$$

with some sequence $(a_n)_{n\in\mathbb{N}} \subset [0,\infty)$, in particular, for $a_n = 1$, $n \in \mathbb{N}$, μ is the *counting measure* (see Examples 3.2 and Remark 3.3), and $A \in \mathscr{P}(\mathbb{N})$,
   (a) $\int_A 0\,d\mu = 0$;
   (b) $\int_A 1\,d\mu = \mu(A) = \sum_{k\in\mathbb{N}:\, k\in A} a_k$, in particular, $\int_{\mathbb{N}} 1\,d\mu = \mu(\mathbb{N}) = \sum_{k=1}^{\infty} a_k$;
   (c) $\int_{\mathbb{N}} \chi_A\,d\lambda = \mu(A) = \sum_{k\in\mathbb{N}:\, k\in A} a_k$, in particular, for any $n \in \mathbb{N}$,

$$\int_{\mathbb{N}} \chi_{\{n\}}\,d\lambda = \mu(\{n\}) = a_n.$$

Exercise 6.3.
(a) Verify.
(b) Show that, for a measure space (X, Σ, μ) and nonnegative Σ-measurable simple functions $s_1, s_2 : X \to [0,\infty)$, such that

$$s_1(x) \le s_2(x),\ x \in X,$$

for any $A \in \Sigma$,

$$\int_A s_1\,d\mu \le \int_A s_2\,d\mu.$$

Now, we extend the notion of Lebesgue integral to arbitrary nonnegative measurable functions.

Definition 6.2 (Lebesgue Integral of a Nonnegative Measurable Function). Let (X, Σ, μ) be a measure space, $f : X \to [0, \infty]$ be a nonnegative Σ-measurable function, and $A \in \Sigma$. For a set $A \in \Sigma$, the *Lebesgue integral of f over A relative to μ* is the nonnegative value

$$\int_A f\,d\mu = \int_A f(x)\,d\mu(x) := \sup_{s \in S(f,A)} \int_A s\,d\mu \in [0, \infty],$$

where

$$S(f, A) := \{s : X \to [0, \infty) \mid s \text{ is } \Sigma\text{-measurable, simple, and } 0 \le s(x) \le f(x),\ x \in A\}.$$

Remarks 6.2.
- If a nonnegative Σ-measurable function $f : X \to [0, \infty]$ is *simple*, Definition 6.2 is consistent with Definition 6.1 (see Exercise 6.3).
- The function f need not be defined on the whole X, but rather on A only. However, since any nonnegative $\Sigma \cap A$-measurable function $f : A \to [0, \infty]$ can be nonnegatively and Σ-measurably extended to X, e. g., as follows:

$$\hat{f}(x) := \begin{cases} f(x), & x \in A, \\ 0 & x \in A^c, \end{cases}$$

without loss of generality, we can always regard f to be defined on the whole X.

Example 6.2. Let the underlying measure space be $([0,1], \Sigma^*([0,1]), \lambda)$, where $\Sigma^*([0,1])$ is the σ-algebra of Lebesgue measurable subsets of $[0,1]$ and λ is the Lebesgue measure, and

$$f(x) := x,\ x \in [0,1].$$

The function f is *continuous* on $[0,1]$, and hence, is Borel, and the more so, Lebesgue measurable (see Section 5.4).

For each $n \in \mathbb{N}$,

$$s_n := \sum_{k=0}^{n-1} \frac{k}{n} \chi_{[\frac{k}{n}, \frac{k+1}{n})} + \chi_{\{1\}} \in S(f, [0,1])$$

and, in view of $\lambda(\{1\}) = 1$,

$$\int_{[0,1]} s_n\,d\lambda = \sum_{k=0}^{n-1} \frac{k}{n} \lambda\left(\left[\frac{k}{n}, \frac{k+1}{n}\right)\right) + 1\lambda(\{1\}) = \sum_{k=0}^{n-1} \frac{k}{n}\frac{1}{n} = \frac{n(n-1)}{2n^2} \to \frac{1}{2},\ n \to \infty.$$

Hence,

$$\int_{[0,1]} f\,d\lambda := \sup_{s \in S(f,[0,1])} \int_{[0,1]} s\,d\lambda \ge \frac{1}{2}. \tag{6.4}$$

Let

$$s = \sum_{j=1}^{m} a_j \chi_{A_j},$$

where $m \in \mathbb{N}$, $a_j \geq 0$, $j = 1, \ldots, m$, and

$$A_j \in \Sigma,\ j = 1, \ldots, m, \quad A_i \cap A_j = \emptyset,\ i \neq j, \quad \bigcup_{j=1}^{m} A_j = [0,1],$$

be an arbitrary Σ-measurable nonnegative simple function in $S(f, [0,1])$.

Since, for an arbitrary $n \in \mathbb{N}$,

$$\bigcup_{k=0}^{n-1} \left[\frac{k}{n}, \frac{k+1}{n}\right) \cup \{1\} = [0,1],$$

for each $j = 1, \ldots, m$ and any $n \in \mathbb{N}$,

$$A_j = \bigcup_{k=0}^{n-1} A_j \cap \left[\frac{k}{n}, \frac{k+1}{n}\right) \cup (A_j \cap \{1\}),$$

and, for $k = 0, 1, \ldots, n-1$,

$$0 \leq a_j < \frac{k+1}{n} \text{ provided } \left[\frac{k}{n}, \frac{k+1}{n}\right) \cap A_j \neq \emptyset \tag{6.5}$$

Exercise 6.4. Explain.

In view of (6.5) and since

$$\bigcup_{j=1}^{m} A_j = [0,1],$$

by the *additivity* and *monotonicity* of λ, for any $n \in \mathbb{N}$,

$$\begin{aligned}
\int_{[0,1]} s\, d\lambda &= \sum_{j=1}^{m} a_j \lambda(A_j) = \sum_{j=1}^{m} a_j \left[\sum_{k=0}^{n-1} \lambda\left(A_j \cap \left[\frac{k}{n}, \frac{k+1}{n}\right)\right) + \lambda(A_j \cap \{1\}) \right] \\
&= \sum_{j=1}^{m} \sum_{k=0}^{n-1} a_j \lambda\left(A_j \cap \left[\frac{k}{n}, \frac{k+1}{n}\right)\right) \\
&< \sum_{j=1}^{m} \sum_{k=0}^{n-1} \frac{k+1}{n} \lambda\left(A_j \cap \left[\frac{k}{n}, \frac{k+1}{n}\right)\right) \\
&= \sum_{k=0}^{n-1} \sum_{j=1}^{m} \frac{k+1}{n} \lambda\left(A_j \cap \left[\frac{k}{n}, \frac{k+1}{n}\right)\right) = \sum_{k=0}^{n-1} \frac{k+1}{n} \sum_{j=1}^{m} \lambda\left(A_j \cap \left[\frac{k}{n}, \frac{k+1}{n}\right)\right)
\end{aligned}$$

$$= \sum_{k=0}^{n-1} \frac{k+1}{n} \lambda\left(\left[\frac{k}{n}, \frac{k+1}{n}\right)\right)$$

$$= \sum_{k=0}^{n-1} \frac{k+1}{n}\frac{1}{n} = \frac{n(n+1)}{2n^2} \to \frac{1}{2}, \; n \to \infty,$$

Hence, for an arbitrary Σ-measurable nonnegative simple function in $S(f, [0,1])$,

$$\int_{[0,1]} s\, d\lambda \le \frac{1}{2},$$

which implies that

$$\int_{[0,1]} f\, d\lambda := \sup_{s \in S(f,[0,1])} \int_{[0,1]} s\, d\lambda \le \frac{1}{2}. \tag{6.6}$$

From estimates (6.4) and (6.6), we conclude that

$$\int_{[0,1]} f\, d\lambda = \int_{[0,1]} x\, d\lambda(x) = \frac{1}{2}.$$

Remarks 6.3.

- To further stretch the notion of Lebesgue integral to arbitrary extended real-valued measurable functions, recall that with an arbitrary function $f : X \to \overline{\mathbb{R}}$ associated are two nonnegative functions

$$f_+(x) := \max(f(x), 0) := \begin{cases} f(x), & x \in X : f(x) \ge 0, \\ 0, & x \in X : f(x) < 0, \end{cases}$$

and

$$f_-(x) := -\min(f(x), 0) := \begin{cases} -f(x), & x \in X : f(x) < 0, \\ 0, & x \in X : f(x) \ge 0, \end{cases}$$

such that

$$f(x) = f_+(x) - f_-(x) \text{ and } |f(x)| = f_+(x) + f_-(x), \; x \in X,$$

we call the *positive* and *antinegative parts* of the function f, respectively, the functions f_+ and f_- being Σ-measurable provided the function f is Σ-measurable (see the *Positive and Antinegative Parts of a Function Corollary* (Corollary 5.4)).
- In particular, if f is *nonnegative*, i. e., $f : X \to [0, \infty]$, $f_- \equiv 0$, and hence, $f \equiv f_+$; if f is *nonpositive*, i. e., $f : X \to [-\infty, 0]$, $f_+ \equiv 0$, and hence, $f \equiv -f_-$.

Definition 6.3 (Lebesgue Integral of an Extended Real-Valued Measurable Function). Let (X, Σ, μ) be a measure space, $f : X \to \overline{\mathbb{R}}$ be a Σ-measurable function. For a set $A \in \Sigma$, the *Lebesgue integral of f over A relative to μ* is the extended value

$$\int_A f\,d\mu = \int_A f(x)\,d\mu(x) := \int_A f_+\,d\mu - \int_A f_-\,d\mu \in \overline{\mathbb{R}},$$

provided the integrals $\int_A f_+\,d\mu$ and $\int_A f_-\,d\mu$ are *not both* infinite.

Remarks 6.4.
- If a Σ-measurable function $f : X \to \overline{\mathbb{R}}$ is *nonnegative*, $f_- \equiv 0$ and $f \equiv f_+$ (see Remarks 6.3), and hence, Definition 6.3 is consistent with Definition 6.2.
- If a Σ-measurable function $f : X \to \mathbb{R}$ is simple, the functions f_+ and f_- are simple as well.
- The function f need not be defined on the whole X, but rather on A only. However, since any $\Sigma \cap A$-measurable function $f : A \to \overline{\mathbb{R}}$ can be Σ-measurably extended to X, e. g., as follows:

$$\hat{f}(x) := \begin{cases} f(x), & x \in A, \\ 0 & x \in A^c, \end{cases}$$

without loss of generality, we can always regard f to be defined on the whole X.

Definition 6.4 (Lebesgue Integrability). Let (X, Σ, μ) be a measure space, $f : X \to \overline{\mathbb{R}}$ be a Σ-measurable function, and $A \in \Sigma$. If

$$\int_A f\,d\mu$$

is finite, i. e., both

$$\int_A f_+\,d\mu \quad \text{and} \quad \int_A f_-\,d\mu$$

are finite, the function f is said to be *Lebesgue integrable on A relative to μ*.

The collection of all Lebesgue integrable extended real-valued functions on a Σ-measurable set A relative to μ is denoted by $L(A, \mu)$.

Remarks 6.5.
- In particular, for a *nonnegative* Σ-measurable function $f : X \to [0, \infty]$, $f_- \equiv 0$ (see Remarks 6.3), and hence,

$$f = f_+ \in L(A, \mu) \Leftrightarrow \int_A f\,d\mu = \int_A f_+\,d\mu < \infty;$$

for a *nonpositive* Σ-measurable function $f : X \to [-\infty, 0]$, $f_+ \equiv 0$ (see Remarks 6.3), and hence,

$$f = -f_- \in L(A,\mu) \Leftrightarrow f_- \in L(A,\mu) \Leftrightarrow -\int_A f\,d\mu = \int_A f_-\,d\mu < \infty;$$

- Thus, in general,

$$f \in L(A,\mu) \Leftrightarrow f_+, f_- \in L(A,\mu).$$

- For any measure space (X,Σ,μ) and arbitrary $A \in \Sigma$, $f \equiv 0 \in L(A,\mu)$.

Examples 6.3. For the Lebesgue measure space $(\mathbb{R},\Sigma^*,\lambda)$ and $A \in \Sigma^*$,

1. $\int_{\mathbb{R}} [\chi_{[0,2]} - \chi_{[3,4]}]\,d\lambda = \int_{\mathbb{R}} \chi_{[0,2]}\,d\lambda - \int_{\mathbb{R}} \chi_{[3,4]}\,d\lambda = \lambda([0,2]) - \lambda([3,4]) = 2 - 1 = 1$, and hence, $\chi_{[0,2]} - \chi_{[3,4]} \in L(\mathbb{R},\lambda)$;
2. $\int_{\mathbb{R}} [\chi_{[1,\infty)} - \chi_{[-1,0]}]\,d\lambda = \int_{\mathbb{R}} \chi_{[1,\infty)}\,d\lambda - \int_{\mathbb{R}} \chi_{[-1,0]}\,d\lambda = \lambda([1,\infty)) - \lambda([-1,0]) = \infty - 1 = \infty$, and hence, $\chi_{[1,\infty)} - \chi_{[-1,0]} \notin L(\mathbb{R},\lambda)$;
3. $\int_{\mathbb{R}} [\chi_{[-1,0]} - \chi_{[1,\infty)}]\,d\lambda = \int_{\mathbb{R}} \chi_{[-1,0]}\,d\lambda - \int_{\mathbb{R}} \chi_{[1,\infty)}\,d\lambda = \lambda([-1,0]) - \lambda([1,\infty)) = 1 - \infty = -\infty$, and hence, $\chi_{[-1,0]} - \chi_{[1,\infty)} \notin L(\mathbb{R},\lambda)$;
4. More generally, for an arbitrary measure space (X,Σ,μ), a simple Σ-measurable function

$$s = \sum_{i=1}^{m} a_i \chi_{A_i},$$

where $m \in \mathbb{N}$, $a_i \in \mathbb{R}$, $i = 1,\dots,m$, and

$$A_i \in \Sigma,\ i = 1,\dots,m, \quad A_i \cap A_j = \emptyset,\ i \neq j, \quad \bigcup_{i=1}^{m} A_i = X$$

(see Remarks 6.1) is Lebesgue integrable on a set $A \in \Sigma$ relative to μ *iff*

$$\int_A f_+\,d\mu := \sum_{i:\,a_i \geq 0} a_i \mu(A_i) < \infty \quad \text{and} \quad \int_A f_-\,d\mu := \sum_{i:\,a_i \leq 0} (-a_i)\mu(A_i) < \infty,$$

in which case

$$\int_A f\,d\mu := \int_A f_+\,d\mu - \int_A f_-\,d\mu = \sum_{i=1}^{m} a_i \mu(A_i).$$

Remark 6.6. The notion of Lebesgue integrability and Lebesgue integral naturally extends to complex-valued measurable functions. Given a measure space (X,Σ,μ), we say that a Σ-measurable function $f : X \to \mathbb{C}$ is Lebesgue integrable on a set $A \in \Sigma$ relative to measure μ and write $f \in L(A,\mu)$ if

$$\operatorname{Re} f, \operatorname{Im} f \in L(A,\mu),$$

in which case

$$\int_A f\,d\mu := \int_A \operatorname{Re} f\,d\mu + i \int_A \operatorname{Im} f\,d\mu.$$

6.2 Properties of Lebesgue Integral

The subsequent theorem establishes certain immediate properties of a Lebesgue integral, with further properties discussed in subsequent sections.

Theorem 6.1 (Properties of Lebesgue Integral. I). *Let (X, Σ, μ) be a noninfinite measure space, $f, g : X \to \overline{\mathbb{R}}$ be Σ-measurable functions, and $A, B \in \Sigma$.*

(1) *If $\mu(A) = 0$, then*

$$\int_A f\,d\mu = 0$$

and $f \in L(A, \mu)$.
In particular, $\int_\emptyset f\,d\mu = 0$.

(2) *If $f(x) = c$, $x \in A$, for some $c \in \mathbb{R}$, then*

$$\int_A f\,d\mu = \int_A c\,d\mu = c\mu(A)$$

and, for $c \neq 0$, $f \in L(A, \mu) \Leftrightarrow \mu(A) < \infty$.

(3) *If $f, g : X \to [0, \infty]$ and*

$$0 \le f(x) \le g(x),\ x \in A,$$

then

$$\int_A f\,d\mu \le \int_A g\,d\mu$$

and

$$g \in L(A, \mu) \Rightarrow f \in L(A, \mu).$$

(4) *If $f, g \in L(A, \mu)$ and*

$$f(x) \le g(x),\ x \in A,$$

then

$$\int_A f\,d\mu \le \int_A g\,d\mu.$$

(5) *If* $A \neq \emptyset$, $\mu(A) < \infty$, *and* f *is bounded on* A, *then* $f \in L(A,\mu)$ *and*

$$\inf_{x\in A} f(x)\cdot\mu(A) \le \int_A f\,d\mu \le \sup_{x\in A} f(x)\cdot\mu(A).$$

(6) *If* $f : X \to [0,\infty]$ *and* $B \subseteq A$, *then*

$$\int_B f\,d\mu \le \int_A f\,d\mu$$

and

$$f \in L(A,\mu) \Rightarrow f \in L(B,\mu).$$

(7) *If* $B \subseteq A$, *then*

$$f \in L(A,\mu) \Rightarrow f \in L(B,\mu).$$

(8) *For* $B \subseteq A$,

$$f \in L(B,\mu) \text{ and } f \in L(A\setminus B,\mu) \ \Leftrightarrow\ f \in L(A,\mu).$$

(9) *(constant factor property) If* $\int_A f\,d\mu$ *exists (finite or infinite), then, for any* $c \in \mathbb{R}$,

$$\int_A cf\,d\mu = c\int_A f\,d\mu \quad \text{with} \quad 0\cdot(\pm\infty) := 0$$

and

$$f \in L(A,\mu) \Rightarrow cf \in L(A,\mu).$$

Proof.
(1) If $f : X \to [0,\infty]$,

$$\int_A f\,d\mu := \sup_{s\in S(f,A)} \int_A s\,d\mu = 0$$

(see Definition 6.2) since, for each $s \in S(f,A)$,

$$\int_A s\,d\mu = 0$$

(see Remarks 6.1).
The general case follows immediately by Definition 6.3.

Exercise 6.5. Fill in the details.

(2)

Exercise 6.6. Prove (2).

(3) For $f, g : X \to [0, \infty]$, the estimate

$$0 \le f(x) \le g(x), \ x \in A,$$

immediately implies the inclusion

$$S(f, A) \subseteq S(g, A)$$

(see Definition 6.2), where the rest follows from the definition.

Exercise 6.7. Fill in the details.

(4) $f, g \in L(A, \mu)$, the estimate

$$f(x) \le g(x), \ x \in A,$$

implies

$$f_+(x) \le g_+(x) \text{ and } g_-(x) \le f_-(x), \ x \in A,$$

where the rest follows from Definition 6.3.

Exercise 6.8. Verify and fill in the details.

(5) Immediately follows from (2) and (4).

Exercise 6.9. Fill in the details.

(6) Since each function

$$s \in S(f, B)$$

can be extended to the function

$$S(f, A) \ni \hat{s}(x) := \begin{cases} s(x), & x \in B, \\ 0, & x \in B^c. \end{cases}$$

$$\int_B f\,d\mu := \sup_{s \in S(f,B)} \int_B s\,d\mu = \sup_{s \in S(f,B)} \int_A \hat{s}\,d\mu \le \sup_{s \in S(f,A)} \int_A s\,d\mu =: \int_A f\,d\mu.$$

(7) For $B \subseteq A$, the fact that

$$f \in L(A)$$

is equivalent to

$$f_+, f_- \in L(A, \mu)$$

(see Remarks 6.5), which, by (6), implies that

$$f_+, f_- \in L(B),$$

the latter being equivalent to

$$f \in L(B, \mu).$$

(8) *"If"* part follows immediately from (7) in view of the inclusions

$$B \subseteq A, \ A \setminus B \subseteq A. \tag{6.7}$$

"Only if" part. Suppose $f \in L(B, \mu)$ and $f \in L(A \setminus B, \mu)$ assuming additionally that $f : X \to [0, \infty]$, i. e., that f is *nonnegative.*
Inclusions (6.7) imply the inclusion

$$S(f, A) \subseteq S(f, B), \ S(f, A) \subseteq S(f, A \setminus B),$$

and hence, by Definition 6.2,

$$\begin{aligned}\int_A f\, d\mu &:= \sup_{s \in S(f,A)} \int_A s\, d\mu \le \sup_{s \in S(f,A)} \left[\int_B s\, d\mu + \int_{A \setminus B} s\, d\mu \right] \\ &\le \sup_{s \in S(f,A)} \int_B s\, d\mu + \sup_{s \in S(f,A)} \int_{A \setminus B} s\, d\mu \\ &\le \sup_{s \in S(f,B)} \int_B s\, d\mu + \sup_{s \in S(f,A \setminus B)} \int_{A \setminus B} s\, d\mu = \int_B f\, d\mu + \int_{A \setminus B} f\, d\mu < \infty,\end{aligned}$$

which implies that $f \in L(A, \mu)$.

Exercise 6.10. Prove the general case of $f : X \to \overline{\mathbb{R}}$.

(9) The statement is, obviously, true for $c = 0$ ($0 \cdot (\pm\infty) := 0$).
Let $c > 0$ and $f : X \to [0, \infty]$. Then

$$s \in S(cf, A) \Leftrightarrow \frac{1}{c} s \in S(f, A),$$

and hence, for each $s \in S(cf)$,

$$\int_A s\, d\mu = c \int_A \frac{1}{c} s\, d\mu \le c \int_A f\, d\mu,$$

which, by Definition 6.2, implies that

$$\int_A cf\,d\mu \le c\int_A f\,d\mu. \tag{6.8}$$

Symmetrically,

$$\int_A f\,d\mu = \int_A \frac{1}{c}(cf)\,d\mu \le \frac{1}{c}\int_A cf\,d\mu. \tag{6.9}$$

Inequalities (6.8) and (6.9) jointly imply

$$\int_A cf\,d\mu = c\int_A f\,d\mu$$

For $c > 0$ and $f : X \to \overline{\mathbb{R}}$, since

$$(cf)_+ = cf_+ \text{ and } (cf)_- = cf_-$$

and $\int_A f_+\,d\mu$, $\int_A f_-\,d\mu$ are not both infinite, by the proven case,

$$\begin{aligned}\int_A cf\,d\mu &:= \int_A (cf)_+\,d\mu - \int_A (cf)_-\,d\mu = \int_A cf_+\,d\mu - \int_A cf_-\,d\mu\\ &= c\int_A f_+\,d\mu - c\int_A f_-\,d\mu = c\left[\int_A f_+\,d\mu - \int_A f_-\,d\mu\right] =: c\int_A f\,d\mu.\end{aligned}$$

For $c < 0$ and $f : X \to \overline{\mathbb{R}}$, since

$$(cf)_+ = (-c)f_- \text{ and } (cf)_- = (-c)f_+$$

and $\int_A f_+\,d\mu$, $\int_A f_-\,d\mu$ are not both infinite, by the proven part,

$$\begin{aligned}\int_A cf\,d\mu &:= \int_A (cf)_+\,d\mu - \int_A (cf)_-\,d\mu = \int_A (-c)f_-\,d\mu - \int_A (-c)f_+\,d\mu\\ &= (-c)\int_A f_-\,d\mu + c\int_A f_+\,d\mu = c\left[\int_A f_+\,d\mu - \int_A f_-\,d\mu\right] =: c\int_A f\,d\mu.\end{aligned}$$

Hence, for any $c \in \mathbb{R}$,

$$\int_A cf\,d\mu = c\int_A f\,d\mu,$$

which immediately implies that

$$f \in L(A,\mu) \Rightarrow cf \in L(A,\mu).$$ □

Remark 6.7. Properties (1), (2), (7)–(9) also hold for a Σ-measurable $f : X \to \mathbb{C}$ and $c \in \mathbb{C}$.

6.3 Countable Additivity

The property of *countable* or *σ-additivity* of the Lebesgue integral truly distinguishes it from the Riemann integral and underlies the subsequent efficient *limit theorems*.

Theorem 6.2 (σ-Additivity of Lebesgue Integral). *Let (X, Σ, μ) be a measure space and $f : X \to [0, \infty]$ be a Σ-measurable function. Then the set function*

$$\Sigma \ni A \mapsto \mu_f(A) := \int_A f\, d\mu \in [0, \infty] \tag{6.10}$$

is nonnegative and σ-additive, i. e., is a measure on Σ.

If, in addition, $f \in L(X, \mu)$, the measure μ_f is finite.

Proof. By Definition 6.2, the set function μ_f in (6.10) is well-defined and *nonnegative*.

Observe that, by the *σ-additivity* of μ, the statement is true for any nonnegative Σ-measurable simple function $s : X \to [0, \infty)$:

$$s = \sum_{k=1}^{m} b_k \chi_{B_k},$$

where $m \in \mathbb{N}$, $b_k \geq 0$, $k = 1, \dots, m$ are *distinct*, and

$$B_k := s^{-1}(\{b_k\}) \in \Sigma,\ k = 1, \dots, m, \quad B_i \cap B_j = \emptyset,\ i \neq j, \quad \bigcup_{k=1}^{m} B_k = X$$

(see Remarks 6.1).

Exercise 6.11. Show.

Let $(A_n)_{n \in \mathbb{N}}$ be an arbitrary *pairwise disjoint* sequence in Σ and

$$A := \bigcup_{n=1}^{\infty} A_n.$$

If, for some $N \in \mathbb{N}$,

$$\mu_f(A_N) := \int_{A_N} f\, d\mu = \infty,$$

since $A_N \subseteq A$, by the *Properties of Lebesgue Integral. I* (Theorem 6.1, part (6)),

$$\int_A f\, d\mu \geq \int_{A_N} f\, d\mu = \infty$$

and we have

$$\int_A f\, d\mu = \infty = \sum_{n=1}^{\infty} \int_{A_n} f\, d\mu.$$

Suppose that, for each $n \in \mathbb{N}$,

$$\int_{A_n} f\,d\mu < \infty. \tag{6.11}$$

By Definition 6.2, for any nonnegative Σ-measurable simple function $s \in S(f,A) \subseteq S(f,A_n)$, $n \in \mathbb{N}$,

$$\int_A s\,d\mu = \sum_{n=1}^{\infty} \int_{A_n} s\,d\mu \le \sum_{n=1}^{\infty} \int_{A_n} f\,d\mu,$$

and hence,

$$\int_A f\,d\mu := \sup_{s\in S(f,A)} \int_A s\,d\mu \le \sum_{n=1}^{\infty} \int_{A_n} f\,d\mu. \tag{6.12}$$

For an arbitrary $\varepsilon > 0$ and any $n \in \mathbb{N}$, by Definition 6.2, there is a

$$s_n \in S\left(f, \bigcup_{k=1}^{n} A_k\right) \subseteq S(f,A_k),\ k = 1,\dots,n,$$

such that

$$\int_{A_k} f\,d\mu - \frac{\varepsilon}{n} < \int_{A_k} s_n\,d\mu \le \int_{A_k} f\,d\mu,\ k = 1,\dots,n,$$

and hence, since $s_n \in S(f,A_k)$, $k = 1,\dots,n$, by Definition 6.2,

$$\sum_{k=1}^{n} \int_{A_k} f\,d\mu < \sum_{k=1}^{n} \left[\int_{A_k} s_n\,d\mu + \frac{\varepsilon}{n}\right] = \sum_{k=1}^{n} \int_{A_k} s_n\,d\mu + \varepsilon = \int_{\bigcup_{k=1}^{n} A_k} s_n\,d\mu + \varepsilon \le \int_{\bigcup_{k=1}^{n} A_k} f\,d\mu + \varepsilon.$$

For each $n \in \mathbb{N}$, passing to the limit as $\varepsilon \to 0+$, we have

$$\sum_{k=1}^{n} \int_{A_k} f\,d\mu \le \int_{\bigcup_{k=1}^{n} A_k} f\,d\mu,$$

where since

$$\bigcup_{k=1}^{n} A_k \subseteq A,\ n \in \mathbb{N},$$

by the *Properties of Lebesgue Integral. I* (Theorem 6.1, part (6)), we infer that, for each $n \in \mathbb{N}$,

$$\sum_{k=1}^{n} \int_{A_k} f\,d\mu \le \int_A f\,d\mu.$$

Passing to the limit as $n \to \infty$, we obtain

$$\sum_{k=1}^{\infty} \int_{A_k} f\,d\mu \le \int_A f\,d\mu. \tag{6.13}$$

Inequalities (6.12) and (6.13) jointly imply that

$$\int_A f\,d\mu = \sum_{n=1}^{\infty} \int_{A_n} f\,d\mu,$$

which completes the proof of the *σ-additivity* of μ_f.

Thus, μ_f is a *measure* on Σ.

If, in addition, $f \in L(X,\mu)$, then

$$\mu(X) := \int_X f\,d\mu < \infty,$$

which implies that the measure μ_f is finite and completes the entire proof. □

Corollary 6.1 (σ-Additivity of Lebesgue Integral). *Let (X,Σ,μ) be a measure space, $f : X \to \overline{\mathbb{R}}$ and $f \in L(X,\mu)$. Then the real-valued set function*

$$\Sigma \ni A \mapsto \nu_f(A) := \int_A f\,d\mu \in \mathbb{R} \tag{6.14}$$

is σ-additive.

Proof. By the *Properties of Lebesgue Integral. I* (Theorem 6.1, part (7)), the real-valued set function ν_f in (6.10) is well-defined.

By Definition 6.3, we have the following representation:

$$\nu_f(A) := \int_A f\,d\mu := \int_A f_+\,d\mu - \int_A f_-\,d\mu =: \mu_{f_+}(A) - \mu_{f_-}(A),\ A \in \Sigma, \tag{6.15}$$

where, since $f \in L(X,\mu)$, $f_+, f_- \in L(X,\mu)$ (see Remarks 6.5). Hence, by the *σ-Additivity of Lebesgue Integral Theorem* (Theorem 6.2), μ_{f_+} and μ_{f_-} are *finite* measures on Σ.

This immediately implies that the set function ν_f is *σ-additive* on Σ.

Exercise 6.12. Fill in the details. □

Remark 6.8. Countable additivity remains in place also for a Σ-measurable function $f : X \to \mathbb{C}, f \in L(X,\mu)$.

6.4 Further Properties

Here, we continue our discourse on the properties of Lebesgue integral.

Theorem 6.3 (Properties of Lebesgue Integral. II). *Let (X, Σ, μ) be a measure space, $f, g : X \to \overline{\mathbb{R}}$ be Σ-measurable functions, and $A, B \in \Sigma$.*

(1) *(additivity) If $f : X \to [0, \infty]$ or $f \in L(X, \mu)$ and $A \cap B = \emptyset$, then*

$$\int_{A \cup B} f \, d\mu = \int_A f \, d\mu + \int_B f \, d\mu.$$

(2) *(continuity) If $f \in L(X, \mu)$ and $(A_n)_{n \in \mathbb{N}} \subseteq \Sigma$ is a monotone sequence with $A := \lim_{n \to \infty} A_n$, then*

$$\lim_{n \to \infty} \int_{A_n} f \, d\mu = \int_A f \, d\mu,$$

for a function $f : X \to [0, \infty]$ and an increasing set sequence $(A_n)_{n \in \mathbb{N}} \subseteq \Sigma$, the condition $f \in L(X, \mu)$ being redundant.

(3) *(independence of values on a null set) If $f : A \to [0, \infty]$ or $f \in L(A, \mu)$ and $A \supseteq N \in \Sigma$ with $\mu(N) = 0$, then*

$$\int_A f \, d\mu = \int_{A \setminus N} f \, d\mu.$$

(4) *(equality of the integrals of equivalent functions) If $f \in L(A, \mu)$ and $f = g \pmod{\mu}$ on A, then $g \in L(A, \mu)$ and*

$$\int_A f \, d\mu = \int_A g \, d\mu.$$

(5) *$f \in L(A, \mu) \Leftrightarrow |f| \in L(A, \mu)$ and*

$$\left| \int_A f \, d\mu \right| \le \int_A |f| \, d\mu.$$

(6) *If $f \in L(A, \mu)$ and $|g(x)| \le f(x)$ a. e. μ on A, then $g \in L(A, \mu)$.*

(7) *(Markov's[2] inequality, also referred to as Chebyshev's[3] inequality) If $f \in L(A, \mu)$, then*

$$\forall \varepsilon > 0 : \mu(\{x \in A \mid |f(x)| \ge \varepsilon\}) \le \frac{1}{\varepsilon} \int_A |f| \, d\mu.$$

(8) *If $f \in L(A, \mu)$, then f is finite a. e. μ on A.*

---

2 Andrey Markov (1856–1922).

3 Pafnuty Chebyshev (1821–1894).

(9) *If* $f : A \to [0, \infty]$, *then*

$$\int_A f\,d\mu = 0 \Leftrightarrow f(x) = 0 \ a.e.\ \mu \ on \ A.$$

Proof. (1) (*additivity*) Immediately follows from the *σ-Additivity of Lebesgue Integral Theorem* (Theorem 6.2) and the *σ-Additivity of Lebesgue Integral Corollary* (Corollary 6.1) (see Exercise 3.1 (b)).

(2) (*continuity*) For $f : X \to [0, \infty]$, (2) follows from the *Continuity of Measure Theorem* (Theorem 3.2).
For $f \in L(X, \mu)$, (2) follows from the above and representation (6.15).

(3) (*independence of values on a null set*) If $f : A \to [0, \infty]$ or $f \in L(A, \mu)$ and $A \supseteq N \in \Sigma$ with $\mu(N) = 0$, then, by the *additivity* and the *Properties of Lebesgue Integral. I* (Theorem 6.1, part 1),

$$\int_A f\,d\mu = \int_{A\setminus N} f\,d\mu + \int_N f\,d\mu = \int_{A\setminus N} f\,d\mu + 0 = \int_{A\setminus N} f\,d\mu.$$

(4) (*equality of the integrals of equivalent functions*) Since $f = g \pmod{\mu}$ on A,

$$N := \{x \in A \,|\, f(x) \neq g(x)\} \in \Sigma \quad \text{and} \quad \mu(N) = 0.$$

Then, by the *Properties of Lebesgue Integral. I* (Theorem 6.1, parts (1) and (7)),

$$g \in L(N, \mu) \quad \text{and} \quad g \in L(A \setminus N, \mu),$$

which, by the *Properties of Lebesgue Integral. I* (Theorem 6.1, part (8)), implies that

$$g \in L(A, \mu).$$

Further, by (3), the *Properties of Lebesgue Integral. I* (Theorem 6.1, part (1)), and the *additivity*

$$\int_A f\,d\mu = \int_{A\setminus N} f\,d\mu = \int_{A\setminus N} g\,d\mu = \int_{A\setminus N} g\,d\mu + \int_N g\,d\mu = \int_A g\,d\mu.$$

(5) *"Only if"* part. Suppose that $f \in L(A, \mu)$, i. e., by Definition 6.3,

$$\int_A f_+\,d\mu < \infty \text{ and } \int_A f_-\,d\mu < \infty.$$

Let

$$A_+ := \{x \in A \,|\, f(x) \geq 0\} \quad \text{and} \quad A_- := \{x \in A \,|\, f(x) < 0\}.$$

By the Σ-*measurability* of f,

$$A_+, A_- \in \Sigma.$$

Also,

$$A_- = A \setminus A_+.$$

In view of the fact that

$$f_+(x) = 0,\ x \in A_-, \quad \text{and} \quad f_-(x) = 0,\ x \in A_+,$$

by the *additivity*,

$$\begin{aligned}
\int_A |f|\, d\mu &= \int_{A_+} |f|\, d\mu + \int_{A_-} |f|\, d\mu = \int_{A_+} f_+\, d\mu + \int_{A_-} f_-\, d\mu \\
&= \int_{A_+} f_+\, d\mu + \int_{A_-} f_+\, d\mu + \int_{A_-} f_-\, d\mu + \int_{A_+} f_-\, d\mu \\
&= \int_A f_+\, d\mu + \int_A f_-\, d\mu < \infty.
\end{aligned} \tag{6.16}$$

"If" part. Suppose $|f| \in L(A,\mu)$.
Since

$$0 \le f_+(x), f_-(x) \le |f|(x),\ x \in A,$$

by the *Properties of Lebesgue Integral. I* (Theorem 6.1, part (3)), $f_+, f_- \in L(A,\mu)$, which is equivalent to $f \in L(A,\mu)$ (see Remarks 6.5).
Furthermore, for $f \in L(A,\mu)$, by (6.16),

$$\int_A |f|\, d\mu = \int_A f_+\, d\mu + \int_A f_-\, d\mu \ge \left| \int_A f_+\, d\mu - \int_A f_-\, d\mu \right| = \left| \int_A f\, d\mu \right|.$$

(6) If $f \in L(A,\mu)$ and $|g(x)| \le f(x)$ a. e. μ on A, then

$$N := \{x \in A \mid |g(x)| > f(x)\} \in \Sigma \quad \text{and} \quad \mu(N) = 0.$$

Hence, by the *Properties of Lebesgue Integral. I* (Theorem 6.1, parts (1), (6), and (3)),

$$|g| \in L(N,\mu) \quad \text{and} \quad |g| \in L(A \setminus N,\mu),$$

which, by the *Properties of Lebesgue Integral. I* (Theorem 6.1, part (8)), implies that

$$|g| \in L(A,\mu).$$

Hence, by (5), $g \in L(A,\mu)$.

(7) (*Markov's Inequality*) Suppose $f \in L(A,\mu)$. Then, by (5), $|f| \in L(A,\mu)$.
Let $\varepsilon > 0$ be arbitrary. Then, by the *Combinations of Measurable Functions Theorem* (Theorem 5.3),

$$A_\varepsilon := \{x \in A \mid |f(x)| \geq \varepsilon\} \in \Sigma.$$

Exercise 6.13. Explain.

By the *Properties of Lebesgue Integral. I* (Theorem 6.1, parts (2), (3), and (6)), $|f| \in L(A_\varepsilon,\mu)$ and

$$\varepsilon\mu(A_\varepsilon) = \int_{A_\varepsilon} \varepsilon \, d\mu \leq \int_{A_\varepsilon} |f| \, d\mu \leq \int_A |f| \, d\mu,$$

where *Markov's inequality* follows immediately.

(8) Suppose $f \in L(A,\mu)$.
The set

$$\{x \in A \mid |f(x)| = \infty\} = \bigcap_{n=1}^{\infty} \{x \in A \mid |f(x)| \geq n\} \in \Sigma$$

is the limit of the *decreasing* set sequence

$$\{x \in A \mid |f(x)| \geq n\} \in \Sigma,\ n \in \mathbb{N}.$$

Exercise 6.14. Verify (cf. Section 5.12, Problem 10).

By *Markov's inequality*,

$$\mu\left(\{x \in A \mid |f(x)| \geq n\}\right) \leq \frac{1}{n} \int_A |f| \, d\mu < \infty,\ n \in \mathbb{N},$$

where by the *Continuity of Measure Theorem* (Theorem 3.2) and the *Squeeze Theorem*, we infer that

$$\mu\left(\{x \in A \mid |f(x)| = \infty\}\right) = \lim_{n\to\infty} \mu\left(\{x \in A \mid |f(x)| \geq n\}\right) = 0,$$

which shows that f is finite a. e. μ on A.

(9) Let $f : A \to [0,\infty]$.
"If" part follows directly from the *additivity* and the *Properties of Lebesgue Integral. I* (Theorem 6.1, part (1)).

Exercise 6.15. Fill in the details.

"Only if" part. Suppose that

$$\int_A f \, d\mu = 0.$$

The set

$$\{x \in A \mid f(x) > 0\} = \bigcup_{n=1}^{\infty} \{x \in A \mid |f(x)| \geq 1/n\} \in \Sigma$$

is the limit of the *increasing* set sequence

$$\{x \in A \mid f(x) \geq 1/n\} \in \Sigma,\ n \in \mathbb{N}.$$

Exercise 6.16. Verify.

By *Markov's inequality*,

$$\mu(\{x \in A \mid f(x) \geq 1/n\}) \leq n \int_A f\, d\mu = 0,\ n \in \mathbb{N},$$

and hence,

$$\mu(\{x \in A \mid f(x) \geq 1/n\}) = 0,\ n \in \mathbb{N}.$$

Therefore, by the *Continuity of Measure Theorem* (Theorem 3.2),

$$\mu(\{x \in A \mid f(x) > 0\}) = \lim_{n \to \infty} \mu(\{x \in A \mid f(x) \geq 1/n\}) = 0,$$

which implies that $f(x) = 0$ a. e. μ on A. □

Remarks 6.9.

- In view of the *Properties of Lebesgue Integral. II* (Theorem 6.3, part (3)), parts (2)–(5) of the *Properties of Lebesgue Integral. I* (Theorem 6.1) hold as soon as their conditions are met a. e. μ on A.
- As follows from the proof of the *Properties of Lebesgue Integral. II* (Theorem 6.3, part (7)), the following sharper version of *Markov' inequality* holds:
  If $f \in L(A, \mu)$, then

$$\forall \varepsilon > 0 : \mu(\{x \in A \mid |f(x)| \geq \varepsilon\}) \leq \frac{1}{\varepsilon} \int_{\{x \in A \mid |f(x)| \geq \varepsilon\}} |f(x)|\, d\mu. \tag{6.17}$$

- Parts (1)–(8) of the *Properties of Lebesgue Integral. II* (Theorem 6.3) also hold for a Σ-measurable function $f : X \to \mathbb{C}$. In particular, for

$$\int_A f\, d\mu = re^{i\theta}$$

  with some $r \geq 0$ and $\theta \in (-\pi, \pi]$,

$$\left|\int_A f\, d\mu\right| = r = e^{-i\theta} \int_A f\, d\mu = \int_A e^{-i\theta} f\, d\mu,$$

and hence, as follows from the *Properties of Lebesgue Integral. II* (Theorem 6.3, part (5)) and the *Properties of Lebesgue Integral. I* (Theorem 6.1, part (3)),

$$\left|\int_A f\,d\mu\right| = \operatorname{Re}\int_A e^{-i\theta} f\,d\mu = \int_A \operatorname{Re}\left(e^{-i\theta}f\right)\,d\mu \le \int_A |\operatorname{Re}\left(e^{-i\theta}f\right)|\,d\mu$$
$$\le \int_A |e^{-i\theta}f|\,d\mu = \int_A |f|\,d\mu.$$

6.5 Monotone Convergence Theorem

The following limit theorem underlies the *linearity* property of the Lebesgue integral and is the source of other important limit theorems.

Theorem 6.4 (Monotone Convergence Theorem). *Let (X, Σ, μ) be a measure space and $(f_n)_{n\in\mathbb{N}}$ be an increasing sequence:*

$$0 \le f_n(x) \le f_{n+1}(x),\ n \in \mathbb{N}, x \in X, \tag{6.18}$$

of nonnegative Σ-measurable functions. Then, for the limit function

$$f(x) := \lim_{n\to\infty} f_n(x),\ x \in X,$$

and arbitrary $A \in \Sigma$,

$$\lim_{n\to\infty}\int_A f_n\,d\mu = \int_A f\,d\mu.$$

Proof. Observe that, by condition (6.18), the limit function $f : X \to [0, \infty]$ is well-defined and

$$0 \le f_n(x) \le f(x),\ n \in \mathbb{N}, x \in X. \tag{6.19}$$

Exercise 6.17. Explain.

By the *Sequences of Measurable Functions Theorem* (Theorem 5.4), the limit function f is Σ-measurable.

Let $A \in \Sigma$ be arbitrary. In view of (6.18), by the *Properties of Lebesgue Integral. I* (Theorem 6.1, part (3)), the nonnegative numeric sequence

$$\left(\int_A f_n\,d\mu\right)_{n\in\mathbb{N}}$$

increases, and hence, there exists

$$\lim_{n\to\infty}\int_A f_n\,d\mu = \sup_{n\in\mathbb{N}}\int_A f_n\,d\mu \in [0, \infty].$$

From (6.19), by the *Properties of Lebesgue Integral. I* (Theorem 6.1, part (3)), we infer that, for each $n \in \mathbb{N}$,

$$\int_A f_n \, d\mu \le \int_A f \, d\mu,$$

and hence, for each $k \in \mathbb{N}$,

$$\int_A f_k \, d\mu \le \lim_{n\to\infty} \int_A f_n \, d\mu \le \int_A f \, d\mu. \tag{6.20}$$

If

$$\lim_{n\to\infty} \int_A f_n \, d\mu = \infty,$$

by (6.20),

$$\lim_{n\to\infty} \int_A f_n \, d\mu = \infty = \int_A f \, d\mu.$$

Suppose that

$$\lim_{n\to\infty} \int_A f_n \, d\mu < \infty.$$

Hence, by (6.20), implies that, for each $n \in \mathbb{N}$,

$$\int_A f_n \, d\mu < \infty.$$

For an arbitrary $s \in S(f, A)$ and any $c \in (0, 1)$, let

$$A_n := \{x \in A \,|\, f_n(x) \ge cs(x)\} \in \Sigma.$$

Exercise 6.18. Explain why $A_n \in \Sigma$, $n \in \mathbb{N}$.

By (6.18),

$$A_n \subseteq A_{n+1}, \; n \in \mathbb{N}, \quad \text{and} \quad \bigcup_{n=1}^{\infty} A_n =: A \in \Sigma.$$

Exercise 6.19. Explain.

Then, by the *Properties of Lebesgue Integral. I* (Theorem 6.1, parts (6), (3), and (9)),

$$\infty > \int_A f_n \, d\mu \ge \int_{A_n} f_n \, d\mu \ge \int_{A_n} cs \, d\mu = c \int_{A_n} s \, d\mu.$$

By the *continuity* of the Lebesgue integral (the *Properties of Lebesgue Integral. II* (Theorem 6.3, part (2))),

$$\infty > \lim_{n\to\infty} \int_A f_n \, d\mu \geq c \lim_{n\to\infty} \int_{A_n} s \, d\mu = c \int_A s \, d\mu.$$

By Definition 6.2,

$$\infty > \lim_{n\to\infty} \int_A f_n \, d\mu \geq c \sup_{s\in S(f,A)} \int_A s \, d\mu = c \int_A f \, d\mu.$$

Passing to the limit as $c \to 1-$, we arrive at

$$\lim_{n\to\infty} \int_A f_n \, d\mu \geq \int_A f \, d\mu,$$

which jointly with (6.20) imply that

$$\lim_{n\to\infty} \int_A f_n \, d\mu = \int_A f \, d\mu$$

completing the proof. □

Remarks 6.10.

- As is seen from the proof, the *Monotone Convergence Theorem* rests upon the *continuity*, and hence, the *σ-additivity* of the Lebesgue integral (Theorem 6.2).
- As the following examples show, each of the conditions of the *Monotone Convergence Theorem* (*nonnegativity* and *monotonicity*) is essential and cannot be dropped.

Examples 6.4.

1. For the Lebesgue measure space $(\mathbb{R}, \Sigma^*, \lambda)$, the sequence

   $$f_n(x) = -1/n, \; n \in \mathbb{N}, x \in \mathbb{R},$$

   of Lebesgue measurable functions, being increasing:

   $$0 \leq f_n(x) \leq f_{n+1}(x), \; n \in \mathbb{N}, x \in \mathbb{R},$$

   is not nonnegative and we have

   $$\lim_{n\to\infty} f_n(x) = 0 =: f(x), \; x \in \mathbb{R},$$

   but

   $$\lim_{n\to\infty} \int_{\mathbb{R}} f_n \, d\lambda = \lim_{n\to\infty} (-1/n)\lambda(\mathbb{R}) = -\infty \neq 0 = \int_{\mathbb{R}} f \, d\lambda.$$

2. For the Lebesgue measure space $(\mathbb{R}, \Sigma^*, \lambda)$, the sequence

$$f_n(x) = n\chi_{(0,1/n)}, \; n \in \mathbb{N}, x \in \mathbb{R},$$

of Lebesgue measurable functions, being nonnegative, is not increasing.

Exercise 6.20. Explain.

We have

$$\lim_{n\to\infty} f_n(x) = 0 =: f(x), \; x \in \mathbb{R},$$

but

$$\lim_{n\to\infty} \int_{\mathbb{R}} f_n \, d\lambda = \lim_{n\to\infty} n\lambda((0,1/n)) = n(1/n) = 1 \neq 0 = \int_{\mathbb{R}} f \, d\lambda.$$

3. For the Lebesgue measure space $([0,1], \Sigma^*([0,1]), \lambda)$, the function sequence

$$f_n(x) := \begin{cases} \frac{1}{x}, & 1/n < x \le 1, \\ n, & 0 \le x \le 1/n, \end{cases} \quad n \in \mathbb{N},$$

with the limit function

$$f(x) := \lim_{n\to\infty} f_n(x) = \begin{cases} \frac{1}{x}, & 0 < x \le 1, \\ \infty, & x = 0, \end{cases} \quad n \in \mathbb{N},$$

satisfies the conditions of the *Monotone Convergence Theorem* (Theorem 6.4). Therefore,

$$\lim_{n\to\infty} \int_{[0,1]} f_n \, d = \int_{[0,1]} f \, d\lambda.$$

The fact that

$$\int_{[0,1]} f \, d\lambda = \infty$$

becomes apparent in Section 6.10.3 (see Examples 6.14).

Exercise 6.21. Give an example demonstrating that the analogue of the *Monotone Convergence Theorem* (Theorem 6.4) does not hold for a *decreasing* sequence:

$$0 \le f_{n+1}(x) \le f_n(x), \; n \in \mathbb{N}, x \in X,$$

of nonnegative Σ-measurable functions and

$$f(x) := \lim_{n\to\infty} f_n(x), \; x \in X.$$

Show that adding the condition:

$$f_n(x) \le g(x),\ n \in \mathbb{N}, x \in X,$$

with some $g \in L(X,\mu)$, would be a remedy for the situation (cf. *Lebesgue's Dominated Convergence Theorem* (Theorem 6.7)).

Hint. Consider the sequence

$$g_n(x) := g(x) - f_n(x),\ n \in \mathbb{N}, x \in X.$$

6.6 Linearity of Lebesgue Integral

The following *additivity* property of the Lebesgue integral relative to the integrand jointly with the *constant factor property* (the *Properties of Lebesgue Integral. I* (Theorem 6.1, part (9))) constitute the essential property of its *linearity*.

Theorem 6.5 (Linearity of Lebesgue Integral (Extended Nonnegative Case)). *Let (X,Σ,μ) be a measure space and $f,g : X \to [0,\infty]$ be Σ-measurable functions. Then, for arbitrary $A \in \Sigma$,*

$$\int_A (f+g)\,d\mu = \int_A f\,d\mu + \int_A g\,d\mu.$$

Proof. Let $A \in \Sigma$ be arbitrary.

First, suppose that f and g are nonnegative Σ-measurable simple functions:

$$f = \sum_{i=1}^{m} a_i \chi_{A_i},\ g = \sum_{j=1}^{n} b_j \chi_{B_j}$$

where $m,n \in \mathbb{N}$, $a_i \ge 0$, $i = 1,\dots,m$, $b_j \ge 0$, $j = 1,\dots,n$,

$$A_i \in \Sigma,\ i = 1,\dots,m,\ B_j \in \Sigma,\ j = 1,\dots,n,$$

and

$$A_i \cap A_j = \emptyset,\ i \ne j,\quad \bigcup_{i=1}^{m} A_i = X = \bigcup_{j=1}^{n} B_j,\quad B_i \cap B_j = \emptyset,\ i \ne j, \tag{6.21}$$

(see Remarks 6.1).

Then

$$f + g = \sum_{i=1}^{m}\sum_{j=1}^{n} (a_i + b_j)\chi_{A_i \cap B_j}$$

is also a nonnegative Σ-measurable simple function and, by Definition 6.1 (see also Remarks 6.1),

$$\begin{aligned}\int_A (f+g)\,d\mu &= \sum_{i=1}^m\sum_{j=1}^n (a_i+b_j)\mu((A_i\cap B_j)\cap A)\\ &= \sum_{i=1}^m\sum_{j=1}^n a_i\mu((A_i\cap B_j)\cap A) + \sum_{i=1}^m\sum_{j=1}^n b_j\mu((A_i\cap B_j)\cap A)\\ &= \sum_{i=1}^m a_i\sum_{j=1}^n \mu((A_i\cap A)\cap B_j) + \sum_{j=1}^n b_j\sum_{i=1}^m \mu(A_i\cap(B_j\cap A))\end{aligned}$$

in view of (6.21), by the *additivity* of μ;

$$= \sum_{i=1}^m a_i\mu(A_i\cap A) + \sum_{j=1}^n b_j\mu(B_j\cap A) = \int_A f\,d\mu + \int_A g\,d\mu.$$

Hence, the statement is true in this case.

Now, suppose that f and g are nonnegative Σ-measurable functions, which need not be simple. Then, by the *Characterization of Measurability* (Theorem 5.5), there exist increasing sequences $(s_n)_{n\in\mathbb{N}}$ and $(t_n)_{n\in\mathbb{N}}$ of nonnegative Σ-measurable simple functions converging pointwise on X to f and g, respectively. Therefore, $(s_n+t_n)_{n\in\mathbb{N}}$ is an increasing sequence of nonnegative Σ-measurable simple functions converging to $f+g$ pointwise on X, and hence, by the *Monotone Convergence Theorem* (Theorem 6.4) and the above,

$$\int_A (f+g)\,d\mu = \lim_{n\to\infty}\int_A (s_n+t_n)\,d\mu = \lim_{n\to\infty}\int_A s_n\,d\mu + \lim_{n\to\infty}\int_A t_n\,d\mu = \int_A f\,d\mu + \int_A g\,d\mu. \qquad \square$$

Corollary 6.2 (Linearity of Lebesgue Integral (Real Case)). *Let* (X,Σ,μ) *and* $f,g : X\to\mathbb{R}$ *be Σ-measurable functions. Then, for an arbitrary* $A\in\Sigma$ *such that* $f,g\in L(A,\mu)$, $f+g\in L(A,\mu)$ *and*

$$\int_A (f+g)\,d\mu = \int_A f\,d\mu + \int_A g\,d\mu.$$

Proof. By the *Properties of Lebesgue Integral. II* (Theorem 6.3, part (5)), $f,g\in L(A,\mu)$ is equivalent to the fact that $|f|,|g|\in L(A,\mu)$, i. e.,

$$\int_A |f|\,d\mu < \infty \quad\text{and}\quad \int_A |g|\,d\mu < \infty.$$

Hence, since

$$|f(x)+g(x)| \le |f(x)| + |g(x)|,\ x\in X,$$

by the *Properties of Lebesgue Integral. II* (Theorem 6.1, part (3)) and the prior theorem,

$$\int_A |f+g|\,d\mu \le \int_A [|f|+|g|]\,d\mu = \int_A |f|\,d\mu + \int_A |g|\,d\mu < \infty.$$

This implies that $|f+g| \in L(A,\mu)$, and hence, by the *Properties of Lebesgue Integral. II* (Theorem 6.3, part (5)), $f+g \in L(A,\mu)$.

Therefore, by Definition 6.4 (see also Remarks 6.5),

$$f_+, f_-, g_+, g_-, (f+g)_+, (f+g)_- \in L(A,\mu).$$

We further have

$$(f+g)_+ - (f+g)_- = f+g = f_+ - f_- + g_+ - g_-,$$

which implies

$$(f+g)_+ + f_- + g_- = f_+ + g_+ + (f+g)_-,$$

where by the prior theorem,

$$\int_A (f+g)_+\,d\mu + \int_A f_-\,d\mu + \int_A g_-\,d\mu = \int_A f_+\,d\mu + \int_A g_+\,d\mu + \int_A (f+g)_-\,d\mu.$$

Rearranging, in view of the fact that all the above integrals are *finite*, we arrive at the desired equality:

$$\int_A (f+g)\,d\mu = \int_A (f+g)_+\,d\mu - \int_A (f+g)_-\,d\mu = \int_A f_+\,d\mu - \int_A f_-\,d\mu$$
$$+ \int_A g_+\,d\mu - \int_A g_-\,d\mu = \int_A f\,d\mu + \int_A g\,d\mu. \qquad \square$$

Remarks 6.11.

- The *linearity* of the Lebesgue integral is also in place for Σ-measurable complex-valued functions $f,g : X \to \mathbb{C}$, $f,g \in L(A,\mu)$.
- Thus, given a measure space (X,Σ,μ), by the *Linearity of Lebesgue Integral Corollary* (Corollary 6.2) and the *constant factor property* (the *Properties of Lebesgue Integral. I* (Theorem 6.1, part (9))), the function set $L(X,\mu)$ is a real or complex *vector space* and, for an arbitrary fixed set $A \in \Sigma$, the mapping

  $$L(X,\mu) \ni f \mapsto \int_A f\,d\mu$$

  is a *linear functional* on it (see, e. g., [14]), i. e.,

  $$\forall f,g \in L(X,\mu)\ \forall c_1,c_2 \in \mathbb{R}(\mathbb{C}) : \ c_1 f + c_2 g \in L(X,\mu) \text{ and}$$
  $$\int_A (c_1 f + c_2 g)\,d\mu = c_1 \int_A f\,d\mu + c_2 \int_A g\,d\mu. \tag{6.22}$$

- By the *Linearity of Lebesgue Integral Theorem* ((Theorem 6.5)) and the *constant factor property* (the *Properties of Lebesgue Integral. I* (Theorem 6.1, part (9))), for nonnegative Σ-measurable functions $f, g : X \to [0, \infty]$ and an arbitrary fixed set $A \in \Sigma$, equality (6.22) holds for any nonnegative $c_1, c_2 \in [0, \infty)$ regardless whether the functions are Lebesgue integrable on A relative to measure μ or not.

Proposition 6.1 (Integrating Series). *Let* (X, Σ, μ) *be a measure space and*

$$f_n : X \to [0, \infty],\ n \in \mathbb{N},$$

be Σ*-measurable functions. Then, for arbitrary* $A \in \Sigma$,

$$\int_A \sum_{k=1}^{\infty} f_k \, d\mu = \sum_{k=1}^{\infty} \int_A f_k \, d\mu.$$

Provided

$$\sum_{k=1}^{\infty} \int_A f_k \, d\mu < \infty,$$

the series

$$\sum_{k=1}^{\infty} f_k(x)$$

converges to a finite sum a. e. on A.

Exercise 6.22. Prove.

Hint. Apply the *Monotone Convergence Theorem* (Theorem 6.4) and the *Linearity of Lebesgue Integral Theorem* (Theorem 6.5) to the sequence of partial sums:

$$s_n(x) := \sum_{k=1}^{n} f_k(x),\ n \in \mathbb{N}, x \in X.$$

6.7 Basic Limit Theorems

Due to its countable additivity, the Lebesgue integral is subject to several momentous *limit theorems* not assuming uniform convergence, one of which, the *Monotone Convergence Theorem* (Theorem 6.4), has already been proved above.

6.7.1 Levi's Theorem

We start with the following direct corollary of the *Monotone Convergence Theorem* (Theorem 6.4), which, in a sense, extends the range of the latter to increasing sequences of measurable functions, which need not be nonnegative.

Corollary 6.3 (Levi's Theorem). *Let (X, Σ, μ) be a measure space, $(f_n)_{n\in\mathbb{N}}$ be an increasing sequence:*

$$f_n(x) \le f_{n+1}(x),\ n \in \mathbb{N}, x \in X, \tag{6.23}$$

of real-valued Σ-measurable functions, and

$$f(x) := \lim_{n\to\infty} f_n(x),\ x \in X,$$

be the limit function. Then, for an arbitrary $A \in \Sigma$ such that

$$\sup_{n\in\mathbb{N}} \int_A |f_n|\, d\mu < \infty, \tag{6.24}$$

the limit function $f \in L(A, \mu)$ and[4]

$$\lim_{n\to\infty} \int_A f_n\, d\mu = \int_A f\, d\mu.$$

Proof. Observe that, by (6.23), the functions

$$g_n(x) := f_n(x) - f_1(x),\ n \in \mathbb{N}, x \in X,$$

and

$$g(x) := f(x) - f_1(x),\ x \in X,$$

satisfy the conditions of the *Monotone Convergence Theorem* (Theorem 6.4).

Exercise 6.23. Verify.

Hence,

$$\int_A (f - f_1)\, d\mu = \lim_{n\to\infty} \int_A (f_n - f_1)\, d\mu, \tag{6.25}$$

where, due to (6.24), the right-hand side is finite, and hence, $f - f_1 \in L(A, \mu)$.

Exercise 6.24. Verify.

Since, by the *Properties of Lebesgue Integral. II* (Theorem 6.3, part (5)), (6.24) implies in particular, that $f_n \in L(A, \mu)$, $n \in \mathbb{N}$, by the *Linearity of Lebesgue Integral Corollary* (Corollary 6.2), we infer that

$$f = (f - f_1) + f_1 \in L(A, \mu)$$

and

$$\int_A f\, d\mu = \lim_{n\to\infty} \int_A f_n\, d\mu. \qquad \square$$

4 Beppo Levi (1875–1961).

6.7.2 Fatou's Lemma

The following essential theorem due to Pierre Fatou[5] is traditionally referred to as a lemma.

Theorem 6.6 (Fatou's Lemma). *Let (X, Σ, μ) be a measure space and*

$$f_n : X \to [0, \infty],\ n \in \mathbb{N},$$

be a sequence of nonnegative Σ-measurable functions. Then, for an arbitrary $A \in \Sigma$,

$$\int_A \varliminf_{n\to\infty} f_n\, d\mu \le \varliminf_{n\to\infty} \int_A f_n\, d\mu.$$

Proof. In view of the *Sequences of Measurable Functions Theorem* (Theorem 5.4), the functions

$$g_n(x) := \inf_{k\ge n} f_k(x),\ n \in \mathbb{N}, x \in X,$$

and

$$f(x) := \varliminf_{n\to\infty} f_n(x) = \lim_{n\to\infty} g_n(x),\ x \in X,$$

satisfy the conditions of the *Monotone Convergence Theorem* (Theorem 6.4).

Exercise 6.25. Verify.

Hence,

$$\int_A f\, d\mu = \lim_{n\to\infty} \int_A g_n\, d\mu.$$

Also,

$$g_n(x) \le f_n(x),\ n \in \mathbb{N}, x \in X,$$

which, by the *Properties of Lebesgue Integral. I* (Theorem 6.1, part (3)), implies that, for any $A \in \Sigma$ and each $n \in \mathbb{N}$,

$$\int_A g_n\, d\mu \le \int_A f_n\, d\mu,$$

and hence,

$$\int_A f\, d\mu = \lim_{n\to\infty} \int_A g_n\, d\mu = \varliminf_{n\to\infty} \int_A g_n\, d\mu \le \varliminf_{n\to\infty} \int_A f_n\, d\mu. \qquad \square$$

5 Pierre Fatou (1878–1929).

Remark 6.12. The analogue of *Fatou's Lemma* does not hold for the upper limit.

Exercise 6.26. Give the corresponding counterexample.

Hint. Use the *Dancing Steps Example* (Example 5.15) or make up a similar one (cf. Section 6.11, Problem 13).

Example 6.5. Let the underlying measure space be $([0,1], \Sigma^*([0,1]), \lambda)$. Then, for

$$f_n(x) := \frac{1}{n}x, \ n \in \mathbb{N}, x \in [0,1],$$

by *Fatou's Lemma*, in view of Example 6.2,

$$0 = \int_{[0,1]} 0 \, d\lambda = \int_{[0,1]} \varliminf_{n\to\infty} f_n \, d\lambda \le \varliminf_{n\to\infty} \int_{[0,1]} f_n \, d\lambda = \lim_{n\to\infty} \frac{1}{n} \int_{[0,1]} x \, d\lambda = \lim_{n\to\infty} \frac{1}{2n} = 0.$$

The following useful corollary represents a typical application of *Fatou's Lemma*.

Corollary 6.4. *Let (X, Σ, μ) be a measure space and $f, f_n : X \to \overline{\mathbb{R}}$, $n \in \mathbb{N}$, be Σ-measurable functions. Then, for an arbitrary $A \in \Sigma$ such that:*
(1) *$f_n(x) \to f(x)$ a. e. μ on A and*
(2) *$\sup_{n\in\mathbb{N}} \int_A |f_n| \, d\mu < \infty$,*

$f \in L(A, \mu)$ and

$$\int_A |f| \, d\mu \le \sup_{n\in\mathbb{N}} \int_A |f_n| \, d\mu.$$

Remarks 6.13.
- For $f, f_n : X \to \mathbb{R}$, $n \in \mathbb{N}$, the statement of the prior corollary remains in place if convergence a. e. μ on A in condition (1) is replaced with convergence in measure: $f_n \xrightarrow{\mu} f$ on A.
- The statement of the prior corollary naturally extends to Σ-measurable functions

$$f, f_n : X \to \mathbb{C}, \ n \in \mathbb{N},$$

via applying separately to the real and imaginary parts.

Exercise 6.27.
(a) Prove the prior corollary.

Hint. Apply *Fatou's Lemma* (Theorem 6.6) to the function sequence $(|f_n|)_{n\in\mathbb{N}}$ and the *Properties of Lebesgue Integral. II* (Theorem 6.3, parts (4) and (5)).

(b) Verify the first of the prior remarks.

Hint. Apply the *Riesz Theorem* (Theorem 5.13).

Example 6.6. For the measure space $([0,1], \Sigma^*([0,1]), \lambda)$ and

$$f_n(x) := (-1)^n n\chi_{[0,1/n]}(x) \to f(x) \equiv 0 \text{ a. e. } \lambda \text{ on } [0,1],$$

$f \in L([0,1], \lambda)$ and

$$\int_{[0,1]} |f|\, d\lambda = 0 \le 1 = \sup_{n\in\mathbb{N}} \int_A |f_n|\, d\mu.$$

6.7.3 Lebesgue's Dominated Convergence Theorem

The following theorem is one of the most important statements in the theory of Lebesgue integration.

Theorem 6.7 (Lebesgue's Dominated Convergence Theorem). *Let (X, Σ, μ) be a measure space and $f, f_n : X \to \overline{\mathbb{R}}$, $n \in \mathbb{N}$, be Σ-measurable functions. Then, for an arbitrary $A \in \Sigma$ such that:*
(1) *$f_n(x) \to f(x)$ a. e. μ on A and*
(2) *$\exists g \in L(A, \mu)\ \forall n \in \mathbb{N} : |f_n(x)| \le g(x)$ a. e. μ on A,*

$f, f_n \in L(A, \mu)$, $n \in \mathbb{N}$, and

$$\lim_{n\to\infty} \int_A f_n\, d\mu = \int_A f\, d\mu.$$

Proof. By condition (1),

$$N_0 := \left\{x \in A \,\middle|\, \lim_{n\to\infty} f_n(x) \ne f(x)\right\} \in \Sigma \quad \text{and} \quad \mu(N_0) = 0.$$

By condition (2), for each $n \in \mathbb{N}$,

$$N_n := \{x \in A \mid |f_n(x)| > g(x)\} \in \Sigma \quad \text{and} \quad \mu(N_n) = 0.$$

Let

$$N_f := \{x \in A \mid |f(x)| > g(x)\} \in \Sigma.$$

Exercise 6.28. Explain why $N_f \in \Sigma$.

We have the inclusion

$$N_f \subseteq \bigcup_{n=1}^{\infty} N_n \subseteq N := \bigcup_{n=0}^{\infty} N_n \in \Sigma.$$

Exercise 6.29. Explain.

Where by the *monotonicity* and *σ-subadditivity* of μ (see the *Properties of Measure* (Theorem 3.1)),

$$\mu(N_f) \le \mu(N) = \mu\left(\bigcup_{n=0}^{\infty} N_n\right) \le \sum_{n=0}^{\infty} \mu(N_n) = 0,$$

and hence, we conclude that

$$|f(x)| \le g(x) \text{ a. e. } \mu \text{ on } A.$$

In view of condition (2), this estimate, by the *Properties of Lebesgue Integral. II* (Theorem 6.3, part (6)), imply that $f, f_n \in L(A, \mu)$, $n \in \mathbb{N}$.

Applying *Fatou's Lemma* (Theorem 6.6) to the sequences

$$(g \pm f_n)_{n\in\mathbb{N}}$$

of nonnegative Σ-measurable functions on the set $A \setminus N$, we have

$$\int_{A\setminus N} (g \pm f)\, d\mu \le \varliminf_{n\to\infty} \int_{A\setminus N} (g \pm f_n)\, d\mu,$$

which, since $\mu(N) = 0$, by the *Properties of Lebesgue Integral. II* (Theorem 6.3, part (3)), implies that

$$\int_A (g \pm f)\, d\mu \le \varliminf_{n\to\infty} \int_A (g \pm f_n)\, d\mu,$$

where by the *Linearity of Lebesgue Integral Corollary* (Corollary 6.2) and in view of the fact that

$$\varliminf_{n\to\infty} \left(-\int_A f_n\, d\mu \right) = -\varlimsup_{n\to\infty} \int_A f_n\, d\mu,$$

we have

$$\int_A g\, d\mu + \int_A f\, d\mu \le \int_A g\, d\mu + \varliminf_{n\to\infty} \int_A f_n\, d\mu,$$

and

$$\int_A g\, d\mu - \int_A f\, d\mu \le \int_A g\, d\mu - \varlimsup_{n\to\infty} \int_A f_n\, d\mu.$$

Subtracting through $\int_A g\, d\mu$ in both of the above inequalities, we infer that

$$\int_A f\, d\mu \le \varliminf_{n\to\infty} \int_A f_n\, d\mu \le \varlimsup_{n\to\infty} \int_A f_n\, d\mu \le \int_A f\, d\mu,$$

which, by the *Characterization of Limit Existence* (Proposition 1.3) implies that

$$\lim_{n\to\infty} \int_A f_n \, d\mu = \int_A f \, d\mu$$

completing the proof. □

Remarks 6.14.

- For $f, f_n : X \to \mathbb{R}$, $n \in \mathbb{N}$, the *Lebesgue Dominated Convergence Theorem* (Theorem 6.7) remains in place if convergence a. e. μ on A in condition (1) is replaced with convergence in measure: $f_n \xrightarrow{\mu} f$ on A.
- The *Lebesgue Dominated Convergence Theorem* (Theorem 6.7) naturally extends to Σ-measurable functions

  $$f, f_n : X \to \mathbb{C}, \; n \in \mathbb{N},$$

  via applying separately to the real and imaginary parts.

Exercise 6.30. Verify the first of the prior remarks.

Hint. Apply the *Characterization of Convergence in Measure* (Proposition 5.13) (Section 5.12, Problem 21) and the *Characterization of Convergence* (Theorem 1.6).

Example 6.7. Let the underlying measure space be $([0,1], \Sigma^*([0,1]), \lambda)$. Then, for

$$f_n(x) := (-1)^n \chi_{[0,1/n]}(x) \to f(x) \equiv 0 \text{ a. e. } \lambda \text{ on } [0,1],$$
$$|f_n(x)| \le g(x) \equiv 1 \in L([0,1], \lambda), \; n \in \mathbb{N}, x \in [0,1],$$

and, by the *Lebesgue Dominated Convergence Theorem*,

$$0 = \int_{[0,1]} f \, d\lambda = \lim_{n\to\infty} \int_A f_n \, d\mu = \lim_{n\to\infty} \frac{(-1)^n}{n} = 0.$$

6.8 Change of Variable Theorem

Here, we are to procure an analogue of the *change of variable* for the Lebesgue integral, for which purpose we need the following notion of *induced measure*.

Definition 6.5 (Induced Measure). Let (X, Σ, μ) be a measure space, (X', Σ') be a measurable space, and

$$T : X \to X'$$

be a Σ-Σ'-measurable mapping. The set function is defined as follows:

$$\Sigma' \ni A' \mapsto \mu'(A') := \mu(T^{-1}A')$$

a measure on Σ' called the *induced measure*.

Exercise 6.31. Verify that μ' is a *measure* on Σ'.

Examples 6.8. Let $X = X' := \mathbb{R}$, $\Sigma = \Sigma' := \mathscr{B}(\mathbb{R})$, and $\mu := \lambda$.

1. For the Borel measurable mapping $Tx := |x|$, $x \in \mathbb{R}$, the induced measure on $\mathscr{B}(\mathbb{R})$ is
$$\lambda'(A') = 2\lambda(A' \cap [0,\infty)),\ A' \in \Sigma.$$
2. For the Borel measurable mapping $Tx := ax + b$ $(a, b \in \mathbb{R},\ a \neq 0)$, $x \in \mathbb{R}$, the induced measure on $\mathscr{B}(\mathbb{R})$ is
$$\lambda'(A') = \frac{1}{|a|}\lambda(A'),\ A' \in \Sigma.$$

Exercise 6.32. Verify.

Theorem 6.8 (Change of Variable Theorem). *Let (X, Σ, μ) be a measure space, (X', Σ') be a measurable space, $T : X \to X'$ be a Σ-Σ'-measurable mapping, and μ' be the induced measure on Σ'. Then, for any Σ'-measurable function $f : X \to \mathbb{R}$,*
$$\int_{X'} f(x')\,d\mu'(x') = \int_X f(Tx)\,d\mu(x) \tag{6.26}$$
provided the left-side integral exists.

Proof. Let
$$\mathscr{F} := \{f : X' \to \mathbb{R}\,|\, f \text{ is } \Sigma'\text{-measurable and equality (6.26) holds}\}.$$

(1) $\mathscr{F} \neq \emptyset$ since $f(x') \equiv 0 \in \mathscr{F}$.
(2) By the *Linearity of Lebesgue Integral Theorem* (Theorem 6.5) (see also Remarks 6.11), the collection $\mathscr{F}$ contains linear combinations with nonnegative coefficients of nonnegative functions from $\mathscr{F}$.
(3) By the *Monotone Convergence Theorem* (Theorem 6.4), $\mathscr{F}$ contains the pointwise on X' limits of increasing sequences of nonnegative Σ'-measurable functions from $\mathscr{F}$.
(4) By the definition of *induced measure*, $\mathscr{F}$ contains all Σ'-measurable characteristic functions, i. e.,
$$\forall A' \in \Sigma' : \chi_{A'} \in \mathscr{F}.$$

Indeed,
$$\int_{X'} \chi_{A'}(x')\,d\mu'(x') = \mu'(A') = \mu(T^{-1}A') = \int_X \chi_{T^{-1}A'}(x)\,d\mu(x) = \int_X \chi_{A'}(Tx)\,d\mu(x).$$

Exercise 6.33. Verify that
$$\chi_{T^{-1}A'}(x) = \chi_{A'}(Tx),\ x \in X.$$

(5) By (2) and (4), $\mathscr{F}$ contains all nonnegative Σ'-measurable simple functions $s : X' \to [0,\infty)$.
(6) By the *Characterization of Measurability* (Theorem 5.5) and (3), $\mathscr{F}$ contains all nonnegative Σ'-measurable functions $f : X' \to [0,\infty]$.
(7) By Definition 6.3, $\mathscr{F}$ contains all Σ'-measurable functions $f : X' \to \mathbb{R}$, for which $\int_{X'} f_+ \, d\mu'$ and $\int_{X'} f_- \, d\mu'$ are not both infinite.

Thus, $\mathscr{F}$ coincides with the collection of all Σ'-measurable functions, for which the left-side integral in *change of variable formula* (6.26) exists. □

Remark 6.15. The *Change of Variable Theorem* is in place also for a Σ-measurable $f : X \to \mathbb{C}$, $f \in L(A,\mu)$.

Examples 6.9. Let $X = X' := \mathbb{R}$, $\Sigma = \Sigma' = \mathscr{B}(\mathbb{R})$, and $\mu := \lambda$.

1. For $Tx := |x|$, $x \in \mathbb{R}$,

$$\lambda'(A') = 2\lambda(A' \cap [0,\infty)), \ A' \in \Sigma,$$

(see Examples 6.8) and, for any Borel measurable function $f : \mathbb{R} \to \mathbb{R}$,

$$\int_{\mathbb{R}} f(x') \, d\lambda'(x') = 2 \int_{[0,\infty)} f(x') \, d\lambda(x') = \int_{\mathbb{R}} f(|x|) \, d\lambda(x)$$

provided the left-side integral exists.

2. For $Tx := ax + b$ $(a, b \in \mathbb{R}, a \neq 0)$, $x \in \mathbb{R}$,

$$\lambda'(A') = \frac{1}{|a|}\lambda(A'), \ A' \in \Sigma,$$

(see Examples 6.8) and, for any Borel measurable function $f : \mathbb{R} \to \mathbb{R}$,

$$\int_{\mathbb{R}} f(x') \, d\lambda'(x') = \frac{1}{|a|} \int_{\mathbb{R}} f(x') \, d\lambda(x') = \int_{\mathbb{R}} f(ax+b) \, d\lambda(x)$$

provided the left-side integral exists.

6.9 Approximation by Continuous Functions

To follow is a useful result on approximation of Borel measurable functions by continuous functions.

Definition 6.6 (Support of a Function). Let (X,ρ) be a metric space. The *support* of a function $f : X \to \mathbb{R}$ is the closure of the set $\{x \in X \mid f(x) \neq 0\}$.

If X is not equipped with a metric, the support of f is simply $\{x \in X \mid f(x) \neq 0\}$.

Notation. $\operatorname{supp} f$.

Examples 6.10. Let $X = \mathbb{R}$ with the regular distance.
1. For $f(x) \equiv 0$, $\operatorname{supp} f = \emptyset$.
2. For $f(x) := x$, $x \in \mathbb{R}$, $\operatorname{supp} f = \mathbb{R}$.
3. For $f(x) := \chi_{[0,1)}(x)$, $x \in \mathbb{R}$, $\operatorname{supp} f = [0,1]$.
4. For $f(x) := \chi_{\mathbb{Q}}(x)$, $x \in \mathbb{R}$, $\operatorname{supp} f = \mathbb{R}$.

Remark 6.16. Linear combinations and products of functions with *compact support* also have compact support.

Exercise 6.34. Verify.

Theorem 6.9 (Approximation by Continuous Functions). *Let $f : \mathbb{R} \to \mathbb{R}$ be a Borel measurable function, $f \in L(\mathbb{R}, \lambda)$. Then, for any $\varepsilon > 0$ such that there exists a continuous function $f : \mathbb{R} \to \mathbb{R}$ with compact support such that*

$$\int_{\mathbb{R}} |f - g| \, d\lambda < \varepsilon.$$

Proof. Since, for a Borel measurable function $f : \mathbb{R} \to [0, \infty)$,

$$f = f_+ - f_-,$$

where $f_+, f_- : \mathbb{R} \to [0, \infty)$ are nonnegative Borel measurable functions, $f_+, f_- \in L(\mathbb{R}, \lambda)$ (see *Positive and Antinegative Parts of a Function Corollary* (Corollary 5.4) and Remarks 6.5), if there exist continuous $g_1, g_2 : \mathbb{R} \to \mathbb{R}$ with compact support such that

$$\int_{\mathbb{R}} |f_+ - g_1| \, d\lambda < \varepsilon/2 \quad \text{and} \quad \int_{\mathbb{R}} |f_- - g_2| \, d\lambda < \varepsilon/2,$$

then, for the continuous function

$$g := g_1 - g_2$$

with compact support, by the *Properties of Lebesgue Integral. I* (Theorem 6.1, part (3)) and the *Linearity of Lebesgue Integral Theorem* (Theorem 6.5),

$$\int_{\mathbb{R}} |f - g| \, d\lambda = \int_{\mathbb{R}} |f_+ - g_1 + g_2 - f_-| \, d\lambda \le \int_{\mathbb{R}} |f_+ - g_1| \, d\lambda + \int_{\mathbb{R}} |g_2 - f_-| \, d\lambda < \varepsilon/2 + \varepsilon/2 = \varepsilon.$$

Hence, it suffices to prove the statement for the case of a nonnegative Borel measurable function $f : \mathbb{R} \to [0, \infty)$.

Suppose, that $f : \mathbb{R} \to [0, \infty)$ and let $\varepsilon >$ be arbitrary.

The function sequence

$$f_n(x) := f(x)\chi_{[-n,n]}(x), \; n \in \mathbb{N}, x \in \mathbb{R},$$

satisfies the conditions of the *Monotone Convergence Theorem* (Theorem 6.4) and

$$\lim_{n\to\infty} f(x)\chi_{[-n,n]}(x) = f(x), \; x \in \mathbb{R}.$$

Exercise 6.35. Explain.

Therefore, we can fix a sufficiently large $n \in \mathbb{N}$ such that

$$\int_{\mathbb{R}} (f - f_n)\, d\lambda = \int_{\mathbb{R}} f\, d\lambda - \int_{\mathbb{R}} f_n\, d\lambda < \varepsilon/2.$$

Thus, if there exists a continuous $g : \mathbb{R} \to \mathbb{R}$ with compact support such that

$$\int_{\mathbb{R}} |f_n - g|\, d\lambda < \varepsilon/2,$$

then, by the *Properties of Lebesgue Integral. I* (Theorem 6.1, part (3)) and the *Linearity of Lebesgue Integral Theorem* (Theorem 6.5),

$$\int_{\mathbb{R}} |f - g|\, d\lambda \le \int_{\mathbb{R}} (f - f_n)\, d\lambda + \int_{\mathbb{R}} |f_n - g|\, d\lambda < \varepsilon/2 + \varepsilon/2 = \varepsilon.$$

Hence, without loss of generality, we can assume that $f(x) = 0$ outside some interval $[-n, n]$, i. e., f has a compact support.

First, suppose that

$$f(x) = \chi_A(x),\ x \in \mathbb{R},$$

where A is a *bounded* Borel set. Then, by the *Approximation of Borel Sets Proposition* (Proposition 4.6) (Section 4.8, Problem 13), for any $\varepsilon > 0$, there exist an open set G and a closed set F such that

$$F \subseteq A \subseteq G \quad \text{and} \quad \lambda(G \setminus A) < \varepsilon/2,\ \lambda(A \setminus F) < \varepsilon/2,$$

and hence, by the *additivity* of λ,

$$\lambda(G \setminus F) \le \lambda(G \setminus A) + \lambda(A \setminus F) < \varepsilon/2 + \varepsilon/2 = \varepsilon.$$

Observe that, without loss of generality, we can regard that G is *bounded*, and hence, by the *Heine–Borel Theorem* (Theorem 1.13), its closure $\overline{G}$ is *compact* in $\mathbb{R}$.

Since for an arbitrary set $C \subseteq \mathbb{R}$, the distance to C function

$$\operatorname{dist}(x, C) := \inf_{y \in C} |x - y|,\ x \in \mathbb{R},$$

is *continuous* on $\mathbb{R}$ and the *closed* sets F and G^c are *disjoint*, the function

$$g(x) := \frac{\operatorname{dist}(x, G^c)}{\operatorname{dist}(x, F) + \operatorname{dist}(x, G^c)},\ x \in \mathbb{R},$$

is

(i) *continuous* on $\mathbb{R}$,
(ii) takes values in $[0, 1]$,
(iii) is equal to 0 on G^c and to 1 on F, and
(iv) $\operatorname{supp} g = \overline{G}$ is *compact*.

We also have

$$|\chi_A(x) - g(x)| \le \chi_G(x) - \chi_F(x),\ x \in \mathbb{R}.$$

Exercise 6.36. Verify.

Hence, by the *Properties of Lebesgue Integral. I* (Theorem 6.1, part (3)),

$$\int_{\mathbb{R}} |\chi_A - g|\, d\lambda \le \int_{\mathbb{R}} (\chi_G - \chi_F)\, d\lambda = \int_{\mathbb{R}} \chi_{G\setminus F}\, d\lambda = \lambda(G \setminus F) < \varepsilon.$$

Thus, the approximation result holds for the characteristic functions of bounded Borel sets.

If f is a nonnegative simple function:

$$f(x) = \sum_{i=1}^{m} a_i \chi_{A_i}(x),\ x \in \mathbb{R},$$

where, $m \in \mathbb{N}$, $a_i \ge 0$, $i = 1, \dots, m$, and A_i, $i = 1, \dots, m$, are *bounded* Borel sets, then, for each $i = 1, \dots, m$, there exists a continuous $g_i : \mathbb{R} \to \mathbb{R}$ such that

$$\int_{\mathbb{R}} |\chi_{A_i} - g_i|\, d\lambda < \frac{\varepsilon}{m(a_i + 1)},\ i = 1, \dots, m,$$

then, for the continuous function

$$g(x) := \sum_{i=1}^{m} g_i(x),\ x \in \mathbb{R},$$

with compact support, by the *Properties of Lebesgue Integral. I* (Theorem 6.1, part (3)) and the *Linearity of Lebesgue Integral Theorem* (Theorem 6.5),

$$\int_{\mathbb{R}} |f - g|\, d\lambda \le \sum_{i=1}^{m} a_i \int_{\mathbb{R}} |\chi_{A_i} - g_i|\, d\lambda < \sum_{i=1}^{m} a_i \frac{\varepsilon}{m(a_i + 1)} < \varepsilon.$$

Thus, the approximation result holds for nonnegative Borel measurable simple functions with compact support.

If $f : \mathbb{R} \to [0, \infty)$ is a nonnegative Borel measurable function with compact support, by the *Characterization of Measurability* (Theorem 5.5), there is an increasing sequence $(s_n)_{n\in\mathbb{N}}$ of nonnegative Borel measurable simple functions with compact support converging to f pointwise on $\mathbb{R}$.

Exercise 6.37. Explain.

Hence, by the *Monotone Convergence Theorem* (Theorem 6.4), we can fix a sufficiently large $n \in \mathbb{N}$ such that

$$\int_{\mathbb{R}} (f - s_n)\, d\lambda = \int_{\mathbb{R}} f\, d\lambda - \int_{\mathbb{R}} s_n\, d\lambda < \varepsilon/2.$$

Choosing a continuous function $g : \mathbb{R} \to \mathbb{R}$ with compact support such that

$$\int_{\mathbb{R}} |s_n - g|\, d\lambda < \varepsilon/2,$$

by the *Properties of Lebesgue Integral. I* (Theorem 6.1, part (3)) and the *Linearity of Lebesgue Integral Theorem* (Theorem 6.5), we have

$$\int_{\mathbb{R}} |f - g|\, d\lambda \le \int_{\mathbb{R}} (f - s_n)\, d\lambda + \int_{\mathbb{R}} |s_n - g|\, d\lambda < \varepsilon/2 + \varepsilon/2 = \varepsilon,$$

which completes the proof. □

Remark 6.17. In Section 7.4.5.3, the *Approximation by Continuous Functions Theorem* is generalized to the *Approximation by Continuous Functions with Compact Support Theorem* (Theorem 7.11).

6.10 Comparison of Riemann and Lebesgue Integrals

We are to compare the particular case of the abstract Lebesgue integral relative to the one-dimensional Lebesgue measure to its familiar counterpart: the *Riemann integral* and establish a *Characterization of Riemann Integrability* (Theorem 6.10), which would not be possible without the notion of *almost everywhere*.

6.10.1 Riemann Integral Basics

To proceed smoothly, we start with outlining certain basics of Riemann integration (see, e. g., [2, 23]).

Let $R[a,b]$ be the vector space of all real-valued functions Riemann integrable on an interval $[a,b]$ $(-\infty < a < b < \infty)$, respectively.

Recall that for a function $f : [a,b] \to \mathbb{R}$ to be Riemann integrable it is *necessary* that f be *bounded* on $[a,b]$, i. e.,

$$\inf_{[a,b]} f \in \mathbb{R} \quad \text{and} \quad \sup_{[a,b]} f \in \mathbb{R},$$

and the Riemann integrability of such a function is equivalent to the following condition:

$$\inf_P U(f,P) = \sup_P L(f,P),$$

where inf and sup are over all *partitions*

$$P := \{a = x_0 < x_1 < \cdots < x_n = b\} \ (n \in \mathbb{N})$$

of $[a,b]$ and

$$U(f,P) := \sum_{k=1}^{n} \left(\sup_{[x_{k-1},x_k]} f \right) (x_k - x_{k-1}), \ L(f,P) := \sum_{k=1}^{n} \left(\inf_{[x_{k-1},x_k]} f \right) (x_k - x_{k-1})$$

are the *upper* and lower *Darboux*[6] *sums* of f over $[a,b]$ corresponding to P, respectively, the *Riemann integral* of f over $[a,b]$ being defined as follows:

$$\int_a^b f(x)\,dx := \inf_P U(f,P) = \sup_P L(f,P).$$

Remark 6.18. For any partitions P and P' of $[a,b]$ such that P' is *finer* than P, i. e., $P \subseteq P'$,

$$L(f,P) \le L(f,P') \le U(f,P') \le U(f,P). \tag{6.27}$$

Exercise 6.38. Verify.

Proposition 6.2 (Characterization of Riemann Integrability). *A bounded function* $f : [a,b] \to \mathbb{R}$ $(-\infty < a < b < \infty)$ *is Riemann integrable on* $[a,b]$ *iff*

$$\forall \varepsilon > 0 \ \exists P \text{ a partition of } [a,b] : \ U(f,P) - L(f,P) < \varepsilon.$$

Exercise 6.39. Prove.

Remarks 6.19.

- If $f \in R[a,b]$ $(-\infty < a < b < \infty)$, then $|f| \in R[a,b]$ (see, e. g., [2, 23]). The converse is not true, i. e., $|f| \in R[a,b]$ does not imply that $f \in R[a,b]$.
  Indeed, for the function

  $$f(x) := \chi_{\mathbb{Q}}(x) - \chi_{\mathbb{Q}^c}(x) = \begin{cases} 1, & x \in \mathbb{Q} \cap [0,1], \\ -1 & x \in \mathbb{Q}^c \cap [0,1], \end{cases}$$

  on $[0,1]$, $|f| \equiv 1 \in R[0,1]$ but $f \notin R[0,1]$.

6 Jean-Gaston Darboux (1842–1917).

Exercise 6.40. Show that $f \notin R[0,1]$.

This fact is one of the fundamental differences between the Riemann and Lebesgue integrals (cf. the *Properties of Lebesgue Integral. II* (Theorem 6.3, part (5))).

- Another essential distinction is that the Riemann integral, although being, in a sense, additive, is *not* countably additive.

6.10.2 Characterization of Riemann Integrability

The following statement characterizes the Riemann integrability of a bounded function $f : [a,b] \to \mathbb{R}$ on $[a,b]$ in terms of the Lebesgue measure and relates the Riemann and the Lebesgue integrals.

Theorem 6.10 (Characterization of Riemann Integrability). *A bounded function $f : [a,b] \to \mathbb{R}$ $(-\infty < a < b < \infty)$ is Riemann integrable on $[a,b]$ iff f is continuous a. e. relative to the Lebesgue measure λ on $[a,b]$, in which case f is also Lebesgue integrable on $[a,b]$ and*

$$\int_a^b f(x)\,dx = \int_{[a,b]} f(x)\,d\lambda(x).$$

Proof. *"Only if"* part. Suppose that $f \in R[a,b]$.

For an arbitrary partition $P := \{a = x_0 < x_1 < \cdots < x_n = b\}$ $(n \in \mathbb{N})$ of $[a,b]$, consider the simple functions

$$u_P(x) := \sum_{k=1}^{n} \left(\sup_{[x_{k-1},x_k]} f \right) \chi_{[x_{k-1},x_k)}(x) + b\chi_{\{b\}}(x),\ x \in [a,b], \tag{6.28}$$

and

$$l_P(x) := \sum_{k=1}^{n} \left(\inf_{[x_{k-1},x_k]} f \right) \chi_{[x_{k-1},x_k)}(x) + b\chi_{\{b\}}(x),\ x \in [a,b], \tag{6.29}$$

Lebesgue integrable on $[a,b]$ with

$$\int_{[a,b]} u_P\,d\lambda = U(f,P) \quad \text{and} \quad \int_{[a,b]} l_P\,d\lambda = L(f,P). \tag{6.30}$$

Exercise 6.41. Verify.

By the premise,

$$\inf_P U(f,P) = \sup_P L(f,P) = \int_a^b f(x)\,dx.$$

Hence, in view of (6.27), there is an *increasing* sequence $(P_n)_{n\in\mathbb{N}}$ of partitions of $[a,b]$:

$$P_n \subseteq P_{n+1},\ n \in \mathbb{N},$$

such that

$$\lim_{n\to\infty} U(f,P_n) = \inf_P U(f,P) = \int_a^b f(x)\,dx = \sup_P L(f,P) = \lim_{n\to\infty} L(f,P_n).$$

Exercise 6.42. Explain.

As is easily seen,

$$l_{P_n}(x) \le l_{P_{n+1}}(x) \le f(x) \le u_{P_{n+1}}(x) \le u_{P_n}(x),\ n \in \mathbb{N}, x \in X. \tag{6.31}$$

Exercise 6.43. Verify.

Hence, by the *Sequences of Measurable Functions Theorem* (Theorem 5.4), the limit functions

$$u(x) := \lim_{n\to\infty} u_{P_n}(x) \quad \text{and} \quad l(x) := \lim_{n\to\infty} l_{P_n}(x),\ x \in [a,b],$$

are Borel measurable and the more so, Lebesgue measurable (see Remark 5.5).

Passing to the limit in (6.31) as $n \to \infty$, we have

$$l(x) \le f(x) \le u(x),\ x \in [a,b]. \tag{6.32}$$

Since

$$0 \le u_{P_n}(x) - l_{P_n}(x) \le g(x) \equiv \sup_{[a,b]} f - \inf_{[a,b]} f \in L([a,b],\lambda),\ n \in \mathbb{N}, x \in [a,b],$$

by the *Lebesgue Dominated Convergence Theorem* (Theorem 6.7), in view of (6.30), we arrive at

$$\int_{[a,b]} (u-l)\,d\lambda = \lim_{n\to\infty} \int_{[a,b]} (u_{P_n} - l_{P_n})\,d\lambda = \lim_{n\to\infty} [U(f,P_n) - L(f,P_n)] = 0.$$

This, in view of (6.32),by the *Properties of Lebesgue Integral. II* (Theorem 6.3, part (9)), implies that

$$l = u = f \text{ a. e. } \lambda \text{ on } [a,b].$$

Let

$$N := \{x \in [a,b] \mid l(x) \ne u(x)\} \cup \bigcup_{n=1}^{\infty} P_n \in \mathscr{B}([a,b]) \subset \Sigma^*([a,b])$$

$$\text{with } \lambda(N) = 0, \tag{6.33}$$

where $\mathscr{B}([a,b])$ and $\Sigma^*([a,b])$ are the σ-algebras of the Borel and Lebesgue measurable subsets of $[a,b]$, respectively.

Exercise 6.44. Verify.

Then, in view of (6.32), for any $x \in [a,b] \setminus N$,

$$l(x) = \lim_{n\to\infty} l_{P_n}(x) = f(x) = \lim_{n\to\infty} u_{P_n}(x) = u(x) \tag{6.34}$$

and f is *continuous* at x. Indeed, let $\varepsilon > 0$ be arbitrary, then, by (6.31),

$$\exists\, n \in \mathbb{N}: \ 0 \le u_{P_n}(x) - l_{P_n}(x) < \varepsilon.$$

Since $x \notin P_n := \{a = x_0 < x_1 < \cdots < x_m = b\}$ $(m \in \mathbb{N})$,

$$\exists\, \delta > 0\ \exists\, 1 \le k \le m: \ x \in (x-\delta, x+\delta) \subset [x_{k-1}, x_k].$$

Then

$$\forall\, x' \in (x-\delta, x+\delta): \ |f(x') - f(x)| \le \sup_{[x_{k-1},x_k]} f - \inf_{[x_{k-1},x_k]} f = u_{P_n}(x) - l_{P_n}(x) < \varepsilon.$$

Hence,

$$D := \{x \in [a,b] \,|\, f \text{ is } \textit{discontinuous} \text{ at } x\} \subseteq N,$$

which, by the *completeness* of the Lebesgue measure λ on $\Sigma^*([a,b])$ (see Section 4.7.1), implies that

$$D \in \Sigma^* \quad \text{and} \quad \lambda(D) = 0,$$

i. e., f is *continuous* a. e. λ on $[a,b]$.

By the *A. E. Continuous is Lebesgue Measurable Proposition* (Proposition 5.10) (Section 5.12, Problem 13 (a)) (see also Example 5.4), we conclude that the function f is *Lebesgue measurable*.

Further, since the function f is *bounded* on $[a,b]$, by the *Properties of Lebesgue Integral. I* (Theorem 6.1, part (5)), we infer that $f \in L([a,b],\lambda)$.

Since, by (6.33) and (6.34), in view of the *completeness* of the Lebesgue measure on $\Sigma^*([a,b])$,

$$l_{P_n}(x) \to f(x) \text{ a. e. } \lambda \text{ on } [a,b]$$

and, for all $n \in \mathbb{N}$ and $x \in [a,b]$,

$$|l_{P_n}(x)|, |u_{P_n}(x)| \le g(x) \equiv \max\left[|\inf_{[a,b]} f|, |\sup_{[a,b]} f| \right] \in L([a,b],\lambda), \tag{6.35}$$

considering (6.30), by the *Lebesgue Dominated Convergence Theorem* (Theorem 6.7), we infer that

$$\int_{[a,b]} f(x)\, d\lambda(x) = \lim_{n\to\infty} \int_{[a,b]} l_{P_n}(x)\, d\lambda(x) = \lim_{n\to\infty} L(f, P_n) = \int_a^b f(x)\, dx.$$

"If" part. Suppose now that a bounded function $f : [a,b] \to \mathbb{R}$ is *continuous* a. e. λ on $[a,b]$, then, by the *A. E. Continuous is Lebesgue Measurable Proposition* (Proposition 5.10), the function f is *Lebesgue measurable*.

Since f is *bounded* on $[a,b]$, by the *Properties of Lebesgue Integral. I* (Theorem 6.1, part (5)), $f \in L([a,b],\lambda)$.

For each $n \in \mathbb{N}$, let

$$P_n := \{a = x_0 < a + (b-a)/n < \cdots < a + (n-1)(b-a)/n < x_m = b\}$$

be the *uniform partition* of $[a,b]$ dividing $[a,b]$ into n subintervals of equal length and

$$N := \{x \in [a,b] \,|\, f \text{ is } \textit{discontinuous} \text{ at } x\} \cup \bigcup_{n=1}^{\infty} P_n \in \Sigma^*([a,b])$$
$$\text{with } \lambda(N) = 0. \tag{6.36}$$

Exercise 6.45. Verify.

Then, for the sequences of Lebesgue integrable simple functions $(u_{P_n})_{n\in\mathbb{N}}$ and $(l_{P_n})_{n\in\mathbb{N}}$ defined by (6.28) and (6.29), respectively,

$$\lim_{n\to\infty} u_{P_n}(x) = f(x) = \lim_{n\to\infty} l_{P_n}(x),\ x \in [a,b] \setminus N.$$

Indeed, let $x \in [a,b] \setminus N$ and $\varepsilon > 0$ be arbitrary, by the *continuity* of f at x, in view of the fact that $x \in (a,b)$,

$$\exists\, \delta > 0\ \forall\, x' \in (x-\delta, x+\delta) \subset (a,b) : |f(x') - f(x)| < \varepsilon. \tag{6.37}$$

Since $x \notin \bigcup_{n=1}^{\infty} P_n$,

$$\exists\, M \in \mathbb{N}\ \forall\, n \geq M\ \exists\, 1 \leq k \leq n : x \in [a + (k-1)(b-a)/n, a + k(b-a)/n] \subset (x-\delta, x+\delta).$$

Exercise 6.46. Explain.

Then, in view of (6.37),

$$f(x) - \varepsilon \leq \inf_{[a+(k-1)(b-a)/n, a+k(b-a)/n]} f \leq \sup_{[a+(k-1)(b-a)/n, a+k(b-a)/n]} f \leq f(x) + \varepsilon,$$

which implies that

$$\forall\, x \in [a,b] \setminus N\ \forall\, \varepsilon > 0\ \exists\, M \in \mathbb{N}\ \forall\, n \geq M : \max\left[u_{P_n}(x) - f(x), f(x) - l_{P_n}(x)\right] < \varepsilon.$$

Therefore, by (6.36), in view of the *completeness* of the Lebesgue measure on $\Sigma^*([a,b])$,

$$u_{P_n}, l_{P_n} \to f \text{ a. e. } \lambda \text{ on } [a,b].$$

Then, in view of (6.30) and estimates (6.35), by the *Lebesgue Dominated Convergence Theorem* (Theorem 6.7),

$$\lim_{n\to\infty} U(f,P_n) = \lim_{n\to\infty} \int_{[a,b]} u_{P_n}\,d\lambda = \int_{[a,b]} f\,d\lambda = \lim_{n\to\infty} \int_{[a,b]} l_{P_n}\,d\lambda = \lim_{n\to\infty} L(f,P_n),$$

which, by the *Characterization of Riemann Integrability* (Proposition 6.2) implies that $f \in R[a,b]$ and

$$\int_a^b f(x)\,dx = \int_{[a,b]} f(x)\,d\lambda(x).$$

Exercise 6.47. Explain. □

Remarks 6.20.

- By the *Characterization of Riemann Integrability* (Theorem 6.10), any real-valued function $f : [a,b] \to \mathbb{R}$ $(-\infty < a < b < \infty)$ Riemann integrable on $[a,b]$ is also Lebesgue integrable on $[a,b]$ and the Riemann and Lebesgue integrals of f over $[a,b]$ coincide:

$$\int_a^b f(x)\,dx = \int_{[a,b]} f(x)\,d\lambda(x).$$

  Thus, we have the inclusion

$$R[a,b] \subset L([a,b],\lambda), \tag{6.38}$$

  which, as the subsequent examples reveal, is *strict*, i. e., one can integrate more functions over $[a,b]$ in the Lebesgue sense than in the Riemann sense.
- Therefore, for each function $f \in R[a,b]$, we can regard the Riemann integral of f over $[a,b]$ to be its Lebesgue integral over $[a,b]$, use the same notation

$$\int_a^b f(x)\,dx$$

  for both integrals, and apply the *limit theorems* of the prior sections.
- The *Characterization of Riemann Integrability* (Theorem 6.10) naturally extends to complex-valued functions via applying separately to the real and imaginary parts.

Examples 6.11.

1. Due to the *Characterization of Riemann Integrability* (Theorem 6.10), finding the Lebesgue integral

$$\int_{[0,1]} x\,d\lambda(x)$$

of the continuous on $[0,1]$ function $f(x) := x$, $x \in [0,1]$, which appears to take such a formidable effort in Example 6.2, is now effortless and straightforward:

$$\int_{[0,1]} x\,d\lambda(x) = \int_0^1 x\,dx = \frac{x^2}{2}\Big|_0^1 = 1/2.$$

2. By the *Characterization of Riemann Integrability* (Theorem 6.10), the bounded function

$$f(x) := \begin{cases} 0, & x = 0, \\ \sin\frac{1}{x} & x \in (0,1], \end{cases}$$

which is continuous on $[0,1]$, except at 0, i. e., a. e. λ on $[0,1]$, is Riemann integrable on $[0,1]$, i. e., $f \in R[0,1]$.
3. The *Dirichlet*[7] *function*

$$f(x) := \chi_{\mathbb{Q}}(x),\ x \in [0,1],$$

is not Riemann integrable on $[0,1]$ since, for any partition P of $[0,1]$,

$$U(f,P) = 1 \text{ and } L(f,P) = 0,$$

which is consistent with the *Characterization of Riemann Integrability* (Theorem 6.10) considering that f is not continuous a. e. λ on $[0,1]$. In fact, f is discontinuous *everywhere* on $[0,1]$.
However, f is Borel measurable and

$$\int_{[0,1]} f\,d\lambda = \int_{[0,1]} \chi_{\mathbb{Q}}\,d\lambda = \lambda(\mathbb{Q} \cap [0,1]) = 0,$$

i. e., $f \in L([0,1],\lambda)$, and hence inclusion (6.38) is strict, indeed.
4. As is noted in Remarks 6.19, the function

$$f(x) := \chi_{\mathbb{Q}}(x) - \chi_{\mathbb{Q}^c}(x),\ x \in [0,1],$$

is not Riemann integrable on $[0,1]$, which is consistent with the *Characterization of Riemann Integrability* (Theorem 6.10) considering that f is discontinuous *everywhere* on $[0,1]$.
However, the function f is Lebesgue integrable on $[0,1]$, i. e., $f \in L([0,1],\lambda)$, since f is Borel measurable and

$$\int_{[0,1]} f\,d\lambda = \int_{[0,1]} \chi_{\mathbb{Q}}\,d\lambda - \int_{[0,1]} \chi_{\mathbb{Q}^c}\,d\lambda$$
$$= \lambda(\mathbb{Q} \cap [0,1]) - \lambda(\mathbb{Q}^c \cap [0,1]) = 0 - 1 = -1.$$

7 Peter Gustav Lejeune Dirichlet (1805–1859).

5. Let A be a Lebesgue measurable subset of the Cantor set C, which is not Borel measurable:

$$\mathscr{B}([0,1]) \ni C \supseteq A \in \Sigma^*([0,1]) \setminus \mathscr{B}([0,1]).$$

Then

$$f(x) := \chi_A(x),\ x \in \mathbb{R},$$

although *not* Borel measurable on $[0,1]$, is continuous a. e. λ on $[0,1]$ (cf. Example 5.9), and hence, by the *Characterization of Riemann Integrability* (Theorem 6.10), $f \in R[a,b]$ and

$$\int_0^1 f(x)\,dx = \int_{[0,1]} f\,d\lambda = \int_{[0,1]} \chi_A\,d\lambda = \lambda(A) = 0.$$

Remark 6.21. As the last example shows, a function $f : [a,b] \to \mathbb{R}$ $(-\infty < a < b < \infty)$ Riemann integrable on $[a,b]$ need not be Borel measurable (cf. Section 5.12, Problem 13 (b)).

Example 6.12. By the *Lebesgue Dominated Convergence Theorem* (Theorem 6.7),
(a) $\lim_{n\to\infty} \int_0^1 x^n\,dx = 0$;
(b) $\lim_{n\to\infty} \int_{-\pi/2}^{\pi/2} \cos^n x\,dx = 0$;
(c) $\lim_{n\to\infty} \int_{-1}^{1} \frac{1-\cos^n x}{x^2+1}\,dx = \frac{\pi}{2}$.

Exercise 6.48. Verify.

6.10.3 Improper Integrals and Lebesgue Integral

Now, we are to parallel Lebesgue integration with improper integration in the Riemann sense (see, e. g., [13]).

Theorem 6.11 (Improper Integral of Type 1). *Let a function $f : [a,\infty) \to \mathbb{R}$ $(a \in \mathbb{R})$ be such that $f \in R[a,b]$ for any $b > a$. Then:*
(1) *f is Lebesgue measurable and*

$$\int_a^\infty |f(x)|\,dx := \lim_{b\to\infty} \int_a^b |f(x)|\,dx = \int_{[a,\infty)} |f(x)|\,d\lambda(x);$$

(2) *the improper integral $\int_a^\infty f(x)\,dx$ converges absolutely, i. e., $\int_a^\infty |f(x)|\,dx < \infty$, iff $\int_{[a,\infty)} |f(x)|\,d\lambda(x) < \infty$, i. e., $f \in L([a,\infty),\lambda)$, in which case*

$$\int_a^\infty f(x)\,dx = \int_{[a,\infty)} f(x)\,d\lambda(x).$$

Proof.

(1) For all $n \in \mathbb{N}$ sufficiently large so that $n > a$, $f \in R[a,n]$, and hence, $|f| \in R[a,n]$ (see Remarks 6.19). By the *Characterization of Riemann Integrability* (Theorem 6.10), $f \in L([a,n],\lambda)$, which implies that the function

$$f_n(x) := f(x)\chi_{[a,n]}(x),\ x \in \mathbb{R},$$

is Lebesgue measurable.
Therefore, by the *Sequences of Measurable Functions Theorem* (Theorem 5.4), the function

$$f(x) = \lim_{n\to\infty} f_n(x),\ x \in \mathbb{R},$$

is Lebesgue measurable.
By the *Monotone Convergence Theorem* (Theorem 2.4) applied to the function sequence $(|f_n|)_{n\in\mathbb{N}}$ and the function $|f|$, we have

$$\lim_{n\to\infty} \int\limits_{[a,n]} |f(x)|\,d\lambda(x) = \lim_{n\to\infty} \int\limits_{[a,\infty)} |f_n(x)|\,d\lambda(x) = \int\limits_{[a,\infty)} |f|(x)\,d\lambda(x).$$

Exercise 6.49. Explain why the *Monotone Convergence Theorem* (Theorem 2.4) applies.

Since, by the *Characterization of Riemann Integrability* (Theorem 6.10), for all $n \in \mathbb{N}$ with $n > a$,

$$\int\limits_{[a,n]} |f(x)|\,d\lambda(x) = \int_a^n |f(x)|\,dx,$$

passing to the limit as $n \to \infty$, we arrive at

$$\int\limits_{[a,\infty)} |f(x)|\,d\lambda(x) = \int_a^\infty |f(x)|\,dx. \tag{6.39}$$

(2) *"Only if"* part. Suppose that the improper integral $\int_a^\infty f(x)\,dx$ converges *absolutely*, i. e.,

$$\int_a^\infty |f(x)|\,dx < \infty,$$

and hence, converges, i. e.,

$$\int_a^\infty f(x)\,dx := \lim_{b\to\infty} \int_a^b f(x)\,dx \in \mathbb{R}$$

(see, e. g., [13]).

In view of proven equality (6.39), this implies that $|f| \in L([a, \infty), \lambda)$, and hence, by the *Properties of Lebesgue Integral. II* (Theorem 6.3, part (5)), $f \in L([a, \infty), \lambda)$.

By the *Lebesgue Dominated Convergence Theorem* (Theorem 6.7) applied to $(f_n)_{n\in\mathbb{N}}$ and f, we have

$$\lim_{n\to\infty} \int_{[a,n]} f\,d\lambda = \lim_{n\to\infty} \int_{[a,\infty)} f_n\,d\lambda = \int_{[a,\infty)} f\,d\lambda.$$

Exercise 6.50. Fill in the details.

Since, by the *Characterization of Riemann Integrability* (Theorem 6.10), for all $n \in \mathbb{N}$ with $n > a$,

$$\int_{[a,n]} f(x)\,d\lambda(x) = \int_a^n f(x)\,dx,$$

passing to the limit as $n \to \infty$, we arrive at

$$\int_{[a,\infty)} f(x)\,d\lambda(x) = \int_a^\infty f(x)\,dx.$$

"If" part. Let us prove this part *by contrapositive* and assume that the improper integral $\int_a^\infty f(x)\,dx$ converges *conditionally*, i. e.,

$$\int_a^\infty |f(x)|\,dx = \infty.$$

In view of proven equality (6.39), this implies that $|f| \notin L([a, \infty), \lambda)$, and hence, by the *Properties of Lebesgue Integral. II* (Theorem 6.3, part (5)), $f \notin L([a, \infty), \lambda)$, which completes the proof. □

Remarks 6.22.

- Thus, whenever the improper integral $\int_a^\infty f(x)\,dx$ converges *absolutely*, $f \in L([a, \infty), \lambda)$ and

$$\int_a^\infty f(x)\,dx = \int_{[a,\infty)} f(x)\,d\lambda(x),$$

  and hence, we can use the same notation $\int_a^\infty f(x)\,dx$ for both integrals, and apply the *limit theorems* of the prior sections.
- The analogue of the *Improper Integral of Type 1 Theorem* (Theorem 6.11) is in place for the improper integral $\int_{-\infty}^b f(x)\,dx$ ($b \in \mathbb{R}$).

- The *Improper Integral of Type 1 Theorem* (Theorem 6.11) and its analogue for the improper integral $\int_{-\infty}^{b} f(x)\,dx$ ($b \in \mathbb{R}$) naturally extend to complex-valued functions via applying separately to the real and imaginary parts.

Examples 6.13. By the *Improper Integral of Type 1 Theorem* (Theorem 6.11) and its analogue for the improper integral $\int_{-\infty}^{b} f(x)\,dx$ ($b \in \mathbb{R}$),

1. $\int_{[0,\infty)} e^{-x}\,d\lambda(x) = \int_0^\infty e^{-x}\,dx = -e^{-x}\big|_0^\infty = 1$, and hence, $f(x) := e^{-x} \in L([0,\infty),\lambda)$;
2. $\int_{[1,\infty)} \frac{1}{x}\,d\lambda(x) = \int_1^\infty \frac{1}{x}\,dx = \ln x\big|_1^\infty = \infty$, and hence, $f(x) := \frac{1}{x} \notin L([1,\infty),\lambda)$.
3. $\int_{-\infty}^{-1} \left|\frac{1}{x^3}\right|\,dx = -\int_{-\infty}^{-1} x^{-3}\,dx = \frac{1}{2}x^{-2}\big|_{-\infty}^{-1} = \frac{1}{2}$, and hence, $f(x) := \frac{1}{x^3} \in L((-\infty,-1],\lambda)$ and

$$\int_{(-\infty,-1]} \frac{1}{x^3}\,d\lambda(x) = \int_{-\infty}^{-1} \frac{1}{x^3}\,dx = -\frac{1}{2}.$$

Similarly, one can prove the following

Theorem 6.12 (Improper Integral of Type 2). *Let an unbounded function $f : [a,b) \to \mathbb{R}$ ($-\infty < a < b < \infty$) be such that $f \in R[a,t]$ for any $a < t < b$. Then*

(1) *f is Lebesgue measurable and*

$$\int_a^b |f(x)|\,dx := \lim_{t\to b-} \int_a^t |f(x)|\,dx = \int_{[a,b)} |f(x)|\,d\lambda(x);$$

(2) *the improper integral $\int_a^b f(x)\,dx$ converges absolutely, i. e., $\int_a^b |f(x)|\,dx < \infty$, iff $\int_{[a,b)} |f(x)|\,d\lambda(x) < \infty$, i. e., $f \in L([a,b),\lambda)$, in which case*

$$\int_a^b f(x)\,dx = \int_{[a,b)} f(x)\,d\lambda(x).$$

Exercise 6.51. Prove.

Remarks 6.23.

- Thus, whenever the improper integral $\int_a^b f(x)\,dx$ converges *absolutely*, $f \in L([a,b),\lambda)$ and

$$\int_a^b f(x)\,dx = \int_{[a,b)} f(x)\,d\lambda(x),$$

and hence, we can use the same notation $\int_a^b f(x)\,dx$ for both integrals, and apply the *limit theorems* of the prior sections.

- If we arbitrarily define the function $f : [a,b) \to \mathbb{R}$ at the right endpoint b of the interval $[a,b)$, then, in view of the fact that $\lambda(\{b\}) = 0$, under the conditions of the prior theorem, by the *Properties of Lebesgue Integral. II* (Theorem 6.3, part (3)), we can replace the interval $[a,b)$ in it (the theorem) with the interval $[a,b]$.
- The analogue of the *Improper Integral of Type 2 Theorem* (Theorem 6.12) is in place for the left-endpoint improper integral of the type 2.
- The *Improper Integral of Type 2 Theorem* (Theorem 6.12) and its analogue for the left-endpoint improper integral of the type 2 naturally extend to complex-valued functions via applying separately to the real and imaginary parts.

Examples 6.14. By the *Improper Integral of Type 2 Theorem* (Theorem 6.12) and its analogue for the left-endpoint improper integral of the type 2,

1. $\int_{[0,1)} \frac{1}{\sqrt{1-x}}\, d\lambda(x) = \int_0^1 (1-x)^{-1/2}\, dx = -2(1-x)^{1/2}\Big|_0^{1-} = 2$, and hence, $f(x) := \frac{1}{\sqrt{1-x}} \in L([0,1),\lambda)$;
2. $\int_{(0,1]} \frac{1}{x}\, d\lambda(x) = \int_0^1 \frac{1}{x}\, dx = \ln x\Big|_{0+}^{1} = \infty$, and hence, $f(x) := \frac{1}{x} \notin L((0,1],\lambda)$.

Example 6.15. For each $n \in \mathbb{N}$, the improper integrals converge absolutely and, by the *Lebesgue Dominated Convergence Theorem* (Theorem 6.7),

(a) $\lim_{n\to\infty} \int_0^\infty e^{-x} \cos^n x\, dx = 0$;

(b) $\lim_{n\to\infty} \int_0^\infty \frac{1-\cos^n x}{x^2+1}\, dx = \frac{\pi}{2}$;

(c) $\lim_{n\to\infty} \int_0^1 \frac{\cos^n x}{\sqrt{x}}\, dx = 0$.

Exercise 6.52. Verify.

6.11 Problems

1. Let (X,Σ) be a measurable space, μ_1, μ_2 be measures on Σ, and $\mu := \mu_1 + \mu_2$. Prove that

$$L(X,\mu) = L(X,\mu_1) \cap L(X,\mu_2)$$

and, for any $f \in L(X,\mu)$,

$$\int_X f\, d\mu = \int_X f\, d\mu_1 + \int_X f\, d\mu_2.$$

2. Let (X,Σ,μ) be a measure space, $f,g : X \to \overline{\mathbb{R}}$ and $f,g \in L(X,\mu)$. Prove that

$$\min(f,g), \max(f,g) \in L(X,\mu).$$

3. For the measure space $(\mathbb{N}, \mathscr{P}(\mathbb{N}), \mu)$, where μ is a *mass distribution measure over* $\mathbb{N}$:

$$\mathscr{P}(X) \ni A \mapsto \mu(A) := \sum_{k\in\mathbb{N}:\, k\in A} a_k$$

with some sequence $(a_n)_{n\in\mathbb{N}} \subset [0,\infty)$ (see Examples 3.2), and $A \in \mathscr{P}(\mathbb{N})$. Prove that, for any real-/complex-valued function f on $\mathbb{N}$, i. e., for any real/complex-termed sequence $(x_n := f(n))_{n\in\mathbb{N}}$,

$$f \in L(\mathbb{N},\mu) \Leftrightarrow \sum_{k=1}^{\infty} a_k|x_k| < \infty,$$

in which case

$$\int_{\mathbb{N}} f\, d\mu = \sum_{k=1}^{\infty} a_k x_k.$$

In particular, for $a_n = 1$, $n \in \mathbb{N}$, μ is the *counting measure* (see Examples 3.2 and Remark 3.3) and

$$f \in L(\mathbb{N},\mu) \Leftrightarrow \sum_{k=1}^{\infty} |x_k| < \infty,$$

in which case

$$\int_{\mathbb{N}} f\, d\mu = \sum_{k=1}^{\infty} x_k.$$

4. Let the underlying measure space be $(\mathbb{R}, \Sigma^*, \lambda)$ and $f \in L(\mathbb{R},\lambda)$. Prove that
   (a) $\int_{[-n,n]} |f|\, d\lambda \to \int_{\mathbb{R}} |f|\, d\lambda$, $n \to \infty$;
   (b) $\int_{\{x\in\mathbb{R}\,|\,|x|>n\}} |f|\, d\lambda \to 0$, $n \to \infty$;
   (c) for any *bounded* set $A \in \Sigma^*$, $\int_{A+n} |f|\, d\lambda \to 0$, $n \to \infty$.
5. Prove

Proposition 6.3 (Characterization of Equality A. E.). *Let* (X,Σ,μ) *be a measure space,* $f,g : X \to \overline{\mathbb{R}}$ *and* $f,g \in L(X,\mu)$. *Then*

$$f(x) = g(x) \text{ a. e. } \mu \text{ on } X \Leftrightarrow \forall A \in \Sigma: \int_A f\, d\mu = \int_A g\, d\mu.$$

Corollary 6.5 (Characterization of Vanishing A. E.). *Let* (X,Σ,μ) *be a measure space,* $f,g : X \to \overline{\mathbb{R}}$ *and* $f \in L(X,\mu)$. *Then*

$$f(x) = 0 \text{ a. e. } \mu \text{ on } X \Leftrightarrow \forall A \in \Sigma: \int_A f\, d\mu = 0.$$

6. Let (X, Σ, μ) be a measure space and $f \in L(X, \mu)$. Prove that the set

$$\{x \in X \mid f(x) \neq 0\}$$

is a countable union of Σ-measurable sets of finite measure μ.

7. Let (X, Σ, μ) be a measure space and $f \in L(X, \mu)$. Prove that

$$n\mu\left(\{x \in X \mid |f(x)| \geq n\}\right) \to 0,\ n \to \infty.$$

Hint. Apply the sharper version of *Markov's inequality* (6.17).

8. Let (X, Σ, μ) be a measure space and $f, f_n : X \to \mathbb{R}$, $n \in \mathbb{N}$, be Σ-measurable functions.
   (a) Prove that

$$\lim_{n\to\infty} \int_X |f_n - f|\, d\mu = 0 \Rightarrow f_n \xrightarrow{\mu} f.$$

   (b) Give an example showing that the converse is not true.

Hint. For (a), apply *Markov's inequality* (the *Properties of Lebesgue Integral. II* (Theorem 6.3, part (7))).

9. Let (X, Σ, μ) be a measure space and $f, f_n : X \to \mathbb{R}$, $n \in \mathbb{N}$, be Σ-measurable functions.
   (a) Prove that

$$\sum_{n=1}^{\infty} \int_X |f_n - f|\, d\mu < \infty \Rightarrow f_n \to f \pmod{\mu}.$$

   (b) Give an example showing that the converse is not true.

Hint. For (a), apply the *Characterization of Convergence A. E.* (Proposition 5.12) (Section 5.12, Problem 18) and *Markov's inequality* (the *Properties of Lebesgue Integral. II* (Theorem 6.3, part (7))).

10. Let (X, Σ, μ) be a measure space with $\mu(X) < \infty$ and for Σ-measurable functions $f, g : X \to \mathbb{R}$,

$$\rho(f, g) := \int_X \min(|f(x) - g(x)|, 1)\, d\mu(x).$$

Prove that

$$f_n \xrightarrow{\mu} f \Leftrightarrow \rho(f_n, f) \to 0,\ n \to \infty.$$

11. Let (X, Σ, μ) be a measure space with $\mu(X) < \infty$ and, for Σ-measurable functions $f, g : X \to \mathbb{R}$,

$$\rho(f, g) := \int_X \frac{|f(x) - g(x)|}{|f(x) - g(x)| + 1}\, d\mu(x).$$

Prove that

$$f_n \xrightarrow{\mu} f \Leftrightarrow \rho(f_n, f) \to 0,\ n \to \infty.$$

12. Let (X, Σ, μ) be a measure space with $\mu(X) < \infty$ and $f : X \to [0, \infty]$ be a Σ-measurable function. Prove that

$$f \in L(X, \mu) \Leftrightarrow \sum_{n=1}^{\infty} n\mu(A_n) < \infty,$$

where

$$A_n := \{x \in X \mid n - 1 \le f(x) < n\},\ n \in \mathbb{N}.$$

Hint. Show that

$$\sum_{n=1}^{\infty} (n-1)\mu(A_n) \le \int_X f\, d\mu = \sum_{n=1}^{\infty} \int_{A_n} f\, d\mu \le \sum_{n=1}^{\infty} n\mu(A_n).$$

13. Let the underlying measure space be $([0,1], \Sigma^*([0,1]), \lambda)$ and

$$f_n := \begin{cases} \chi_{[0,1/3]}, & n \in \mathbb{N} \text{ is odd}, \\ \chi_{(1/3,1]}, & n \in \mathbb{N} \text{ is even}. \end{cases}$$

Find
(a) $\int_{[0,1]} \underline{\lim}_{n\to\infty} f_n\, d\lambda$, $\underline{\lim}_{n\to\infty} \int_{[0,1]} f_n\, d\lambda$;
(b) $\int_{[0,1]} \overline{\lim}_{n\to\infty} f_n\, d\lambda$, $\overline{\lim}_{n\to\infty} \int_{[0,1]} f_n\, d\lambda$.

14. Let (X, Σ, μ) be a measure space and $f : X \to \overline{\mathbb{R}}$ be a Σ-measurable function. Prove that, if

$$\Sigma \ni A_n \uparrow \bigcup_{k=1}^{\infty} A_k = X$$

and

$$\sup_{n\in\mathbb{N}} \int_{A_n} |f|\, d\mu < \infty,$$

then $f \in L(X, \mu)$ and

$$\lim_{n\to\infty} \int_{A_n} f\, d\mu = \int_X f\, d\mu.$$

Hint. Apply *Levi's Theorem* (Corollary 6.3) to the function sequence

$$f_n(x) := |f(x)|\chi_{A_n}(x),\ n \in \mathbb{N}, x \in X.$$

15. Let (X, Σ, μ) be a measure space with $\mu(X) < \infty$ and $f, f_n : X \to \mathbb{R}$, $n \in \mathbb{N}$, be Σ-measurable functions. Prove that, if

$$f_n \to f \pmod{\mu},$$

then
(a) $\lim_{n\to\infty} \int_X \sin f_n(x)\, d\mu(x) = \int_X \sin f(x)\, d\mu(x)$;
(b) $\lim_{n\to\infty} \int_X e^{-f_n^2(x)}\, d\mu(x) = \int_X e^{-f^2(x)}\, d\mu(x)$.

16. Let (X, Σ, μ) be a measure space with $\mu(X) < \infty$ and $f : X \to \mathbb{R}$ be a Σ-measurable function. Verify that
(a) $\lim_{n\to\infty} \int_X e^{-nf^2(x)}\, d\mu(x) = \mu(\{x \in X \mid f(x) = 0\})$.

Hint. Use the *additivity* of the Lebesgue integral (the *Properties of Lebesgue Integral. II* (Theorem 6.3, part (1))) and find the limit over the sets $\{x \in X \mid f(x) = 0\}$ and $\{x \in X \mid f(x) \neq 0\}$ separately.

(b) $\lim_{n\to\infty} \int_X \cos^{2n} f(x)\, d\mu(x) = \mu(\{x \in X \mid f(x) = n\pi,\ n \in \mathbb{Z}\})$.

17. Let (X, Σ, μ) be a measure space and $f, f_n : X \to \mathbb{R}$, $n \in \mathbb{N}$, be Σ-measurable functions. Prove that, if

$$f_n \to f \pmod{\mu}$$

and

$$\exists\, c \geq 0 : |f_n(x)| \leq c,\ n \in \mathbb{N}, x \in X.$$

Prove that, for any $g : X \to \mathbb{R}$ such that $g \in L(X, \mu)$,

$$\lim_{n\to\infty} \int_X f_n g\, d\mu = \int_X fg\, d\mu.$$

18. Evaluate
(a) $\lim_{n\to\infty} \int_0^1 e^{-n\sin x}\, dx$;
(b) $\lim_{n\to\infty} \int_0^\pi e^{-n\sin^2 \frac{1}{x}}\, dx$.

Remark 6.24. In (b), at 0, the integrand can be defined arbitrarily.

19. Evaluate

$$\lim_{n\to\infty} \int_D \sin^n(x^2 + y^2)\, dxdy,$$

where

$$D := \left\{(x, y) \in \mathbb{R}^2 \,\middle|\, x^2 + y^2 \leq 1\right\}.$$

20. Prove that, for each $n \in \mathbb{N}$, the improper integrals converge absolutely and evaluate
 (a) $\lim_{n\to\infty} \int_0^\infty e^{-x} \sin^n x\, dx$;
 (b) $\lim_{n\to\infty} \int_0^\infty e^{-x} \sin^n x^2\, dx$;
 (c) $\lim_{n\to\infty} \int_0^\infty e^{-x} \sin^n \frac{1}{x}\, dx$.
21. Prove that, for each $n \in \mathbb{N}$, the improper integral converges absolutely and evaluate

$$\lim_{n\to\infty} \int_0^\infty ne^{-nx} \sin x^3\, dx.$$

22. Prove that, for each $n \in \mathbb{N}$, the improper integral converges absolutely and evaluate

$$\sum_{n=1}^{\infty} \int_1^\infty \frac{1}{(x^2+1)^n}\, dx.$$

23. Prove that, for each $t > 0$, the improper integral converges absolutely and evaluate

$$\lim_{t\to 0+} \int_0^\infty \frac{1}{x^2+1+t}\, dx.$$

24. Let a function $f : [0,\infty) \to [0,\infty)$ be *continuous* on $[0,\infty)$ and

$$\sup_{0<t<1} \int_0^\infty \frac{1-e^{-tx}}{t} f(x)\, dx < \infty.$$

 Prove that

$$\int_0^\infty xf(x)\, dx < \infty.$$

25. Prove

 Theorem 6.13 (Parametric Differentiation). *Let (X,Σ,μ) be a measure space, $A \in \Sigma$, and $I \subseteq \mathbb{R}$ be an interval, and a function $f : A \times I \to \mathbb{R}$ be such that:*
 (a) $\forall\, t \in I : f(\cdot,t) \in L(A,\mu)$;
 (b) $\exists N \in \Sigma, \mu(N) = 0 : \forall\, x \in A \setminus N\ f_t(x,\cdot)$ *exists on* I;
 (c) $\exists g \in L(A,\mu) : \forall\,(x,t) \in (A\setminus N)\times I : |f_t(x,t)| \le g(x)$.
 Then the function

$$F(t) := \int_A f(x,t)\, d\mu(x),\ t \in I,$$

is differentiable on I and

$$F'(t) := \int_A f_t(x,t)\,d\mu(x),\ t \in I.$$

Hint. Apply the *Mean Value Theorem* and the *Lebesgue Dominated Convergence Theorem* (Theorem 6.7).

26. Let a function $f : [0,\infty) \to \mathbb{R}$ be *continuous* on $[0,\infty)$ and

$$\int_0^\infty x|f(x)|\,dx < \infty.$$

Prove that

$$\frac{d}{dt}\int_0^\infty e^{-tx}f(x)\,dx = \int_0^\infty e^{-tx}(-x)f(x)\,dx,\ t \geq 0.$$

7 L_p Spaces

This chapter, we study L_p *spaces*, whose importance in contemporary analysis cannot be overestimated, and outline the basics of abstract *normed vector spaces* (see, e. g., [14]).

7.1 Hölder's and Minkowski's Inequalities

Hölder's[1] and *Minkowski's*[2] *inequalities*, being fundamental inequalities between integrals, are indispensable for studying L_p spaces.

7.1.1 Conjugate Indices

Definition 7.1 (Conjugate Indices). We call $1 \le p, q \le \infty$ *conjugate indices* if they are related as follows:

$$\begin{aligned} &\frac{1}{p} + \frac{1}{q} = 1 \text{ for } 1 < p, q < \infty, \\ &q = \infty \text{ for } p = 1, \\ &q = 1 \text{ for } p = \infty. \end{aligned}$$

Examples 7.1. In particular, $p = 2$ and $q = 2$ are conjugate as well as $p = 3$ and $q = 3/2$.

Remark 7.1. Thus, for $1 < p, q < \infty$,

$$q = \frac{p}{p-1} = 1 + \frac{1}{p-1} \to \begin{cases} \infty, & p \to 1+, \\ 1, & p \to \infty, \end{cases}$$

with $q > 2$ if $1 < p < 2$ and $1 < q < 2$ if $p > 2$, and the following relationships hold:

$$\begin{aligned} &p + q = pq, \\ &pq - p - q + 1 = 1 \Rightarrow (p-1)(q-1) = 1, \\ &(p-1)q = p \Rightarrow p - 1 = \frac{p}{q}, \\ &(q-1)p = q \Rightarrow q - 1 = \frac{q}{p}. \end{aligned} \tag{7.1}$$

1 Otto Ludwig Hölder (1859–1937).
2 Hermann Minkowski (1864–1909).

https://doi.org/10.1515/9783110600995-007

7.1.2 Two Important Inequalities

To proceed, we are to prove the following important inequalities.

Theorem 7.1 (Young's Inequality). *Let $1 < p, q < \infty$ be conjugate indices. Then, for any $a, b \geq 0$,*[3]

$$ab \leq \frac{a^p}{p} + \frac{b^q}{q}.$$

Proof. The inequality is, obviously, true if $a = 0$ or $b = 0$ and for $p = q = 2$.

Exercise 7.1. Verify.

Suppose that $a, b > 0$ and $1 < p < 2$ or $p > 2$ and recall that

$$(p-1)(q-1) = 1 \quad \text{and} \quad (p-1)q = p.$$

(see (7.1)).

Comparing the areas in Figure 7.1, which corresponds to the case of $p > 2$, the case of $1 < p < 2$ being symmetric, we conclude that

$$A \leq A_1 + A_2,$$

where A is the area of the rectangle $[0, a] \times [0, b]$, the equality being the case *iff* $b = a^{p-1} = a^{p/q}$.

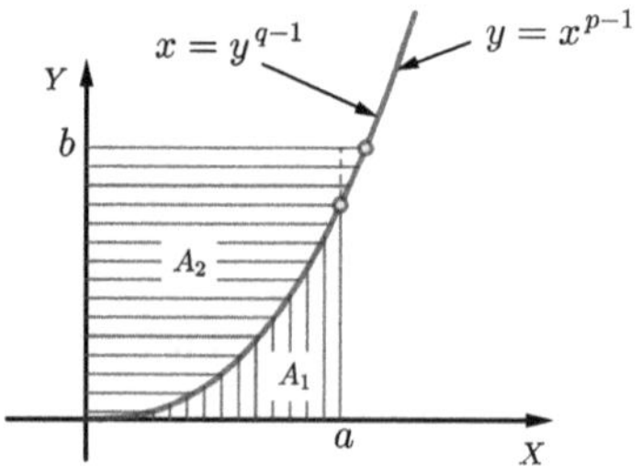

Figure 7.1: The case of $p > 2$.

Hence,

$$ab \leq \int_0^a x^{p-1}\,dx + \int_0^b y^{q-1}\,dy = \frac{a^p}{p} + \frac{b^q}{q}. \qquad \square$$

3 William Henry Young (1863–1942).

Remark 7.2. As Figure 7.1 shows, equality in *Young's inequality* (Theorem 7.1) holds *iff* $a^p = b^q$.

Proposition 7.1 (Important Inequality). *Let* $1 \le p < \infty$. *Then, for any* $a, b \in \mathbb{R}$,

$$|a+b|^p \le (|a|+|b|)^p \le 2^{p-1}\left(|a|^p + |b|^p\right). \tag{7.2}$$

Proof. Obviously, the inequality is true for $p = 1$.

Suppose that $p > 1$. Then the function

$$f(x) := x^p,\ x \ge 0,$$

is *concave up*.

Exercise 7.2. Verify.

Hence, for any $a, b \in \mathbb{R}$,

$$\left(\frac{|a|+|b|}{2}\right)^p = f\left(\frac{|a|+|b|}{2}\right) \le \frac{f(|a|)+f(|b|)}{2} = \frac{|a|^p+|b|^p}{2}.$$

Whence, multiplying through by 2^p, we arrive at desired inequality (7.2). □

7.1.3 Essential Supremum

Definition 7.2 (Essential Upper Bounds and Essential Supremum). Let (X, Σ, μ) be a measure space and $f : X \to \mathbb{R}$ be a Σ-measurable function.

A number $a \in \mathbb{R}$ is called an *essential upper bound for the function f relative to measure* μ *on X* if

$$\mu\left(f^{-1}((a,\infty))\right) = \mu(\{x \in X \mid f(x) > a\}) = 0,$$

i. e., $f(x) \le a$ a. e. μ on X.

Let

$$U_f^{\text{ess}} := \{a \in \mathbb{R} \mid \mu(\{x \in X \mid f(x) > a\}) = 0\}$$

be the set of all essential upper bounds. The value

$$[-\infty, \infty] \ni \operatorname*{ess\,sup}_{x\in X} f(x) := \begin{cases} \inf U_f^{\text{ess}} & \text{if } U_f^{\text{ess}} \ne \emptyset, \\ \infty & \text{if } U_f^{\text{ess}} = \emptyset, \end{cases}$$

is called the *essential supremum for the function f relative to measure* μ *on X*.

If $\operatorname{ess\,sup}_{x\in X} f(x) < \infty$, the function f is said to be *essentially bounded above relative to measure* μ *on X* or μ*-essentially bounded above on X*.

If $\operatorname{ess\,sup}_{x\in X} |f(x)| < \infty$, the function f is said to be *essentially bounded relative to measure* μ *on X* or μ*-essentially bounded on X*.

Remarks 7.3.

- If $a \in \mathbb{R}$ is an essential upper bound for the function f relative to measure μ on X, then any $b > a$ is also an essential upper bound for f relative to μ on X, i. e.,

$$a \in U_f^{\text{ess}} \Rightarrow \forall b > a : \; b \in U_f^{\text{ess}}.$$

- Provided $U_f^{\text{ess}} \neq \emptyset$, the *essential supremum* of the function f relative to measure μ on X is the *smallest* in $\overline{\mathbb{R}}$ essential upper bound for f relative to μ on X, i. e.,

$$\inf U_f^{\text{ess}} = \min U_f^{\text{ess}}.$$

- The *essential supremum* for the function f relative to measure μ on X can be equivalently defined as follows:

$$\operatorname*{ess\,sup}_{x\in X} f(x) := \inf \left\{ \sup_{x \in X\setminus N} f(x) \,\middle|\, N \in \Sigma,\ \mu(N) = 0 \right\}, \tag{7.3}$$

and hence, for $\operatorname{ess\,sup}_{x\in X} f(x)$, the values of f on any μ-null set are superfluous.
- For an arbitrary *noninfinite* measure space (X, Σ, μ), every bounded (above) on X Σ-measurable function $f : X \to \mathbb{R}$ is, obviously, essentially bounded (above) relative to measure μ on X. As the following examples demonstrate, the converse is not true, i. e. an essentially bounded (above) function need not be bounded. They also show that sup and ess sup need not be equal.

Exercise 7.3. Verify.

Examples 7.2. Let the underlying measure space be $(\mathbb{R}, \Sigma^*, \lambda)$.

1. For the function

$$f(x) := \begin{cases} 0, & x < 0, \\ 2, & x = 0, \\ 1, & x > 0, \end{cases}$$

$$\sup_{x\in\mathbb{R}} f(x) = 2 \quad \text{but} \quad \operatorname*{ess\,sup}_{x\in\mathbb{R}} f(x) = 1.$$

2. For the function

$$f(x) := x\chi_{\mathbb{Q}}(x) = \begin{cases} x, & x \in \mathbb{Q}, \\ 0, & x \in \mathbb{Q}^c, \end{cases}$$

$$\sup_{x\in\mathbb{R}} |f(x)| = \infty \quad \text{but} \quad \operatorname*{ess\,sup}_{x\in\mathbb{R}} |f(x)| = 0,$$

i. e., f, being *unbounded* on $\mathbb{R}$, is *essentially bounded* relative to the Lebesgue measure λ on $\mathbb{R}$.

3. For the function

$$f(x) := x\chi_{\mathbb{Q}^c}(x) = \begin{cases} x, & x \in \mathbb{Q}^c, \\ 0, & x \in \mathbb{Q}, \end{cases}$$

$$\sup_{x\in\mathbb{R}} |f(x)| = \operatorname*{ess\,sup}_{x\in\mathbb{R}} |f(x)| = \infty,$$

   i. e., f is both *unbounded* and *essentially unbounded* relative to the Lebesgue measure λ on $\mathbb{R}$.
4. For the function

$$f(x) := \begin{cases} 1/x, & x \neq 0, \\ 0, & x = 0, \end{cases}$$

   similarly,

$$\sup_{x\in\mathbb{R}} |f(x)| = \operatorname*{ess\,sup}_{x\in\mathbb{R}} |f(x)| = \infty.$$

Exercise 7.4. Verify.

Remarks 7.4.

- For a measure space (X, Σ, μ) such that $\emptyset$ is the only μ-null set, in particular for $(X, \mathscr{P}(X), \mu)$ with the *counting measure* μ (see Examples 3.2), for any real- or complex-valued Σ-measurable function f on X, every essential upper bound is a regular upper bound for f, and hence,

$$\operatorname*{ess\,sup}_{x\in X} f(x) = \sup_{x\in X} f(x).$$

- One can similarly define *essential lower bounds* and the *essential infimum.*

7.1.4 *p*-Norms

Here, for convenience, we introduce the notion of *p-norm* ($1 \le p \le \infty$) for a function, which, getting ahead of the rigorous definition of the general concept of *norm* (see Definition 7.5), has a certain inherent deficiency, which becomes apparent and is taken care of in Section 7.4.

Definition 7.3 (p-Norm). Let (X, Σ, μ) be a measure space and f be a real- or complex-valued Σ-measurable function on X.

For $1 \le p < \infty$, if $|f|^p \in L(X, \mu)$, i. e.,

$$\int_X |f|^p \, d\mu < \infty,$$

then the *p-norm* of f is the nonnegative number

$$\|f\|_p := \left[\int_X |f|^p \, d\mu \right]^{1/p} \in [0, \infty).$$

For $p = \infty$, if f is essentially bounded relative to μ on X, i. e.,

$$\operatorname*{ess\,sup}_{x \in X} |f(x)| < \infty,$$

then the ∞-*norm* of f is the nonnegative number

$$\|f\|_\infty := \operatorname*{ess\,sup}_{x \in X} |f(x)| \in [0, \infty).$$

Exercise 7.5. Prove that, given a measure space (X, Σ, μ), for a real- or complex-valued Σ-measurable function f on X,

$$\|f\|_p = 0 \Leftrightarrow f(x) = 0 \text{ a. e. } \mu \text{ on } X.$$

Examples 7.3.

1. For $X := \{1, \dots, n\}$ ($n \in \mathbb{N}$), $\Sigma := \mathscr{P}(X)$, and the *counting measure* μ (see Examples 3.2), any real- or complex-valued function f on X, i. e., a real- or complex-component ordered n-tuple $(x_1, \dots, x_n)$ with $x_k = f(k)$, $k = 1, \dots, n$, is Σ-measurable (see Examples 5.2) and

$$\|f\|_p = \begin{cases} \left[\sum_{i=1}^n |x_i|^p\right]^{1/p} & \text{if } 1 \le p < \infty, \\ \max_{1 \le i \le n} |x_i| & \text{if } p = \infty. \end{cases}$$

2. For $X := \mathbb{N}$, $\Sigma := \mathscr{P}(X)$, and the *counting measure* μ, any real- or complex-valued function f on X, i. e., a real- or complex-termed sequence $(x_n)_{n \in \mathbb{N}}$ with $x_n = f(n)$, $n \in \mathbb{N}$, is Σ-measurable and, for $1 \le p < \infty$, if $|f|^p \in L(X, \mu)$, i. e.,

$$\int_X |f|^p \, d\mu = \sum_{n=1}^\infty |x_n|^p < \infty$$

(see Section 6.11, Problem 3), then

$$\|f\|_p = \left[\sum_{n=1}^\infty |x_n|^p \right]^{1/p};$$

if

$$\operatorname*{ess\,sup}_{n \in \mathbb{N}} |f| = \sup_{n \in \mathbb{N}} |x_n| < \infty$$

(Remarks 7.4), then

$$\|f\|_\infty = \sup_{n \in \mathbb{N}} |x_n|.$$

Exercise 7.6. Verify.

7.1.5 Hölder's Inequality

Theorem 7.2 (Hölder's Inequality). *Let (X, Σ, μ) be a measure space and $1 \le p, q \le \infty$ be conjugate indices. Then, for any real- or complex-valued Σ-measurable functions f and g on X such that:*

(i) $|f|^p, |g|^q \in L(X, \mu)$ *if* $1 < p, q < \infty$;

(ii) $f \in L(X, \mu)$ *and* $\operatorname{ess\,sup}_{x \in X} |g(x)| < \infty$ *if* $p = 1$, $q = \infty$; *or*

(iii) $\operatorname{ess\,sup}_{x \in X} |f(x)| < \infty$ *and* $g \in L(X, \mu)$ *if* $p = \infty$, $q = 1$,

$fg \in L(X, \mu)$ and Hölder's inequality holds:

$$\int_X |fg| \, d\mu \le \|f\|_p \|g\|_q. \tag{7.4}$$

Proof. The cases of $p = 1$, $q = \infty$, and $p = \infty$, $q = 1$ follow immediately from the *Properties of Lebesgue Integral. I* (Theorem 6.1, parts (3) and (9)) and the *Properties of Lebesgue Integral. II* (Theorem 6.3, part (3)).

Exercise 7.7. Explain.

Suppose that $1 < p, q < \infty$.
If

$$\|f\|_p = 0 \quad \text{or} \quad \|g\|_q = 0,$$

then, by the *Properties of Lebesgue Integral. II* (Theorem 6.3, part (9)),

$$f(x) = 0 \text{ or } g(x) = 0 \text{ a. e. } \mu \text{ on } X, \text{ respectively,}$$

(cf. Exercise 7.5), and hence, by the *Properties of Lebesgue Integral. II* (Theorem 6.3, part (3)), *Hölder's inequality* (7.4) holds.

Exercise 7.8. Explain.

Suppose now that

$$\|f\|_p, \|g\|_q > 0.$$

Then, applying *Young's inequality* (Theorem 7.1), for each $x \in X$, to

$$a = \frac{|f(x)|}{\|f\|_p} \quad \text{and} \quad b = \frac{|g(x)|}{\|g\|_q},$$

we have

$$\frac{|f(x)g(x)|}{\|f\|_p \|g\|_q} \le \frac{|f(x)|^p}{p\|f\|_p^p} + \frac{|g(x)|^q}{q\|g\|_q^q}, \quad x \in X.$$

Whence, integrating over X, by the *Properties of Lebesgue Integral. I* (Theorem 6.1, parts (3) and (9)) and the *Linearity of Lebesgue Integral Theorem* (Theorem 6.5), we obtain

$$\frac{1}{\|f\|_p\|g\|_q}\int_X |fg|\,d\mu \le \frac{1}{p\|f\|_p^p}\int_X |f|^p\,d\mu + \frac{1}{q\|g\|_q^q}\int_X |g|^q\,d\mu = \frac{1}{p}+\frac{1}{q} = 1.$$

Where multiplying through by $\|f\|_p\|g\|_q$, we arrive at *Hölder's inequality* (7.4), which, by the *Properties of Lebesgue Integral. II* (Theorem 6.3, part (5)), in particular, implies that $fg \in L(X,\mu)$ (see Remarks 6.5). □

We immediately obtain *Hölder's inequality* for ordered n-tuples ($n \in \mathbb{N}$) and sequences (see Examples 7.3).

Corollary 7.1 (Hölder's Inequality for n-Tuples). *Let $n \in \mathbb{N}$ and $1 \le p,q \le \infty$ be conjugate indices. Then, for any real- or complex-component ordered n-tuples*

$$x := (x_1,\dots,x_n) \text{ and } y := (y_1,\dots,y_n),$$

Hölder's inequality holds:

$$\sum_{k=1}^{n} |x_k y_k| \le \|x\|_p\|y\|_q.$$

Corollary 7.2 (Hölder's Inequality for Sequences). *Let $1 \le p,q \le \infty$ be conjugate indices. Then, for any real- or complex-termed sequences*

$$x := (x_n)_{n\in\mathbb{N}} \text{ and } y := (y_n)_{n\in\mathbb{N}}$$

such that:

(i) $\sum_{n=1}^{\infty}|x_n|^p, \sum_{n=1}^{\infty}|y_n|^q < \infty$ *if* $1 < p,q < \infty$;
(ii) $\sum_{n=1}^{\infty}|x_n| < \infty$ *and* $\sup_{n\in\mathbb{N}}|y_n| < \infty$ *if* $p = 1$, $q = \infty$; *or*
(iii) $\sup_{n\in\mathbb{N}}|x_n| < \infty$ *and* $\sum_{n=1}^{\infty}|y_n| < \infty$ *if* $p = \infty$, $q = 1$,

Hölder's inequality holds:

$$\sum_{n=1}^{\infty} |x_n y_n| \le \|x\|_p\|y\|_q.$$

Remark 7.5. The important special case of *Hölder's inequality* for $p = q = 2$:

$$\int_X |fg|\,d\mu \le \left[\int_X |f|^2\,d\mu\right]^{1/2}\left[\int_X |g|^2\,d\mu\right]^{1/2} \tag{7.5}$$

is known as the *Cauchy–Schwarz*[4] *inequality*. For real- or complex-component ordered n-tuples and real- or complex-termed sequences, it acquires the following form:

$$\sum_{k=1}^{n}|x_k y_k| \leq \left[\sum_{k=1}^{n}|x_k|^2\right]^{1/2}\left[\sum_{k=1}^{n}|y_k|^2\right]^{1/2} \quad (n \in \mathbb{N}),$$
$$\sum_{k=1}^{\infty}|x_k y_k| \leq \left[\sum_{k=1}^{\infty}|x_k|^2\right]^{1/2}\left[\sum_{k=1}^{\infty}|y_k|^2\right]^{1/2}, \tag{7.6}$$

respectively.

7.1.6 Minkowski's Inequality

Theorem 7.3 (Minkowski's Inequality). *Let (X, Σ, μ) be a measure space and $1 \leq p \leq \infty$. Then, for any real- or complex-valued Σ-measurable functions f and g on X such that:*
(i) $|f|^p, |g|^p \in L(X, \mu)$ *if* $1 \leq p < \infty$ *or*
(ii) $\operatorname{ess\,sup}_{x\in X}|f(x)|, \operatorname{ess\,sup}_{x\in X}|g(x)| < \infty$ *if* $p = \infty$,

$|f+g|^p \in L(X, \mu)$ *or* $\operatorname{ess\,sup}_{x\in X}|f(x)+g(x)| < \infty$, *respectively, and Minkowski's inequality holds:*

$$\|f+g\|_p \leq \|f\|_p + \|g\|_p. \tag{7.7}$$

Proof. Let $1 \leq q \leq p$ be conjugate to p index (see Definition 7.1).

The cases of $p = 1$ or $p = \infty$ are rather trivial as well as the case of $1 < p < \infty$ with

$$\int_X |f+g|^p \, d\mu = 0.$$

Exercise 7.9. Consider.

Suppose $1 < p < \infty$ and

$$\int_X |f+g|^p \, d\mu > 0.$$

By the *important inequality* (Proposition 7.1), the *Properties of Lebesgue Integral. I* (Theorem 6.1, part (3)), and the *Linearity of Lebesgue Integral Theorem* (Theorem 6.5), we infer that

$$\int_X |f+g|^p \, d\mu < \infty,$$

i. e., $|f+g|^p \in L(X, \mu)$.

4 Karl Hermann Amandus Schwarz (1843–1921).

Exercise 7.10. Fill in the details.

For each $x \in X$,

$$|f(x)+g(x)|^p = |f(x)+g(x)||f(x)+g(x)|^{p-1} \qquad \text{since } |f(x)+g(x)| \le |f(x)|+|g(x)|;$$
$$\le |f(x)||f(x)+g(x)|^{p-1} + |g(x)||f(x)+g(x)|^{p-1}.$$

Integrating over X, in view of the *Properties of Lebesgue Integral. I* (Theorem 6.1, part (3)) and the *Linearity of Lebesgue Integral Theorem* (Theorem 6.5), we obtain

$$\int_X |f(x)+g(x)|^p\,d\mu \le \int_X |f(x)||f(x)+g(x)|^{p-1}\,d\mu + \int_X |g(x)||f(x)+g(x)|^{p-1}\,d\mu$$

by *Hölder's inequality* (Theorem 7.2);

$$\le \left[\int_X |f(x)|^p\,d\mu\right]^{1/p}\left[\int_X |f(x)+g(x)|^{(p-1)q}\,d\mu\right]^{1/q}$$
$$+ \left[\int_X |g(x)|^p\,d\mu\right]^{1/p}\left[\int_X |f(x)+g(x)|^{(p-1)q}\,d\mu\right]^{1/q}$$

since $(p-1)q = p$ (see (7.1));

$$= \left[\int_X |f(x)|^p\,d\mu\right]^{1/p}\left[\int_X |f(x)+g(x)|^{p}\,d\mu\right]^{1/q}$$
$$+ \left[\int_X |g(x)|^p\,d\mu\right]^{1/p}\left[\int_X |f(x)+g(x)|^{p}\,d\mu\right]^{1/q}$$
$$= \left[\int_X |f(x)+g(x)|^p\,d\mu\right]^{1/q}\left(\left[\int_X |f(x)|^p\,d\mu\right]^{1/p} + \left[\int_X |g(x)|^p\,d\mu\right]^{1/p}\right).$$

Dividing through by

$$0 < \left[\int_X |f(x)+g(x)|^p\,d\mu\right]^{1/q} < \infty,$$

in view of $1 - \frac{1}{q} = \frac{1}{p}$ (see Definition 7.1), we arrive at *Minkowski's inequality* (7.7). □

Remark 7.6. The vital implication of *Minkowski's inequality* is the fact that, given a measure space (X, Σ, μ) and $1 \le p < \infty$, the sets of Σ-measurable real or complex-valued functions

$$\mathcal{L}_p(X,\Sigma,\mu) := \{f : X \to \mathbb{R} \text{ (or } \mathbb{C}) \mid f \text{ is } \Sigma\text{-measurable and } |f|^p \in L(X,\mu)\}$$

(p-*integrable on* X *functions*), in particular $\mathcal{L}_1(X, \Sigma, \mu) = L(X, \mu)$, and

$$\mathcal{L}_\infty(X, \Sigma, \mu) := \{f : X \to \mathbb{R} \text{ (or } \mathbb{C}) \mid f \text{ is } \Sigma\text{-measurable and } \mu\text{-essentially bounded on } X\}$$

are real or complex *vector spaces*, respectively (cf. Remarks 6.11).

Exercise 7.11. Explain.

Minkowski's inequality for ordered n-tuples ($n \in \mathbb{N}$) and sequences (see Examples 7.3) follow immediately.

Corollary 7.3 (Minkowski's Inequality for n-Tuples). *Let $n \in \mathbb{N}$ and $1 \le p \le \infty$. Then, for any real- or complex-component ordered n-tuples $x := (x_1, \dots, x_n)$ and $y := (y_1, \dots, y_n)$, Minkowski's inequality holds:*

$$\|x + y\|_p \le \|x\|_p + \|y\|_p.$$

Corollary 7.4 (Minkowski's Inequality for Sequences). *Let $1 \le p \le \infty$. Then, for any real- or complex-termed sequences $x := (x_n)_{n\in\mathbb{N}}$ and $y := (y_n)_{n\in\mathbb{N}}$ such that:*

(i) $\sum_{n=1}^{\infty} |x_n|^p, \sum_{n=1}^{\infty} |y_n|^p < \infty$ *if* $1 \le p < \infty$ *or*

(ii) $\sup_{n\in\mathbb{N}} |x_n|, \sup_{n\in\mathbb{N}} |y_n| < \infty$ *if* $p = \infty$,

$$\sum_{n=1}^{\infty} |x_n + y_n|^p < \infty,$$

or

$$\sup_{n\in\mathbb{N}} |x_n + y_n| < \infty,$$

respectively, and Minkowski's inequality holds:

$$\|x + y\|_p \le \|x\|_p + \|y\|_p.$$

7.2 Convergence in p-Norm

Along with *convergence almost everywhere* and in *convergence measure*, *convergence in p-norm* ($1 \le p \le \infty$) is yet another type of convergence for sequences of measurable functions, which is based on the idea of measure.

7.2.1 Definitions and Examples

Definition 7.4 (Convergence in p-Norm). Let (X, Σ, μ) be a measure space, f, f_n, $n \in \mathbb{N}$, be real- or complex-valued Σ-measurable functions on X, and $1 \le p \le \infty$.

The function sequence $(f_n)_{n\in\mathbb{N}}$ is said to *converge to the function f in p-norm on X* if:

(i) for $1 \le p < \infty$, $|f|^p, |f_n|^p \in L(X,\mu)$, $n \in \mathbb{N}$, i. e.,

$$\int_X |f(x)|^p \, d\mu(x) < \infty, \int_X |f_n(x)|^p \, d\mu(x) < \infty, \ n \in \mathbb{N},$$

and

$$\|f_n - f\|_p := \left[\int_X |f_n(x) - f(x)|^p \, d\mu(x)\right]^{1/p} \to 0, \ n \to \infty;$$

(ii) for $p = \infty$, f, f_n, $n \in \mathbb{N}$, are μ-essentially bounded on X, i. e.,

$$\operatorname*{ess\,sup}_{x\in X} |f(x)| < \infty, \ \operatorname*{ess\,sup}_{x\in X} |f_n(x)| < \infty, \ n \in \mathbb{N},$$

and

$$\|f_n - f\|_\infty := \operatorname*{ess\,sup}_{x\in X} |f_n(x) - f(x)| \to 0, \ n \to \infty.$$

Remark 7.7. Due to *Minkowski's inequality* (Theorem 7.3), convergence in p-norm is well-defined (see Remark 7.6).

Exercise 7.12. Explain.

Examples 7.4.

1. Let the underlying measure space be $(\mathbb{R}, \Sigma^*, \lambda)$.
   (a) The function sequence

   $$f_n(x) := \chi_{[0,1/n]}(x), \ n \in \mathbb{N}, x \in \mathbb{R},$$

   converges to the function $f(x) := 0$, $x \in \mathbb{R}$, in p-norm on $\mathbb{R}$ for $1 \le p < \infty$, but not in ∞-norm.

   (b) The function sequence

   $$f_n(x) := \chi_{[n,n+1]}(x), \ n \in \mathbb{N}, x \in \mathbb{R},$$

   does not converge to the function $f(x) := 0$, $x \in \mathbb{R}$, in p-norm on $\mathbb{R}$ for any $1 \le p \le \infty$.

   (c) The function sequence

   $$f_n(x) := \chi_{[1,n]}(x)\frac{1}{x}, \ n \in \mathbb{N}, x \ge 1,$$

   converges to the function $f(x) := \frac{1}{x}$, $x \ge 1$, in p-norm on $\mathbb{R}$ for $1 < p \le \infty$, but not in 1-norm.

2. For $X := \{1, \dots, n\}$ ($n \in \mathbb{N}$), $\Sigma := \mathscr{P}(X)$, and the *counting measure* μ, a sequence of real or complex-component ordered n-tuples $\left(x_k := (x_1^{(k)}, \dots, x_n^{(k)})\right)_{k\in\mathbb{N}}$ converges to a real or complex-component ordered n-tuple $x := (x_1, \dots, x_n)$ in p-norm ($1 \le p \le \infty$) *iff*

$$\|x_k - x\|_p = \begin{cases} \left[\sum_{i=1}^n |x_i^{(k)} - x_i|^p\right]^{1/p} & \text{if } 1 \le p < \infty, \\ \max_{1\le i\le n} |x_i^{(k)} - x_i| & \text{if } p = \infty, \end{cases} \to 0,\ k \to \infty.$$

3. For $X := \mathbb{N}$, $\Sigma := \mathscr{P}(X)$, and the *counting measure* μ, a sequence of real- or complex-termed sequences $\left(x_k := (x_n^{(k)})_{n\in\mathbb{N}}\right)_{k\in\mathbb{N}}$ converges to a real- or complex-termed sequence $x := (x_n)_{n\in\mathbb{N}}$ in p-norm ($1 \le p \le \infty$) if
   (i) for $1 \le p < \infty$,

$$\sum_{n=1}^{\infty} |x_n|^p < \infty,\ \sum_{n=1}^{\infty} |x_n^{(k)}|^p < \infty,\ k \in \mathbb{N},$$

   and

$$\|x_n - x\|_p := \left[\sum_{n=1}^{\infty} |x_n^{(k)} - x_n|^p\right]^{1/p} \to 0,\ k \to \infty;$$

   (ii) for $p = \infty$,

$$\sup_{n\in\mathbb{N}} |x_n| < \infty,\ \sup_{n\in\mathbb{N}} |x_n^{(k)}| < \infty,\ k \in \mathbb{N},$$

   and

$$\|x_n - x\|_\infty := \sup_{n\in\mathbb{N}} |x_n^{(k)} - x_n| \to 0,\ k \to \infty.$$

Exercise 7.13. Verify.

Remarks 7.8.

- For a sequence $(x_k := (x_1^{(k)}, \dots, x_n^{(k)}))_{k\in\mathbb{N}}$ of ordered n-tuples ($n \in \mathbb{N}$), p-norm convergence ($1 \le p \le \infty$) to an ordered n-tuple $x := (x_1, \dots, x_n)$ is equivalent to *componentwise convergence*, i. e.,

$$\|x_k - x\|_p \to 0,\ k \to \infty \Leftrightarrow \forall\, i = 1, \dots, n :\ x_i^{(k)} \to x_i,\ k \to \infty.$$

- For a sequence $(x_k := (x_n^{(k)})_{n\in\mathbb{N}})_{k\in\mathbb{N}}$ of sequences, p-norm convergence ($1 \le p \le \infty$) to a sequence $x := (x_n)_{n\in\mathbb{N}}$ implies *termwise convergence*, i. e.,

$$\|x_k - x\|_p \to 0,\ k \to \infty \Rightarrow \forall\, n \in \mathbb{N} :\ x_n^{(k)} \to x_n,\ k \to \infty,$$

  but not vice versa.

Exercise 7.14. Verify (see Examples 1.7).

7.2.2 Convergence in ∞-Norm

The following statement illuminates the nature of convergence in ∞-norm.

Theorem 7.4 (Convergence in ∞-Norm). *Let (X, Σ, μ) be a measure space and f, f_n, $n \in \mathbb{N}$, be real- or complex-valued Σ-measurable functions on X. Then*

$$\|f_n - f\|_\infty \to 0,\ n \to \infty \Leftrightarrow \exists N \in \Sigma,\ \mu(N) = 0 :\ \sup_{x \in X \setminus N} |f_n(x) - f(x)| \to 0,\ n \to \infty,$$

i. e., convergence in ∞-norm on X is equivalent to the uniform convergence a. e. μ on X.

Proof. *"If"* part follows immediately from the fact that

$$\operatorname*{ess\,sup}_{x \in X} |f_n(x) - f(x)| = \inf\left\{ \sup_{x \in X \setminus N} |f_n(x) - f(x)| \,\middle|\, N \in \Sigma,\ \mu(N) = 0 \right\}. \tag{7.8}$$

(see Remarks 7.3).

Exercise 7.15. Fill in the details.

"Only if" part. Suppose that

$$\|f_n - f\|_\infty \to 0,\ n \to \infty.$$

Then, in view of (7.8),

$$\forall n \in \mathbb{N}\ \exists N_n \in \Sigma,\ \mu(N_n) = 0 :\ \sup_{x \in X \setminus N_n} |f_n(x) - f(x)| < \|f_n - f\|_\infty + 1/n.$$

Then

$$N := \bigcup_{n=1}^{\infty} N_n$$

is a *μ-null set.*

Exercise 7.16. Explain.

Furthermore, for each $n \in \mathbb{N}$, in view of the inclusion $N_n \subseteq N$, we have

$$\sup_{x \in X \setminus N} |f_n(x) - f(x)| \le \sup_{x \in X \setminus N_n} |f_n(x) - f(x)| < \|f_n - f\|_\infty + 1/n \to 0,\ n \to \infty,$$

which completes the proof. □

Remark 7.9. For a measure space (X, Σ, μ) such that $\emptyset$ is the only μ-null set, in particular for $(X, \mathscr{P}(X), \mu)$ with the *counting measure* μ (see Examples 3.2), a. e. μ on X

being merely everywhere on X (see Remarks 5.15), for any real- or complex-valued Σ-measurable function f on X,

$$\operatorname*{ess\,sup}_{x\in X} f(x) = \sup_{x\in X} f(x).$$

(see Remarks 7.4), and hence, convergence in ∞-norm on X is *uniform convergence on X*.

7.2.3 Uniqueness A. E. of Limit in p-Norm

The following statement is the natural analogue of the a. e. uniqueness of the limit function for convergence almost everywhere (Theorem 5.8) and convergence in measure (Theorem 5.11).

Theorem 7.5 (Uniqueness A. E. of Limit in p-Norm). *Let (X, Σ, μ) be a measure space, f, g, f_n, $n \in \mathbb{N}$, be real- or complex-valued Σ-measurable functions on X, and $1 \le p \le \infty$. If,*

$$\|f_n - f\|_p \to 0 \text{ and } \|f_n - g\|_p \to 0,\ n \to \infty,$$

then $f = g \pmod{\mu}$.

Exercise 7.17. Prove.

Hint. Use *Minkowski's inequality* (Theorem 7.3) to show that $|f(x) - g(x)| = 0$ a. e. μ on X.

7.2.4 Relationships Between Different Types of Convergence

Theorem 7.6 (Relationships Between Different Types of Convergence). *Let (X, Σ, μ) be a measure space, f, g, f_n, $n \in \mathbb{N}$, be real- or complex-valued Σ-measurable functions on X, and $1 \le p \le \infty$.*

(1) *For $1 \le p \le \infty$, if*

$$\|f_n - f\|_p \to 0,\ n \to \infty,$$

then $f_n \xrightarrow{\mu} f$, but not vice versa.

(2) *For $1 \le p < \infty$, convergence in p-norm and convergence a. e. μ are independent of each other.*
Convergence in ∞-norm implies convergence a. e. μ on X, but not vice versa.

Proof. Here, we prove part (1) only, proving part (2) being left as an exercise.

The implication

$$\|f_n - f\|_p \to 0,\ n \to \infty \Rightarrow f \xrightarrow{\mu} f$$

for $1 \le p < \infty$ immediately follows from *Markov's inequality* (the *Properties of Lebesgue Integral. II* (Theorem 6.3, part (7))). Indeed, for any $\varepsilon > 0$,

$$\mu(\{x \in A \mid |f_n(x) - f(x)| \ge \varepsilon\}) = \mu(\{x \in A \mid |f_n(x) - f(x)|^p \ge \varepsilon^p\})$$
$$\le \frac{1}{\varepsilon^p} \int_A |f_n(x) - f(x)|^p \, d\mu(x) = \frac{1}{\varepsilon^p} \|f_n - f\|^p \to 0,\ n \to \infty.$$

For $p = \infty$, by the *Convergence in* ∞*-Norm Theorem* (Theorem 7.4),

$$\|f_n - f\|_\infty \to 0,\ n \to \infty,$$

implies that

$$\exists N \in \Sigma \text{ with } \mu(N) = 0\ \forall \varepsilon > 0\ \exists K \in \mathbb{N}\ \forall n \ge K : \sup_{x \in X \setminus N} |f_n(x) - f(x)| < \varepsilon,$$

and hence, in view of the Σ-measurability of f and f_n, $n \in \mathbb{N}$,

$$\forall \varepsilon > 0\ \exists K \in \mathbb{N}\ \forall n \ge K : \Sigma \ni \{x \in X \mid |f_n(x) - f(x)| \ge \varepsilon\} \subseteq N.$$

The latter, by the *monotonicity* of μ (see Theorem 3.1), implies that

$$\forall \varepsilon > 0\ \exists K \in \mathbb{N}\ \forall n \ge K : \mu(\{x \in X \mid |f_n(x) - f(x)| \ge \varepsilon\}) = 0,$$

and hence,

$$f_n \xrightarrow{\mu} f.$$

For the Lebesgue measure space $(\mathbb{R}, \Sigma^*, \mu)$, the function sequence

$$f_n(x) := n\chi_{[0,1/n]}(x),\ n \in \mathbb{N}, x \in \mathbb{R},$$

converges to $f(x) := 0$, $x \in \mathbb{R}$, in the Lebesgue measure λ on $\mathbb{R}$ but not in p-norm for any $1 \le p \le \infty$. □

Exercise 7.18. Provide corresponding counterexamples and a proof for (2).

7.3 Fundamentals of Normed Vector Spaces

7.3.1 Definitions and Examples

Let $\mathbb{F}$ designate either $\mathbb{R}$ or $\mathbb{C}$.

Definition 7.5 (Normed Vector Space). A *normed vector space* over $\mathbb{F}$ is a *vector space* X over $\mathbb{F}$ equipped with a *norm*, i. e., a mapping

$$\|\cdot\| : X \to \mathbb{R}$$

subject to the following *norm axioms*:

1. $\|x\| \geq 0$, $x \in X$. *Nonnegativity*
2. $\|x\| = 0$ iff $x = 0$. *Separation*
3. $\|\lambda x\| = |\lambda| \|x\|$, $\lambda \in \mathbb{F}$, $x \in X$. *Absolute Homogeneity/Scalability*
4. $\|x + y\| \leq \|x\| + \|y\|$, $x, y \in X$. *Subadditivity/Triangle Inequality*

The space is said to be *real* if $\mathbb{F} = \mathbb{R}$ and *complex* if $\mathbb{F} = \mathbb{C}$.

Notation. $(X, \|\cdot\|)$.

See, e. g., [13].

Remarks 7.10.

- A function $\|\cdot\| : X \to \mathbb{R}$ satisfying the norm axioms of *absolute scalability* and *subadditivity* only, which immediately imply the following weaker version of the *separation axiom*:

  2w. $\|x\| = 0$ if $x = 0$,

  and hence, also the axiom of *nonnegativity*, is called a *seminorm* on X and $(X, \|\cdot\|)$ is called a *seminormed vector space* (see the examples to follow).
- A norm $\|\cdot\|$ on a vector space X generates a metric on X, called the *norm metric*, as follows:

  $$X \times X \ni (x, y) \mapsto \rho(x, y) := \|x - y\|, \tag{7.9}$$

  which turns X into a *metric space*, endows it with the *norm metric topology*, and brings to life all the relevant concepts: *openness, closedness, denseness, category, boundedness*, and *compactness* for sets,*continuity* for functions, *fundamentality* and *convergence* for sequences, *separability* and *completeness* for spaces.
- Due to the *axioms* of *subadditivity* and *absolute scalability*, the linear operations of *vector addition*

  $$X \times X \ni (x, y) \mapsto x + y \in X$$

  and *scalar multiplication*

  $$\mathbb{F} \times X \ni (\lambda, x) \mapsto \lambda x \in X$$

  are *jointly continuous*.
- The following immediate implication of *subadditivity*

  $$\big|\|x\| - \|y\|\big| \leq \|x - y\|, \quad x, y \in X, \tag{7.10}$$

  showing that the norm is *continuous* on X, holds.
  Observe that the inequality applies to *seminorms* as well.

Exercise 7.19. Verify.

Examples 7.5. The following are examples of normed vector spaces.

1. The spaces $\mathbb{R}$ and $\mathbb{C}$ relative to the *absolute-value norm*:

$$\mathbb{F} \ni x \mapsto |x|.$$

2. The (real or complex) space $l_p^{(n)}$ ($n \in \mathbb{N}, 1 \le p \le \infty$) (see Examples 1.6 and 7.3) with componentwise linear operations relative to *p-norm*

$$l_p^{(n)} \ni x := (x_1, \dots, x_n) \mapsto \|x\|_p = \begin{cases} \left[\sum_{k=1}^{n} |x_k|^p\right]^{1/p} & \text{if } 1 \le p < \infty, \\ \max_{1\le k\le n} |x_k| & \text{if } p = \infty. \end{cases}$$

Remarks 7.11.

- For $n = 1$, all these norms coincide with the *absolute-value norm* $|\cdot|$.
- For $n = 2, 3$ and $p = 2$, we have the usual *Euclidean norm* $\|\cdot\|_2$.
- $(\mathbb{C}, |\cdot|) = (\mathbb{R}^2, \|\cdot\|_2)$.

3. The (real or complex) space l_p ($1 \le p \le \infty$) (see Examples 1.6 and 7.3) with termwise linear operations relative to *p-norm*

$$l_p \ni x := (x_n)_{n\in\mathbb{N}} \mapsto \|x\|_p = \begin{cases} \left[\sum_{k=1}^{\infty} |x_k|^p\right]^{1/p} & \text{if } 1 \le p < \infty, \\ \sup_{k\in\mathbb{N}} |x_k| & \text{if } p = \infty. \end{cases}$$

4. The spaces of real- or complex-termed sequences

$$\begin{aligned} c_{00} &:= \{(x_n)_{n\in\mathbb{N}} \mid \exists N \in \mathbb{N} : x_n = 0, \ n \ge N\} \quad \textit{(eventually zero sequences)}, \\ c_0 &:= \left\{(x_n)_{n\in\mathbb{N}} \,\middle|\, \lim_{n\to\infty} x_n = 0\right\} \quad \textit{(vanishing sequences)}, \\ c &:= \left\{(x_n)_{n\in\mathbb{N}} \,\middle|\, \lim_{n\to\infty} x_n \in \mathbb{R} \text{ (or } \mathbb{C})\right\} \quad \textit{(convergent sequences)} \end{aligned}$$

with termwise linear operations relative to the *supremum norm*

$$x := (x_n)_{n\in\mathbb{N}} \mapsto \|x\|_\infty = \sup_{n\in\mathbb{N}} |x_n|.$$

Each of the spaces in the following chain of strict inclusions

$$c_{00} \subset c_0 \subset c \subset l_\infty$$

is called a *subspace* of larger one.

5. The space $C[a, b]$ of real- or complex-valued functions *continuous* on $[a, b]$ ($-\infty < a < b < \infty$) (see Examples 1.6) with pointwise linear operations relative to the *maximum norm*

$$C[a, b] \ni f \mapsto \|f\|_\infty := \max_{a\le x\le b} |f(x)|.$$

6. The space $C[a,b]$ $(-\infty < a < b < \infty)$ relative to the *integral norm*

$$C[a,b] \ni f \mapsto \|f\|_1 := \int_a^b |f(x)|\,dx.$$

Exercise 7.20. Verify.

Hint. For examples 2 and 3, apply *Minkowski's Inequality for n-Tuples* (Corollary 7.3) and *Minkowski's Inequality for Sequences* (Corollary 7.4), respectively.

Remark 7.12. *Minkowski's Inequality for Sequences* (Corollary 7.4) implies, in particular, that l_p $(1 \le p \le \infty)$ is a *vector space* (see Remark 7.6).

Definition 7.6 (Banach Space). A *Banach space* is a normed vector space $(X, \|\cdot\|)$ *complete* relative to the norm metric ρ defined by (7.9), i. e., such that every *Cauchy sequence* (or *fundamental sequence*)

$$\{x_n\}_{n=1}^{\infty} \subseteq X : \rho(x_n, x_m) = \|x_n - x_m\| \to 0,\ n, m \to \infty,$$

converges to an element $x \in X$:

$$\lim_{n\to\infty} x_n = x \Leftrightarrow \rho(x_n, x) = \|x_n - x\| \to 0,\ n \to \infty$$

(see Definition 1.10).[5]

Examples 7.6.

1. The spaces $\mathbb{R}$ and $\mathbb{C}$ are Banach spaces relative to the *absolute-value norm*, which generates the usual metric (see Examples 7.5 and 1.9).
2. As we see in Section 7.4.4, the (real or complex) space $l_p^{(n)}$ $(n \in \mathbb{N},\ 1 \le p \le \infty)$ is a Banach space relative to *p-norm*, which generates *p*-metric (see Examples 7.5 and 1.6).
3. As we see in Section 7.4.4, the (real or complex) space l_p $(1 \le p \le \infty)$ is a Banach space relative to *p-norm*, which generates *p*-metric (see Examples 7.5 and 1.6).
4. The (real or complex) spaces $(c, \|\cdot\|_\infty)$, $(c_0, \|\cdot\|_\infty)$ are Banach spaces and the space $(c_{00}, \|\cdot\|_\infty)$ is an *incomplete* normed vector space (see, e. g., [14]).
5. The (real or complex) space $C[a,b]$ $(-\infty < a < b < \infty)$ is a Banach space relative to the *maximum norm* $\|\cdot\|_\infty$, which generates the maximum metric ρ_∞ (see Examples 7.5 and 1.9) and is an *incomplete* normed vector space relative to the *integral norm* $\|\cdot\|_1$ generating the integral metric ρ_1 (see Examples 7.5 and 1.6).

5 Stefan Banach (1892–1945).

7.3.2 Incompleteness of $R[a, b]$

On the vector space $R[a,b]$ $(-\infty < a < b < \infty)$ of real- or complex-valued functions *Riemann integrable* on $[a,b]$ with pointwise linear operations, the mapping

$$R[a,b] \ni f \mapsto \|f\|_1 := \int_a^b |f(x)|\,dx \tag{7.11}$$

is *not* a norm but a *seminorm.*

Exercise 7.21. Verify norm axioms 1, 3, and 4.

Indeed, by the *Characterization of Riemann Integrability* (Theorem 6.10) and the *Properties of Lebesgue Integral. II* (Theorem 6.3, part (9)), for any $f \in R[a,b]$, $f \in L([a,b],\lambda)$ and

$$\int_{[a,b]} |f(x)|\,d\lambda(x) = \int_a^b |f(x)|\,dx = \|f\|_1 = 0 \Leftrightarrow f(x) = 0 \text{ a. e. } \lambda \text{ on } [a,b],$$

e. g., for the nonzero function

$$f(x) := \begin{cases} 0, & 0 \le x < 1, \\ 1, & x = 1 \end{cases} \in R[0,1],$$

$$\|f\|_1 = \int_0^1 f(x)\,dx = 0.$$

For the *separation axiom* to hold, we are not to distinguish Riemann integrable functions equal a. e. λ on $[a,b]$, i. e., *equivalent relative to the Lebesgue measure* λ *on* $[a,b]$ (see Definition 5.9).

This *equivalence relation* (see Remarks 5.16) partitions the vector space $R[a,b]$ into *equivalence classes*, each class f being represented by a function $f(\cdot) \in R[a,b]$, i. e.,

$$f := \{g(\cdot) \in R[a,b] \,|\, g(\cdot) \text{ is equivalent to } f(\cdot) \text{ relative to } \lambda \text{ on } [a,b]\}$$

Remark 7.13. Henceforth, we use notation f and $f(\cdot)$ to distinguish between an *equivalence class* f and its *representative function* $f(\cdot)$.

The set of all such equivalence classes is a *vector space* relative to the linear operations:

$$f + g \text{ is the class represented by } f(\cdot) + g(\cdot)$$

and, for any scalar λ,

$$\lambda f \text{ is the class represented by } \lambda f(\cdot),$$

with the *zero element* being the equivalence class represented by $f(x) = 0$, $x \in [a,b]$, i. e., the *zero subspace*

$$0 := \{g(\cdot) \in R[a,b] \,|\, g(x) = 0 \text{ a. e. } \lambda \text{ on } [a,b]\}.$$

The obtained vector space, for which we adopt the same notation $R[a,b]$, i. e., the *quotient space* of the original space $R[a,b]$ *modulo the zero subspace* (see, e. g., [14]), is a normed vector space relative to the *integral norm* defined by (7.11).

Exercise 7.22. Verify.

Let us prove the *incompleteness* of the quotient space $(R[a,b], \|\cdot\|_1)$.

Proposition 7.2 (Incompleteness of $(R[a,b], \|\cdot\|_1)$). *The quotient space $(R[a,b], \|\cdot\|_1)$ is an incomplete normed vector space.*

Proof. Without loss of generality, suppose that $[a,b] = [0,1]$, consider the *fundamental sequence* of equivalence classes $(f_n)_{n\in\mathbb{N}}$ in $(R[0,1], \|\cdot\|_1)$ represented by the functions

$$f_n(x) := \begin{cases} n^{1/2}, & 0 \le x \le 1/n, \\ x^{-1/2}, & 1/n < x \le 1, \end{cases} \quad n \in \mathbb{N},$$

and let

$$f(x) := \begin{cases} 0, & x = 0, \\ x^{-1/2}, & 0 < x \le 1. \end{cases}$$

For each $x \in (0,1]$,

$$0 \le f_n(x) \le f_{n+1}(x) \le f(x),\ n \in \mathbb{N},$$

and

$$f_n(x) \to f(x),\ n \to \infty.$$

Exercise 7.23. Verify.

In view of the fact that

$$\int_0^1 f(x)\,dx = \int_0^1 x^{-1/2}\,dx = 2x^{1/2}\Big|_{0-}^{1} = 2 < \infty,$$

by the *Improper Integral of Type 2 Theorem* (Theorem 6.12), $f \in L([0,1],\lambda)$, and hence, by the *Properties of Lebesgue Integral. I* (Theorem 6.1, part (3)) and the *Properties of Lebesgue Integral. II* (Theorem 6.3, part (3)), $f_n \in L([0,1],\lambda)$.

By the *Improper Integral of Type 2 Theorem*,

$$\int_0^1 [f(x) - f_n(x)]\,dx = \int_{[0,1]} [f(x) - f_n(x)]\,d\lambda(x).$$

As is easily verified, the sequence $(f - f_n)_{n\in\mathbb{N}}$ satisfies the conditions of the *Lebesgue Dominated Convergence Theorem* (Theorem 6.7).

Exercise 7.24. Verify.

Hence,

$$\lim_{n\to\infty} \int_{[0,1]} [f(x) - f_n(x)]\,d\lambda(x) = \int_{[0,1]} 0\,d\lambda(x) = 0. \tag{7.12}$$

Assume that there exists an equivalence class $g \in R[0,1]$ such that

$$f_n \to g,\ n \to \infty,\ \text{in } (R[0,1], \|\cdot\|_1),$$

i. e.,

$$\int_0^1 |f_n(x) - g(x)|\,dx =: \|f_n - g\|_1 \to 0,\ n \to \infty. \tag{7.13}$$

By the *Characterization of Riemann Integrability* (Theorem 6.10),

$$g(\cdot) \in L([0,1],\lambda) \text{ and } \quad f_n(\cdot) - g(\cdot) \in L([0,1],\lambda),\ n \in \mathbb{N},$$

and, for each $n \in \mathbb{N}$,

$$\int_0^1 |f_n(x) - g(x)|\,dx = \int_{[0,1]} |f_n(x) - g(x)|\,d\lambda(x)$$

Hence, by the *Properties of Lebesgue Integral. I* (Theorem 6.1, part (3)) and the *Linearity of Lebesgue Integral Theorem* (Theorem 6.5), for each $n \in \mathbb{N}$,

$$\int_{[0,1]} |f(x) - g(x)|\,d\lambda(x) \le \int_{[0,1]} |f(x) - f_n(x)|\,d\lambda(x) + \int_{[0,1]} |f_n(x) - g(x)|\,d\lambda(x).$$

Passing to the limit as $n \to \infty$ in view of (7.12) and (7.13), we arrive at

$$\int_{[0,1]} |f(x) - g(x)|\,d\lambda(x) = 0,$$

which, by the *Properties of Lebesgue Integral. II* (Theorem 6.3, part (9)), implies that

$$|f(x) - g(x)| = 0 \text{ a. e. } \lambda \text{ on } [0,1],$$

and hence,

$$f(x) = g(x) \text{ a. e. } \lambda \text{ on } [0,1],$$

where we infer that the Riemann integrable on $[0,1]$ representative function $g(\cdot)$ is *unbounded* on $[0,1]$.

Exercise 7.25. Explain.

Hint. Show that, for each, $n \in \mathbb{N}$

$$\forall n \in \mathbb{N}\, \exists x \in (0,1/n] : f(x) = g(x).$$

The unboundedness of the function $g(\cdot)$ on $[0,1]$ *contradicts* its Riemann integrability on $[0,1]$ (see Section 6.10.1). Thus, we conclude that the sequence $(f_n)_{n\in\mathbb{N}}$, although fundamental, *does not* converge in the quotient space $(R[0,1], \|\cdot\|_1)$, and hence, the latter is *incomplete*. □

Remark 7.14. As becomes apparent from the following section, the incompleteness of the quotient space $(R[a,b], \|\cdot\|_1)$ is yet another deficiency of the Riemann integral compared to its Lebesgue counterpart.

7.4 L_p Spaces

Here, we are to study L_p *spaces*, which are the answer to the lack of completeness of the spaces based on the Riemann and whose construct is based on the idea of the quotient space discussed in the prior section.

7.4.1 Definition

For a measure space (X,Σ,μ) and $1 \le p < \infty$, let

$$\mathcal{L}_p(X,\Sigma,\mu) := \{f : X \to \mathbb{R} \text{ (or } \mathbb{C}) \,|\, f \text{ is } \Sigma\text{-measurable and } |f|^p \in L(X,\mu)\}$$

(*p-integrable on X functions*), in particular $\mathcal{L}_1(X,\Sigma,\mu) = L(X,\mu)$, and

$$\mathcal{L}_\infty(X,\Sigma,\mu) := \{f : X \to \mathbb{R} \text{ (or } \mathbb{C}) \,|\, f \text{ is } \Sigma\text{-measurable and } \mu\text{-essentially bounded on } X\}$$

(see Definition 7.2).

Due to *Minkowski's inequality* (Theorem 7.3), for each $1 \le p \le \infty$, $\mathcal{L}_p(X,\Sigma,\mu)$ is a *vector space* (real or complex, respectively) (see Remark 7.6) and, as is easily verified, on $\mathcal{L}_p(X,\Sigma,\mu)$, p-norm satisfies all the *norm axioms*, except *separation*:

$$\forall f \in \mathcal{L}_p(X,\Sigma,\mu) : \|f\|_p = 0 \Leftrightarrow f(x) = 0 \text{ a. e. } \mu \text{ on } X.$$

(see Exercise 7.5).

Exercise 7.26. Verify norm axioms 1, 3, and 4.

Hint. *Minkowski's inequality* (Theorem 7.3) implies *subadditivity*.

Thus, for the *separation axiom* to hold, we are not to distinguish functions in $\mathcal{L}_p(X,\Sigma,\mu)$ equal a. e. μ on X, i. e., *equivalent relative to measure μ on X* (see Definition 5.9).

Definition 7.7 (L_p **Spaces).** Let (X,Σ,μ) be a measure space and $1 \le p \le \infty$.

The equivalence relation of *equality a. e. μ on X* partitions the vector space $\mathcal{L}_p(X,\Sigma,\mu)$ into equivalence classes, each class f being represented by a function $f(\cdot) \in \mathcal{L}_p(X,\Sigma,\mu)$, i. e.,

$$f := \{g(\cdot) \in \mathcal{L}_p(X,\Sigma,\mu) \,|\, g(\cdot) \text{ is equivalent to } f(\cdot) \text{ relative to } \mu \text{ on } X\}.$$

The set $L_p(X,\Sigma,\mu)$ of all such equivalence classes is a *vector space* relative to the linear operations:

$$f + g \text{ is the class represented by } f(\cdot) + g(\cdot)$$

and, for any scalar λ,

$$\lambda f \text{ is the class represented by } \lambda f(\cdot),$$

the *zero element* being the class represented by $f(x) = 0$, $x \in X$, i. e., the *zero subspace*

$$0 := \{g(\cdot) \in L_p(X,\Sigma,\mu) \,|\, g(x) = 0 \text{ a. e. } \mu \text{ on } X\}.$$

The vector space $L_p(X,\Sigma,\mu)$, i. e., the *quotient space* of the original space $\mathcal{L}_p(X,\Sigma,\mu)$ *modulo the zero subspace*, is a normed vector space relative to p-norm

$$L_p(X,\Sigma,\mu) \ni f \mapsto \|f\|_p := \|f(\cdot)\|_p.$$

Exercise 7.27.

(a) Verify that the linear operations are well defined, i. e., are independent of the choice of representative function.

(b) Verify that p-norm is well-defined, i. e., is independent of the choice of a representative function and satisfies all the norm axioms (see Definition 7.5).

Remarks 7.15.

- For a measure space (X,Σ,μ) such that $\emptyset$ is the only μ-null set, in particular for $(X,\mathscr{P}(X),\mu)$ with the *counting measure* μ (see Examples 3.2), the equivalence of functions relative to measure μ on X, i. e., their equality a. e. μ on X, is merely equality everywhere on X (see Remarks 5.16), which implies that the equivalence classes are *singletons*, and hence, in this case, the spaces $L_p(X,\Sigma,\mu)$ and $\mathcal{L}_p(X,\Sigma,\mu)$ are identical.
- The convergence of a sequence of equivalence classes in $L_p(X,\Sigma,\mu)$ is the convergence in p-norm of an arbitrary sequence of their representatives to a representative of the limit equivalence class.

7.4.2 Important Particular Cases

7.4.2.1 $L_p(a,b)$ Spaces

For $X := I$, where an interval I consists of an open interval (a,b) $(-\infty \le a < b \le \infty)$ and, possibly, one or both of its finite endpoints, a or b, if any, $\Sigma := \Sigma^*(I)$ is the σ-algebra of the Lebesgue measurable subsets of I, and $\mu := \lambda$ is the Lebesgue measure, the normed vector space $L_p(X,\Sigma,\mu)$ $(1 \le p \le \infty)$ is denoted by $L_p(a,b)$.

Remarks 7.16.

- The above notation is void of indicating whether any of the endpoints, if finite, is included, this fact being superfluous since singletons are Lebesgue measure-null sets.
- By the *Lebesgue Measurable versus Borel Measurable Proposition* (Proposition 5.11) (Section 5.12, Problem 14), without loss of generality, we can regard each equivalence class $f \in L_p(a,b)$ to be represented by a p-integrable on X *Borel measurable* function $f(\cdot)$.

Examples 7.7.

1. For $f(x) = x^{-1}$, $x \in [1,\infty)$, $f \in L_p(1,\infty)$ for $1 < p \le \infty$, but $f \notin L_1(1,\infty)$.
2. For

$$f(x) := \begin{cases} x^{-1/2}, & 0 < x < 1, \\ x^{-2}, & x \ge 1, \end{cases}$$

$f \in L_p(0,\infty)$ for $1 \le p < 2$, but $f \notin L_p(0,\infty)$ for $2 \le p \le \infty$.

Exercise 7.28. Show that, if a class $f \in L_p(a,b)$ $(1 \le p \le \infty, -\infty \le a < b \le \infty)$ is represented by a function $f(\cdot) \in C(a,b)$, then such a continuous representation of f is unique.

7.4.2.2 $l_p^{(n)}$ Spaces

For $X := \{1,\dots,n\}$ $(n \in \mathbb{N})$, $\Sigma := \mathscr{P}(X)$, and the *counting measure* μ,

$$L_p(X,\Sigma,\mu) = l_p^{(n)} \quad (1 \le p \le \infty)$$

(see Examples 7.3 and 7.5).

7.4.2.3 l_p Spaces

For $X := \mathbb{N}$, $\Sigma := \mathscr{P}(X)$, and the *counting measure* μ, $L_p(X,\Sigma,\mu)$,

$$L_p(X,\Sigma,\mu) = l_p \quad (1 \le p \le \infty)$$

(see Examples 7.3 and 7.5).

Examples 7.8.
1. $(1/n)_{n\in\mathbb{N}} \in l_p$ for $1 < p \leq \infty$, but $(1/n)_{n\in\mathbb{N}} \notin l_1$.
2. $((-1)^n)_{n\in\mathbb{N}} \in l_\infty$, but $((-1)^n)_{n\in\mathbb{N}} \notin l_p$ for $1 \leq p < \infty$.

7.4.3 Hölder's and Minkowski's Inequalities in L_p Spaces

In the L_p-space setting, *Hölder's* and *Minkowski's inequalities* for functions and sequences acquire the following simpler formulations.

Theorem 7.7 (Hölder's Inequality for L_p Spaces). *Let (X, Σ, μ) be a measure space and $1 \leq p, q \leq \infty$ be conjugate indices. Then, for any $f \in L_p(X, \Sigma, \mu)$ and $g \in L_q(X, \Sigma, \mu)$, $fg \in L_1(X, \Sigma, \mu)$ and*

$$\|fg\|_1 \leq \|f\|_p \|g\|_q.$$

Corollary 7.5 (Hölder's Inequality for Sequences). *Let $1 \leq p, q \leq \infty$ be conjugate indices. Then, for any real- or complex-termed sequences $x := (x_n)_{n\in\mathbb{N}} \in l_p$ and $y := (y_n)_{n\in\mathbb{N}} \in l_q$, $xy := (x_n y_n)_{n\in\mathbb{N}} \in l_1$ and*

$$\sum_{n=1}^{\infty} |x_n y_n| = \|xy\|_1 \leq \|x\|_p \|y\|_q.$$

Theorem 7.8 (Minkowski's Inequality for L_p Spaces). *Let (X, Σ, μ) be a measure space and $1 \leq p \leq \infty$. Then, for any $f, g \in L_p(X, \Sigma, \mu)$, $f + g \in L_p(X, \Sigma, \mu)$ and*

$$\|f + g\|_p \leq \|f\|_p + \|g\|_p.$$

Corollary 7.6 (Minkowski's Inequality for Sequences). *Let $1 \leq p \leq \infty$. Then, for any real- or complex-termed sequences $x := (x_n)_{n\in\mathbb{N}}, y := (y_n)_{n\in\mathbb{N}} \in l_p$, $x + y := (x_n + y_n)_{n\in\mathbb{N}} \in l_p$ and*

$$\|x + y\|_p \leq \|x\|_p + \|y\|_p.$$

7.4.4 Completeness of L_p Spaces

Now, we are ready to prove completeness for L_p spaces (cf. the *Incompleteness of* $(R[a, b], \|\cdot\|_1)$ *Proposition* (Proposition 7.2)).

Theorem 7.9 (Completeness of L_p Spaces). *Let (X, Σ, μ) be a measure space. Then, for any $1 \leq p \leq \infty$, $L_p(X, \Sigma, \mu)$ is a Banach space.*

Proof. Suppose that $1 \leq p < \infty$ and let $(f_n)_{n\in\mathbb{N}}$ be an arbitrary fundamental sequence in $L_p(X, \Sigma, \mu)$. Then there exists a sequence $(n(k))_{k\in\mathbb{N}} \subset \mathbb{N}$ such that

$$n(k) < n(k + 1), \ k \in \mathbb{N},$$

and

$$\forall k \in \mathbb{N}\ \exists n(k) \in \mathbb{N}\ \forall m, n \geq n(k) : \|f_n - f_m\|_p < 2^{-k}.$$

Exercise 7.29. Explain.

Since, in particular,

$$\forall k \in \mathbb{N} : \|f_{n(k+1)} - f_{n(k)}\|_p < 2^{-k}, \tag{7.14}$$

by the *Comparison Test*, the telescoping series

$$\sum_{k=0}^{\infty} (f_{n(k+1)} - f_{n(k)})$$

with $f_{n(0)}(x) := 0$, $x \in X$, converges *absolutely* in $L_p(X, \Sigma, \mu)$, i. e.,

$$\sum_{k=0}^{\infty} \left\|f_{n(k+1)} - f_{n(k)}\right\|_p < \infty.$$

Exercise 7.30. Verify (cf. Section 7.5, Problem 9).

Consider the sequence

$$g_m(x) := \sum_{k=0}^{m-1} \left|f_{n(k+1)}(x) - f_{n(k)}(x)\right|, \quad m \in \mathbb{N}, x \in X.$$

The functions $g_m : X \to [0, \infty)$, $m \in \mathbb{N}$, are Σ-measurable by the *properties of measurable functions* (Theorems 5.2 and 5.3) and, since $L_p(X, \Sigma, \mu)$ is a vector space,

$$g_m = \sum_{k=0}^{m-1} \left|f_{n(k+1)} - f_{n(k)}\right| \in L_p(X, \Sigma, \mu), \quad m \in \mathbb{N},$$

where g_m is the equivalence class represented by the function $g_m(\cdot) \in \mathcal{L}_p(X, \Sigma, \mu)$, $m \in \mathbb{N}$, (see Definition 7.7).

Further, by the *subadditivity* of p-norm and in view of (7.14),

$$\begin{aligned}\|g_m\|_p &= \sum_{k=0}^{m-1} \left\|f_{n(k+1)} - f_{n(k)}\right\|_p \leq \|f_{n(1)}\|_p + \sum_{k=1}^{\infty} 2^{-k} \\ &= \|f_{n(1)}\|_p + 1.\end{aligned} \tag{7.15}$$

Since

$$g_m(x) \leq g_{m+1}(x), \quad m \in \mathbb{N}, x \in X,$$

let

$$g(x) := \lim_{m\to\infty} g_m(x) = \sum_{k=0}^{\infty} \left|f_{n(k+1)}(x) - f_{n(k)}(x)\right|, \ x \in X.$$

By the *Sequences of Measurable Functions* (Theorem 5.4), the function $g : X \to [0, \infty]$ is Σ-measurable and, by *Fatou's Lemma* (Theorem 6.6), in view of (7.15),

$$\int_X |g(x)|^p \, d\mu(x) \le \varliminf_{m\to\infty} \int_X |g_m(x)|^p \, d\mu(x) = \varliminf_{m\to\infty} \|g_m\|_p^p \le [\|f_{n(1)}\|_p + 1]^p .$$

Hence, by the *Properties of Lebesgue Integral. II* (Theorem 6.3, part (8)),

$$g(x) = \sum_{k=0}^{\infty} |f_{n(k+1)}(x) - f_{n(k)}(x)| < \infty \text{ a. e. } \mu \text{ on } X.$$

Therefore, by the *Sequences of Measurable Functions Theorem* (Theorem 5.4), the real- or complex-valued function $f(\cdot)$ defined on X as follows:

$$f(x) := \begin{cases} \sum_{k=0}^{\infty} (f_{n(k+1)}(x) - f_{n(k)}(x)) & \text{if } g(x) < \infty, \\ 0 & \text{otherwise,} \end{cases}$$

is Σ-measurable.

Exercise 7.31. Explain.

Since, for each $x \in X$ such that $g(x) < \infty$,

$$f(x) = \lim_{m\to\infty} \sum_{k=0}^{m-1} (f_{n(k+1)}(x) - f_{n(k)}(x)) = \lim_{m\to\infty} f_{n(m)}(x),$$

by *Fatou's Lemma* (Theorem 6.6), the *Properties of Lebesgue Integral. II* (Theorem 6.3, part (3)), and in view of (7.14), for each $k \in \mathbb{N}$,

$$\begin{aligned} \int_X |f_{n(k)}(x) - f(x)|^p \, d\mu(x) &\le \varliminf_{m\to\infty} \int_X |f_{n(k)}(x) - f_{n(m)}(x)|^p \, d\mu(x) \\ &= \varliminf_{m\to\infty} \|f_{n(k)} - f_{n(m)}\|_p^p \le 2^{-kp}, \end{aligned}$$

where we conclude that $f \in L_p(X, \Sigma, \mu)$, where f is the equivalence class represented by the function $f(\cdot) \in \mathcal{L}_p(X, \Sigma, \mu)$, (see Definition 7.7), and

$$\|f_{n(k)} - f\|_p \le 2^{-k} \to 0, \ k \to \infty,$$

which implies that

$$f_{n(k)} \to f, \ k \to \infty,$$

in $L_p(X, \Sigma, \mu)$.

Exercise 7.32. Explain.

Thus, an arbitrary fundamental sequence $(f_n)_{n\in\mathbb{N}}$ in $L_p(X,\Sigma,\mu)$ contains a subsequence $(f_{n(k)})_{k\in\mathbb{N}}$ convergent to an equivalence class f in $L_p(X,\Sigma,\mu)$, which, by the *Fundamental Sequence with Convergent Subsequence Proposition* (Proposition 7.7) (Section 7.5, Problem 8), implies that

$$f_n \to f,\ n \to \infty,$$

in $L_p(X,\Sigma,\mu)$ and concludes the proof of the completeness of $L_p(X,\Sigma,\mu)$ for $1 \le p < \infty$.

Now suppose that $p = \infty$ and let $(f_n)_{n\in\mathbb{N}}$ be an arbitrary fundamental sequence in $L_\infty(X,\Sigma,\mu)$, i. e.,

$$\|f_n - f_m\|_\infty \to 0,\ m,n \to \infty. \tag{7.16}$$

By the *Boundedness of Fundamental Sequences Theorem* (Theorem 1.8), $(f_n)_{n\in\mathbb{N}}$ is *bounded*, i. e.,

$$\exists\, c > 0 : \ \|f_n\|_\infty < c,\ n \in \mathbb{N},$$

and hence, by the definition of essential supremum (see (7.3)),

$$\forall\, n \in \mathbb{N}\ \exists\, N_n \in \Sigma \text{ with } \mu(N_n) = 0 : \ \sup_{x\in X\setminus N_n} |f_n(x)| < c.$$

Then

$$N := \bigcup_{n=1}^{\infty} N_n$$

is a *μ-null set.*

Exercise 7.33. Explain.

Furthermore, for any $n \in \mathbb{N}$, in view of the inclusion $N_n \subseteq N$, we have

$$\sup_{x\in X\setminus N} |f_n(x)| \le \sup_{x\in X\setminus N_n} |f_n(x)| < c. \tag{7.17}$$

By the definition of essential supremum (see (7.3)), (7.16) implies that

$$\forall\, m,n \in \mathbb{N}\ \exists\, M_{m,n} \in \Sigma,\ \mu(M_{m,n}) = 0 :$$
$$\sup_{x\in X\setminus M_{m,n}} |f_n(x) - f_m(x)| < \|f_n - f_m\|_\infty + 1/(mn).$$

Then

$$M := \bigcup_{m,n=1}^{\infty} M_{m,n}$$

is a *μ-null set.*

Exercise 7.34. Explain.

Furthermore, for any $m, n \in \mathbb{N}$, in view of the inclusion $M_{m,n} \subseteq M$, we have

$$\sup_{x \in X \setminus M} |f_n(x) - f_m(x)| \le \sup_{x \in X \setminus M_{m,n}} |f_n(x) - f_m(x)|$$
$$< \|f_n - f_m\|_\infty + 1/(mn) \to 0, \; m, n \to \infty,$$

which implies that the sequence of representative functions $(f_n(\cdot))_{n \in \mathbb{N}}$ is *uniformly fundamental* on $X \setminus M$, i. e.,

$$\forall \varepsilon > 0 \; \exists K \in \mathbb{N} \; \forall m, n \ge K : \; \sup_{x \in X \setminus M} |f_n(x) - f_m(x)| < \varepsilon. \tag{7.18}$$

Hence, for each $x \in X \setminus M$, there exists a *finite* limit $\lim_{n\to\infty} f_n(x) \in \mathbb{R}$, where $\mathbb{F} = \mathbb{R}$ or $\mathbb{F} = \mathbb{C}$.

Therefore, by the *Sequences of Measurable Functions Theorem* (Theorem 5.4), the real- or complex-valued function $f(\cdot)$ defined on X as follows:

$$f(x) := \begin{cases} \lim_{n\to\infty} f_n(x), & x \in X \setminus M, \\ 0, & x \in M, \end{cases}$$

is Σ-measurable.

Exercise 7.35. Explain.

Further, by the *subadditivity* of μ (see Theorem 3.1), $M \cup N$ is a *μ-null set* and, for each $x \in X \setminus (M \cup N)$, in view of (7.17),

$$|f(x)| = \lim_{n\to\infty} |f_n(x)| \le c,$$

which implies that the function $f(\cdot)$ is *μ-essentially bounded on X*, i. e., $f(\cdot)$ represents an equivalence class $f \in L_\infty(X, \Sigma, \mu)$.

Exercise 7.36. Explain.

In (7.18), fixing an arbitrary $m \ge K$ in and passing to the limit as $n \to \infty$, we arrive at

$$\forall \varepsilon > 0 \; \exists K \in \mathbb{N} \; \forall m \ge K : \; \sup_{x \in X \setminus M} |f(x) - f_m(x)| \le \varepsilon,$$

which, in view of $\mu(M) = 0$, by the *Convergence in ∞-Norm Theorem* (Theorem 7.4), implies that the sequence of representative functions $(f_n(\cdot))_{n \in \mathbb{N}}$ converges in ∞-norm to the function $f(\cdot)$, and hence,

$$f_n \to f, \; n \to \infty,$$

in $L_\infty(X, \Sigma, \mu)$.

Thus, we conclude that the space $L_\infty(X, \Sigma, \mu)$ is *complete* and so is the proof. □

Directly follows the subsequent corollary for the important particular cases discussed in Section 7.4.2.

Corollary 7.7 (Completeness of $L_p(a,b)$, $l_p^{(n)}$, and l_p). *For any $1 \le p \le \infty$, $L_p(X,\Sigma,\mu)$, $L_p(a,b)$ $(-\infty \le a < b \le \infty)$, $l_p^{(n)}$ $(n \in \mathbb{N})$, and l_p are Banach spaces.*

7.4.5 Approximation in L_p Spaces

This section contains a number of useful statements on approximation in L_p spaces $(1 \le p < \infty)$.

7.4.5.1 Approximation by Simple Functions

Theorem 7.10 (Approximation by Simple Functions). *Let (X,Σ,μ) be a measure space and $1 \le p < \infty$. Then the set of equivalence classes represented by p-integrable simple functions is dense in $L_p(X,\Sigma,\mu)$, i. e., for an arbitrary equivalence class $f \in L_p(X,\Sigma,\mu)$ and any $\varepsilon > 0$, there exists an equivalence class $s \in L_p(X,\Sigma,\mu)$ represented by a simple function $s(\cdot) \in \mathcal{L}_p(X,\Sigma,\mu)$ such that*

$$\|f - s\|_p < \varepsilon.$$

Proof. Let $f \in L_p(X,\Sigma,\mu)$ represented by a function $f(\cdot) \in \mathcal{L}_p(X,\Sigma,\mu)$ be arbitrary. We are to prove that, there exists a sequence $(s_n)_{n\in\mathbb{N}}$ represented by simple functions $s_n(\cdot) \in \mathcal{L}_p(X,\Sigma,\mu)$, $n \in \mathbb{N}$, such that

$$\|f - s_n\|_p \to 0, \ n \to \infty.$$

Since the complex case directly follows from the real case by separating the real and imaginary parts, without loss of generality, we can regard the representative function $f(\cdot)$ to be *real-valued*. Then, by the *Positive and Antinegative Parts of a Function Corollary* (Corollary 5.4),

$$f(x) = f_+(x) - f_-(x), \ x \in X,$$

where $f_+(\cdot)$ and $f_-(\cdot)$ are nonnegative Σ-measurable functions.

Since $f(\cdot)$ *p-integrable*, so are the functions $f_+(\cdot)$ and $f_-(\cdot)$.

Exercise 7.37. Explain.

Hence, by the *subadditivity* of p-norm, without loss of generality, we can regard the representative function $f(\cdot)$ to be *nonnegative*.

Exercise 7.38. Explain.

By the *Characterization of Measurability* (Theorem 5.5), there exists an increasing sequence $(s_n(\cdot))_{n\in\mathbb{N}}$:

$$s_n(x) \le s_{n+1}(x) \le f(x),\ n \in \mathbb{N}, x \in X,$$

of nonnegative Σ-measurable simple functions on X converging to f pointwise on X:

$$\forall x \in X : s_n(x) \to f(x),\ n \to \infty.$$

Therefore,

$$|s_n(x)|^p \le |f(x)|^p,\ n \in \mathbb{N}, x \in X,$$

which, by the *Properties of Lebesgue Integral. I* (Theorem 6.1, part (3)), implies that $s_n(\cdot) \in \mathcal{L}_p(X,\Sigma,\mu)$, $n \in \mathbb{N}$, i. e., $s_n \in L_p(X,\Sigma,\mu)$, $n \in \mathbb{N}$.

Further, by the *Lebesgue Dominated Convergence Theorem* (Theorem 6.7), applied to the function sequence $(|f(\cdot) - s_n(\cdot)|^p)_{n\in\mathbb{N}}$

$$\|f - s_n\|_p^p = \int_X |f(x) - s_n(x)|^p\, d\mu(x) \to 0,\ n \to \infty.$$

Exercise 7.39. Verify that the *Lebesgue Dominated Convergence Theorem* (Theorem 6.7) applies.

Hint. Use the *important inequality* (Proposition 7.1).

This completes the proof. □

Proposition 7.3 (Approximation by Special Simple Functions). *Let (X,Σ,μ), where μ is the unique extension of a σ-finite measure on a semiring $\mathscr{S}$ on X to the generated σ-algebra $\Sigma := \sigma a(\mathscr{S})$ and $1 \le p < \infty$. Then, for an arbitrary equivalence class $f \in L_p(X,\Sigma,\mu)$ and any $\varepsilon > 0$, there exists an equivalence class $s \in L_p(X,\Sigma,\mu)$ represented by a simple function $s(\cdot) \in \mathcal{L}_p(X,\Sigma,\mu)$ such that*

$$s(x) = \sum_{k=1}^{n} a_k \chi_{A_k}(x),\ x \in X,$$

with some $n \in \mathbb{N}$, $A_k \in \mathscr{S}$ and $a_k \in \mathbb{F}$ ($\mathbb{F} = \mathbb{R}$ or $\mathbb{F} = \mathbb{C}$), $k = 1,\dots,n$, and

$$\|f - s\|_p < \varepsilon.$$

Proof. Observe that, by the *Measure Extension from a Semiring Theorem* (Theorem 4.1), there is a unique σ-finite measure extension μ of the σ-finite measure μ on the semiring $\mathscr{S}$ to the generated ring $\mathscr{R} := r(\mathscr{S})$.

Exercise 7.40. Explain why the extension is σ-finite.

By the *Carathéodory's Extension Theorem* (Theorem 4.5) and the *Generated Ring Theorem* (Theorem 2.2), there is a unique extension μ of the measure μ on $\mathscr{R}$ to the generated σ-algebra

$$\Sigma := \sigma a(\mathscr{S}) = \sigma a(\mathscr{R}).$$

By the prior theorem, without loss of generality, it suffices to prove the statement for an arbitrary equivalence class $s \in L_p(X, \Sigma, \mu)$ represented by a simple function $s(\cdot) \in \mathcal{L}_p(X, \Sigma, \mu)$.

Exercise 7.41. Explain.

Further, by the *subadditivity* and *absolute scalability* of p-norm, without loss of generality, it suffices to prove the statement for an arbitrary equivalence class $\chi_A \in L_p(X, \Sigma, \mu)$ represented by a characteristic function $\chi_A(\cdot) \in \mathcal{L}_p(X, \Sigma, \mu)$ with $A \in \Sigma$.

Exercise 7.42. Explain.

Since

$$\int_X |\chi_A(x)|^p \, d\mu(x) = \mu(A),$$

$$\chi_A(\cdot) \in \mathcal{L}_p(X, \Sigma, \mu) \iff \mu(A) < \infty.$$

Therefore, by the *Approximation Theorem* (Theorem 4.6) and the *Generated Ring Theorem* (Theorem 2.3), for any $\varepsilon > 0$ there exist *pairwise disjoint* sets $A_1, \dots, A_n \in \mathscr{S}$ with some $n \in \mathbb{N}$ such that, for $B := \bigcup_{k=1}^n A_k \in \mathscr{R}$,

$$\mu(A \triangle B) = \mu(A \setminus B) + \mu(B \setminus A) < \varepsilon,$$

and hence, in view of the pairwise disjointness of the sets A_k, $k = 1, \dots, n$,

$$\begin{aligned}\|\chi_A - \chi_B\|_p^p &= \|\chi_A - \chi_B\|_p^p = \left\|\chi_A - \sum_{k=1}^n \chi_{A_k}\right\|_p^p \\ &= \int_X \left|\chi_A(x) - \sum_{k=1}^n \chi_{A_k}(x)\right|^p d\mu(x) = [\mu(A \triangle B)]^p < \varepsilon^p.\end{aligned}$$

Exercise 7.43. Verify.

This completes the proof. □

7.4.5.2 Separability of $L_p(a, b)$

As a direct corollary of the above, we now derive the following important statement.

Corollary 7.8 (Separability of $L_p(a,b)$). *Let $1 \le p < \infty$ and $-\infty \le a < b \le \infty$. The Banach space $L_p(a,b)$ is separable, the countable dense subset being the set of equivalence classes represented by p-integrable simple functions of the form*

$$s(x) = \sum_{k=1}^{n} a_k \chi_{A_k}(x),\ x \in X,$$

where $n \in \mathbb{N}$, $A_k \in \mathscr{S}$, a_k, $k = 1, \dots, n$, are real or complex rational numbers, and

$$\mathscr{S} := \{(p,q] \mid a < p < q < b,\ p,q \in \mathbb{Q}\} \cup \{\emptyset\}$$

is a countable semiring on (a,b).

Exercise 7.44. Prove.

Hint. Use the prior theorem and the fact

$$\mathscr{B}((a,b)) = \sigma a(\mathscr{S}),$$

considering that, without loss of generality, we can regard each equivalence class $f \in L_p(a,b)$ to be represented by a p-integrable on (a,b) *Borel measurable function* $f(\cdot)$ (see Remarks 7.16).

Remark 7.17. Generally, the space $L_\infty(X, \Sigma, \mu)$ is not separable. For instance, the spaces l_∞ and $L_\infty(a,b)$ (see Examples 1.15 and Section 7.5, Problem 15).

7.4.5.3 Approximation by Continuous Functions with Compact Support

Theorem 7.11 (Approximation by Continuous Functions with Compact Support). *Let $1 \le p < \infty$ and $-\infty \le a < b \le \infty$. Then the set of equivalence classes represented by continuous on (a,b) functions with compact support is dense in $L_p(a,b)$, i. e., for an arbitrary equivalence class $f \in L_p(a,b)$ and any $\varepsilon > 0$, there exists an equivalence class $g \in L_p(a,b)$ represented by a continuous on (a,b) function $g(\cdot)$ with compact support such that*

$$\|f - g\|_p < \varepsilon.$$

Proof. Consider the case of $(-\infty, \infty)$, all other cases being proved similarly.

Let $f \in L_p(\mathbb{R})$ represented by a p-integrable *Borel measurable* function $f(\cdot)$ (see Remarks 7.16), be arbitrary.

Then, by the *Lebesgue Dominated Convergence Theorem* (Theorem 6.7) applied to the function sequence $(|f(\cdot) - f(\cdot)\chi_{[-n,n]}(\cdot)|^p)_{n\in\mathbb{N}}$,

$$\|f - f\chi_{[-n,n]}\|_p^p = \int_{\mathbb{R}} |f(x) - f(x)\chi_{[-n,n]}(x)|^p \, d\lambda(x) \to 0,\ n \to \infty.$$

Exercise 7.45. Verify that the *Lebesgue Dominated Convergence Theorem* (Theorem 6.7) applies.

Hint. Use the *important inequality* (Proposition 7.1).

Hence, without loss of generality, we can regard the p-integrable Borel measurable representative function $f(\cdot)$ to have a *compact support* $\operatorname{supp} f(\cdot) \subseteq [-n, n]$ with some $n \in \mathbb{N}$.

Exercise 7.46. Explain (cf. the proof of the *Approximation by Continuous Functions Theorem* (Theorem 6.9)).

Further, by the *Approximation by Simple Functions Theorem* (Theorem 7.10), without loss of generality, we can regard the p-integrable Borel measurable representative function $f(\cdot)$ with compact support to be *simple*.

Exercise 7.47. Explain.

Hence, by the *subadditivity* and *absolute scalability* of p-norm, we can further regard the p-integrable Borel measurable representative function $f(\cdot)$ with compact support to be a *characteristic function* $\chi_A(\cdot)$ with a *bounded Borel* set A, for which the statement is true as is shown in the proof of the *Approximation by Continuous Functions* (Theorem 6.9).

This completes the proof. □

Remark 7.18. The *Approximation by Continuous Functions with Compact Support Theorem* generalizes the *Approximation by Continuous Functions Theorem* (Theorem 6.9) (cf. Remark 6.17).

Corollary 7.9 (Denseness of $C[a,b]$ in $L_p(a,b)$). *Let $1 \le p < \infty$ and $-\infty < a < b < \infty$. Then the set of equivalence classes represented by continuous on $[a,b]$ functions is dense in $L_p(a,b)$, i. e., for an arbitrary equivalence class $f \in L_p(a,b)$ and any $\varepsilon > 0$, there exists an equivalence class $g \in L_p(a,b)$ represented by a function $g(\cdot) \in C[a,b]$ such that*

$$\|f - g\|_p < \varepsilon.$$

Exercise 7.48. Prove.

7.5 Problems

1. Let the underlying measure space be $(\mathbb{R}, \Sigma^*, \lambda)$. For

$$f(x) := \begin{cases} x^2, & x \in \mathbb{Q}, \\ 2\arctan x, & x \in \mathbb{Q}^c, \end{cases}$$

find $\sup_{x\in\mathbb{R}} f(x)$ and $\operatorname{ess\,sup}_{x\in\mathbb{R}} f(x)$.

2. Prove

   Proposition 7.4 (Equality in Hölder's Inequality). *For arbitrary conjugate indices $1 < p, q < \infty$, equality in Hölder's inequality (Theorem 7.2) holds iff*

   $$\|g\|_q^q |f(x)|^p = \|f\|_p^p |g(x)|^q \text{ a. e. } \mu \text{ on } X.$$

   Hint. For the case of $\|f\|_p, \|g\|_q > 0$, reduce to the question of equality in *Young's inequality* (Theorem 7.1) for

   $$a = \frac{|f(x)|}{\|f\|_p} \text{ and } b = \frac{|g(x)|}{\|g\|_q}, \quad x \in X,$$

   (see Remark 7.2).

3. Prove

   Proposition 7.5 (Equality in Minkowski's Inequality). *For arbitrary $1 < p < \infty$, equality in Minkowski's inequality (Theorem 7.3) holds iff*

   $$\exists \lambda, \mu \geq 0, \lambda + \mu > 0 : \lambda f(x) = \mu g(x) \text{ a. e. } \mu \text{ on } X.$$

   Hint. Use the proposition from the prior problem and show that the equality implies that

   $$|f(x) + g(x)| = |f(x)| + |g(x)| \text{ a. e. } \mu \text{ on } X,$$

   which means that the *signs* (or the principal values of *arguments*, in the complex case) of $f(x)$ and $g(x)$ are the same a. e. μ on the set of all $x \in X$, for which $f(x), g(x) \neq 0$.

4. Let (X, Σ, μ) be a measure space and f, f_n, $n \in \mathbb{N}$, be real- or complex-valued Σ-measurable functions on X. Prove that, if:
   (1) $f_n \xrightarrow{\mu} f$ and
   (2) $\exists g \in \mathcal{L}_1(X, \Sigma, \mu) \, \forall n \in \mathbb{N} : |f_n(x)| \leq g(x)$ a. e. μ on X,
   then $f, f_n \in \mathcal{L}_1(X, \Sigma, \mu)$, $n \in \mathbb{N}$, and

   $$\|f_n - f\|_1 \to 0, \; n \to \infty.$$

5. Let (X, Σ, μ) be a measure space, $f, f_n \in \mathcal{L}_1(X, \Sigma, \mu)$, $n \in \mathbb{N}$, and $g \in \mathcal{L}_\infty(X, \Sigma, \mu)$. Prove that, if

   $$\|f_n - f\|_1 \to 0, \; n \to \infty,$$

   then

   $$\|f_n g - fg\|_1 \to 0, \; n \to \infty.$$

6. Let (X, Σ, μ) be a *complete* measure space and $f, f_n \in \mathcal{L}_p(X, \Sigma, \mu)$, $n \in \mathbb{N}$, for some $1 \le p \le \infty$. Prove that, if

$$\|f_n - f\|_p \to 0,\ n \to \infty, \quad \text{and} \quad f_n \to g \pmod{\mu},$$

then $f = g \pmod{\mu}$.

Hint. Apply the *Riesz Theorem* (Theorem 5.13).

7. Prove

Proposition 7.6 (Incompleteness of $(C[a, b], \|\cdot\|_1)$). *The space* $C[a, b]$ $(-\infty < a < b < \infty)$ *is an incomplete normed vector space relative to the integral norm:*

$$C[a, b] \ni f \mapsto \|f\|_1 := \int_a^b |f(x)|\,dx.$$

Hint. Without loss of generality, consider the case of $C[-1, 1]$ and the function sequence

$$f_n(x) := \begin{cases} 0 & \text{for } -1 \le x \le -1/n, \\ nx + 1 & \text{for } -1/n < x < 0, \\ 1 & \text{for } 0 \le x \le 1 \end{cases} \quad n \in \mathbb{N},$$

provide another example of a similar function sequence exploiting the same idea.

8. Prove

Proposition 7.7 (Fundamental Sequence with Convergent Subsequence). *If a fundamental sequence* $(x_n)_{n\in\mathbb{N}}$ *in a normed vector space* $(X, \|\cdot\|)$ *contains a subsequence* $(x_{n(k)})_{k\in\mathbb{N}}$ *such that*

$$\exists\, x \in X : x_{n(k)} \to x,\ k \to \infty,$$

then

$$x_n \to x,\ n \to \infty.$$

9. Let $(x_n)_{n\in\mathbb{N}}$ be a fundamental sequence in a normed vector space $(X, \|\cdot\|)$.
   (a) Prove that there exists a subsequence $(x_{n(k)})_{k\in\mathbb{N}}$ such that the telescoping series

$$\sum_{k=1}^{\infty} \left(x_{n(k+1)} - x_{n(k)}\right) \tag{7.19}$$

   converges absolutely, i. e.,

$$\sum_{k=1}^{\infty} \left\|x_{n(k+1)} - x_{n(k)}\right\| < \infty.$$

(b) Show that, provided $(X, \|\cdot\|)$ is a *Banach space*, the latter implies that series (7.19) converges.

10. Let (X, Σ, μ) be a finite measure space (i. e., $\mu(X) < \infty$). Prove that, for each $1 \le p < \infty$,

$$L_\infty(X, \Sigma, \mu) \subseteq L_p(X, \Sigma, \mu)$$

and

$$\forall f \in L_\infty(X, \Sigma, \mu) : \ \|f\|_\infty = \lim_{p\to\infty} \|f\|_p.$$

11. Let (X, Σ, μ) be a finite measure space (i. e., $\mu(X) < \infty$). Prove the inclusions

$$L_p(X, \Sigma, \mu) \supseteq L_q(X, \Sigma, \mu) \supseteq L_\infty(X, \Sigma, \mu), \tag{7.20}$$

where $1 \le p < q < \infty$.

Hint. Apply *Hölder's inequality* (Theorem 7.2).

12. Show that, if $\mu(X) = \infty$, the inclusions (7.20) or their inverses need not hold.
13. Prove the *proper* inclusions

$$c_{00} \subset l_p \subset l_q \subset c_0 \subset c \subset l_\infty,$$

where $1 \le p < q < \infty$.

14. Let $n \in \mathbb{N}$. Show that the space $l_p^{(n)}$ is *separable* for any $1 \le p \le \infty$ (see Examples 1.15).
15. Show that the space l_p is *separable* for any $1 \le p < \infty$ and is *not* separable for $p = \infty$ (see Examples 1.15).

8 Differentiation and Integration

In this chapter, we are to see that a novel approach based on the Lebesgue measure and integration theory, as in the case of the Riemann integration (see Section 6.10), allows us to better understand differentiation and extend the classical *total change formula* linking differentiation with integration to a substantially wider class of functions.

8.1 Derivative Numbers

We start our discourse with the following notion, which generalizes the concept of *derivative* of a function at a point.

Definition 8.1 (Derivative Number). Let $f : I \to \mathbb{R}$ be a real-valued function on an interval $I \subseteq \mathbb{R}$ containing a point x_0. A (finite or infinite) value $\lambda \in \overline{\mathbb{R}}$ is called a *derivative number of f at x_0* if there exists a sequence $(x_n)_{n\in\mathbb{N}} \subset I \setminus \{x_0\}$ such that

$$\lim_{n\to\infty} x_n = x_0 \quad \text{and} \quad \lim_{n\to\infty} \frac{f(x_n) - f(x_0)}{x_n - x_0} = \lambda.$$

Notation. $\mathrm{D}f(x_0)$.

Examples 8.1.

1. The set of all derivative numbers of the absolute-value function $f(x) := |x|$, $x \in \mathbb{R}$, is
   (a) $\{\pm 1\}$ at $x_0 = 0$,
   (b) $\{-1\}$ at any point $x_0 < 0$, and
   (c) $\{1\}$ at any point $x_0 > 0$.
2. The set of all derivative numbers of the *Dirichlet function* $f(x) := \chi_{\mathbb{Q}}(x)$, $x \in \mathbb{R}$, is $\{0, \pm\infty\}$ at any point $x_0 \in \mathbb{R}$.
3. The set of all derivative numbers of the function

$$f(x) := \begin{cases} x, & x \in \mathbb{Q}, \\ x^2, & x \in \mathbb{Q}^c, \end{cases}$$

   is
   (a) $\{1, 2x_0\}$ at $x_0 = 0, 1$,
   (b) $\{1, \pm\infty\}$ at any point $x_0 \in \mathbb{Q} \setminus \{0, 1\}$, and
   (c) $\{2x_0, \pm\infty\}$ at any point $x_0 \in \mathbb{Q}^c$.

Exercise 8.1. Verify.

The following two statements shed the light on the *existence* and *uniqueness* of derivative numbers.

https://doi.org/10.1515/9783110600995-008

Theorem 8.1 (Existence of Derivative Numbers). *Let $f : I \to \mathbb{R}$ be a real-valued function on an interval $I \subseteq \mathbb{R}$. Then, at each point of I there exist (finite or infinite) derivative numbers.*

Proof. Let $x_0 \in I$ be arbitrary and $(x_n)_{n\in\mathbb{N}} \subset I \setminus \{x_0\}$ be a sequence such that

$$\lim_{n\to\infty} x_n = x_0.$$

If the sequence

$$y_n := \frac{f(x_n) - f(x_0)}{x_n - x_0}, \quad n \in \mathbb{N}, \tag{8.1}$$

is *bounded*, by the *Bolzano–Weierstrass Theorem* (Theorem 1.15), there exists a subsequence $(y_{n(k)})_{k\in\mathbb{N}}$ convergent to a finite limit $\lambda \in \mathbb{R}$, which hence is a derivative number of f at x_0.

If the sequence defined by (8.1) is *unbounded*, there exists a subsequence $(y_{n(k)})_{k\in\mathbb{N}}$ such that

$$\lim_{k\to\infty} y_{n(k)} \to -\infty \quad \text{or} \quad \lim_{k\to\infty} y_{n(k)} \to \infty.$$

Exercise 8.2. Explain.

Hence, in this case, $-\infty$ or ∞ is a derivative number of f at x_0. □

Theorem 8.2 (Uniqueness of Derivative Numbers). *Let $f : I \to \mathbb{R}$ be a real-valued function on an interval $I \subseteq \mathbb{R}$ containing a point x_0. The function f has a (finite or infinite) derivative $f'(x_0)$ at x_0 iff there exists a unique derivative number $\mathrm{Df}(x_0)$ of f at x_0, in which case $\mathrm{Df}(x_0) = f'(x_0)$.*

Proof. Here, we are to prove the *"if"* part leaving the proof of the *"only if"* part as an exercise.

"If" part. Suppose that $\lambda \in \mathbb{R}$ is the unique derivative number of f at x_0 and assume that there exists a sequence $(x_n)_{n\in\mathbb{N}} \subset I \setminus \{x_0\}$ such that

$$\lim_{n\to\infty} x_n = x_0, \quad \text{but} \quad \lim_{n\to\infty} \frac{f(x_n) - f(x_0)}{x_n - x_0} \neq \lambda.$$

Then the sequence

$$y_n := \frac{f(x_n) - f(x_0)}{x_n - x_0}, \quad n \in \mathbb{N},$$

has a subsequence $(y_{n(k)})_{k\in\mathbb{N}}$ no subsequence of which has limit λ.

As is shown in the proof of the prior theorem, the subsequence $(y_{n(k)})_{k\in\mathbb{N}}$ has a subsequence $(y_{n(k(j))})_{j\in\mathbb{N}}$ such that

$$\lim_{j\to\infty} y_{n(k(j))} = \mu \in \overline{\mathbb{R}} \neq \lambda,$$

which implies that the derivative number of f at x_0 is not unique *contradicting* the premise, and hence, making the assumption false.

Hence, for each sequence $(x_n)_{n\in\mathbb{N}} \subset I \setminus \{x_0\}$ such that

$$\lim_{n\to\infty} x_n = x_0,$$

we have

$$\lim_{n\to\infty} \frac{f(x_n) - f(x_0)}{x_n - x_0} = \lambda,$$

which means that the function f at x_0 has the derivative $f'(x_0) = \lambda$. □

Exercise 8.3. Prove the *"only if"* part.

Remark 8.1. The prior theorem explains the fact why the absolute-value function $f(x) := |x|$, $x \in \mathbb{R}$, has no derivative at $x_0 = 0$ and is *differentiable* (i. e., has a *finite derivative*) at any $x_0 \in \mathbb{R} \setminus \{0\}$ (see Examples 8.1).

8.2 Vitali Covers and Vitali Covering Lemma

Just as *Fatou's Lemma* (Theorem 6.6) in the Lebesgue integration theory, the *Vitali Covering Lemma* (Theorem 8.3) is a fundamental statement in the theory of functions, also traditionally referred to as a lemma, indispensable for our subsequent discourse.

Definition 8.2 (Vitali Cover). A nonempty collection

$$\mathscr{C} := \{[a_i, b_i]\}_{i\in I}, \quad -\infty < a_i < b_i < \infty,\ i \in I,$$

where I is a nonempty indexing set, of nondegenerate ($a_i < b_i$, $i \in I$) closed bounded intervals (to be referred to as *segments*) is said to be a *Vitali cover* of a set $A \subseteq \mathbb{R}$ if

$$\forall\, x \in A\ \forall\, \varepsilon > 0\ \exists\, i \in I :\ x \in [a_i, b_i] \quad \text{and} \quad \lambda([a_i, b_i]) = b_i - a_i < \varepsilon$$

(see Remark 4.10).

Remark 8.2. As the following example shows, a Vitali cover exists for any set $A \subseteq \mathbb{R}$.

Examples 8.2.

1. For an arbitrary nonempty set $A \subseteq \mathbb{R}$, the segment collection

$$\mathscr{C} := \{[x, x + 1/n] \mid x \in A, n \in \mathbb{N}\}$$

(in this case $I = A \times \mathbb{N}$) is a Vitali cover of A.

A Vitali cover of any nonempty set is, trivially, a Vitali cover of $\emptyset$.

2. The segment collection

$$\mathscr{C}_1 := \{[x, x + 1/n] \mid x \in \mathbb{Q}, n \in \mathbb{N}\}$$

(in this case, $I = \mathbb{Q} \times \mathbb{N}$) is a Vitali cover of $\mathbb{R}$, whereas the segment collection

$$\mathscr{C}_2 := \{[n, n + 1]\}_{n\in\mathbb{N}}$$

is a cover for $\mathbb{R}$ (see Definition 1.26) but *not* a Vitali cover.

Exercise 8.4.
(a) Verify.
(b) Give some more examples.

Theorem 8.3 (Vitali Covering Lemma). *For any set $A \subseteq \mathbb{R}$ with $\lambda^*(A) < \infty$ and an arbitrary Vitali cover $\mathscr{C}$ of A, there exists a countable subcollection $\{[a_i, b_i]\}_{i\in I}$ of $\mathscr{C}$ ($I \subseteq \mathbb{N}$) consisting of pairwise disjoint segments such that*

$$\lambda^*\left(A \setminus \bigcup_{i\in I}[a_i, b_i]\right) = 0,$$

where λ^* is the Lebesgue outer measure.

Proof. Let a set $A \subseteq \mathbb{R}$ with $\lambda^*(A) < \infty$ be arbitrary.

Since $\lambda^*(A) < \infty$, by the *Approximation of Sets with Finite Outer Measure Corollary* (Corollary 4.1) (Section 4.8, Problem 14), for any $\varepsilon > 0$, there exists an *open* set $G \subseteq \mathbb{R}$ such that

$$A \subseteq G \quad \text{and} \quad \lambda(G) < \lambda^*(A) + \varepsilon < \infty. \tag{8.2}$$

Let $\mathscr{C}$ be an arbitrary *Vitali cover* of A. Then, without loss of generality, we can regard that

$$\forall\, [a, b] \in \mathscr{C} : \ [a, b] \subset G. \tag{8.3}$$

Exercise 8.5. Explain.

Let

$$l_0 := \sup\{\lambda([a, b]) = b - a \mid [a, b] \in \mathscr{C}\}$$

From inclusion (8.3) in view of (8.2), by the *monotonicity* of the Lebesgue measure λ (Theorem 3.1), we infer that

$$0 < l_0 \leq \lambda(G) < \infty,$$

and hence,

$$\exists\,[a_1, b_1] \in \mathscr{C} : \lambda([a_1, b_1]) = b_1 - a_1 > l_0/2.$$

If

$$A \subseteq [a_1, b_1],$$

then $\{[a_1, b_1]\}$ is the desired finite subcollection of $\mathscr{C}$.

Exercise 8.6. Explain.

Otherwise,

$$\exists\,[a_2, b_2] \in \mathscr{C} : [a_1, b_1] \cap [a_2, b_2] = \emptyset \quad \text{and} \quad \lambda([a_2, b_2]) = b_2 - a_2 > l_1/2,$$

where

$$l_1 := \sup\{\lambda([a, b]) = b - a \mid [a, b] \in \mathscr{C} \textit{ disjoint from } [a_1, b_1]\} \in (0, \lambda(G)].$$

Exercise 8.7. Explain.

Hint. Use the fact that, in this case,

$$\exists x \in G \setminus [a_1, b_1].$$

If

$$A \subseteq [a_1, b_1] \cup [a_2, b_2],$$

then $\{[a_1, b_1], [a_2, b_2]\}$ is the desired finite subcollection of $\mathscr{C}$. Otherwise, continuing inductively in this fashion, we choose on the $(n+1)$th step ($n \in \mathbb{N}$) a segment $[a_{n+1}, b_{n+1}] \in \mathscr{C}$ *disjoint from* the chosen segments $[a_k, b_k] \in \mathscr{C}$, $k = 1, \dots, n$, and such that

$$\lambda([a_{n+1}, b_{n+1}]) = b_{n+1} - a_{n+1} > l_n/2, \tag{8.4}$$

where

$$\begin{aligned} l_n := \sup\{\lambda([a, b]) = b - a \mid [a, b] \in \mathscr{C} \textit{ disjoint from } [a_k, b_k] \in \mathscr{C}, k = 1, \dots, n\} \\ \in (0, \lambda(G)]. \end{aligned} \tag{8.5}$$

If the process terminates for some $n \in \mathbb{N}$, i. e., we arrive at the inclusion

$$A \subseteq \bigcup_{k=1}^{n} [a_k, b_k],$$

then $\{[a_1, b_1], \dots, [a_n, b_n]\}$ is the desired finite subcollection of $\mathscr{C}$. Otherwise, we obtain a countably infinite subcollection $\{[a_k, b_k]\}_{k \in \mathbb{N}}$ of $\mathscr{C}$ consisting of *pairwise disjoint* segments.

Let us show that

$$\lambda^*\left(A \setminus \bigcup_{k=1}^{\infty}[a_k, b_k]\right) = 0. \tag{8.6}$$

This is, obviously, true if

$$A \subseteq \bigcup_{k=1}^{\infty}[a_k, b_k].$$

Suppose that

$$A \setminus \bigcup_{k=1}^{\infty}[a_k, b_k] \neq \emptyset.$$

Since

$$\bigcup_{k=1}^{\infty}[a_k, b_k] \subseteq G,$$

by the *σ-additivity* and *monotonicity* of λ (Theorem 3.1),

$$\sum_{k=1}^{\infty} \lambda([a_k, b_k]) = \lambda\left(\bigcup_{k=1}^{\infty}[a_k, b_k]\right) \leq \lambda(G) < \infty. \tag{8.7}$$

For each $k \in \mathbb{N}$ if $n = \infty$, let $[c_k, d_k]$ be the segment having the *same midpoint* as $[a_k, b_k]$ and *five times longer* than $[a_k, b_k]$, i. e.,

$$\lambda([c_k, d_k]) = d_k - c_k = 5\lambda([a_k, b_k]) = 5(b_k - a_k).$$

Such a peculiar choice of the segments $[c_k, d_k]$, $k \in \mathbb{N}$, is amply justified below.

As follows from (8.7),

$$\sum_{k=1}^{\infty} \lambda([c_k, d_k]) = 5\sum_{k=1}^{\infty} \lambda([a_k, b_k]) < \infty.$$

Hence, to prove (8.6), it suffices to show that

$$\forall\, n \in \mathbb{N} : A \setminus \bigcup_{k=1}^{n}[a_k, b_k] \subseteq \bigcup_{k=n}^{\infty}[c_k, d_k]. \tag{8.8}$$

Exercise 8.8. Explain.

Let $x \in A \setminus \bigcup_{k=1}^{\infty}[a_k, b_k]$ be arbitrary, then

$$\forall\, n \in \mathbb{N} : x \in G_n := G \setminus \bigcup_{k=1}^{n}[a_k, b_k],$$

which, since G_n, $n \in \mathbb{N}$, is an *open set*, implies that

$$\forall\, n \in \mathbb{N}\, \exists\, [e_n, f_n] \in \mathscr{C} : x \in [e_n, f_n] \subset G_n. \tag{8.9}$$

Exercise 8.9. Explain.

However, it is impossible that, for the same $n \in \mathbb{N}$,

$$[e_n, f_n] \subset G_m \text{ for all } m \in \mathbb{N},$$

since, by (8.5) and (8.4), this would imply that

$$\lambda([e_n, f_n]) \le l_m < 2\lambda([a_{m+1}, b_{m+1}]) \text{ for all } m \in \mathbb{N},$$

and hence, since, by (8.7),

$$\lambda([a_{m+1}, b_{m+1}]) \to 0, \ m \to \infty$$

the segment $[e_n, f_n]$ *degenerates* into a singleton, i. e., $e_n = f_n$ which *contradicts* the fact that the segment $[e_n, f_n]$ belongs to the *Vitali cover* $\mathscr{C}$ of A.

Hence,

$$\forall\, n \in \mathbb{N}\ \exists\, m \in \mathbb{N} : \ [e_n, f_n] \not\subset G_m, \text{ i. e., } [e_n, f_n] \cap \bigcup_{k=1}^{m} [a_k, b_k] \ne \emptyset$$

Let, for each $n \in \mathbb{N}$, $m(n) \in \mathbb{N}$ be the *smallest* such number.

In view of the fact that

$$\bigcup_{k=1}^{m} [a_k, b_k] \subset \bigcup_{k=1}^{m+1} [a_k, b_k], \ m \in \mathbb{N},$$

and (8.8), we infer that $m(n) \ge n + 1$, $n \in \mathbb{N}$, an hence,

$$\forall\, n \in \mathbb{N} : \ [e_n, f_n] \cap \bigcup_{k=1}^{m(n)-1} [a_k, b_k] = \emptyset,$$

which immediately implies that

$$\forall\, n \in \mathbb{N} : \ [e_n, f_n] \subset G_{m(n)-1} \quad \text{and} \quad [e_n, f_n] \cap [a_{m(n)}, a_{m(n)}] \ne \emptyset. \tag{8.10}$$

Therefore, in view of (8.4),

$$\forall\, n \in \mathbb{N} : \ [e_n, f_n] : \ \lambda([e_n, f_n]) \le l_{m(n)-1} < 2\lambda([a_{m(n)}, a_{m(n)}]). \tag{8.11}$$

As follows from (8.10), (8.11), and the definition of $[c_{m(n)}, d_{m(n)}]$, which now makes sense,

$$\forall\, n \in \mathbb{N} : \ [e_n, f_n] \subset [c_{m(n)}, d_{m(n)}].$$

Exercise 8.10. Explain.

Moreover, in view of (8.9),

$$\forall n \in \mathbb{N} : x \in [e_n, f_n] \subseteq [c_{m(n)}, d_{m(n)}] \subseteq \bigcup_{k=n}^{\infty} [c_k, d_k],$$

whence desired inclusion (8.8) follows, which completes the proof. □

Theorem 8.4 (Vitali Covering Lemma (Alternative Version)). *Let $A \subseteq \mathbb{R}$ an arbitrary set such that $\lambda^*(A) < \infty$ and $\mathscr{C}$ be its Vitali cover. Then, for any $\varepsilon > 0$, there exists a finite subcollection $\{[a_1, b_1], \dots, [a_n, b_n]\}$ ($n \in \mathbb{N}$) of $\mathscr{C}$ consisting of pairwise disjoint segments such that*

$$\lambda^*\left(A \setminus \bigcup_{i=1}^{n} [a_i, b_i]\right) < \varepsilon,$$

where λ^* is the Lebesgue outer measure.

Proof. By choosing a countable subcollection $\{[a_i, b_i]\}_{i \in I}$ ($I \subseteq \mathbb{N}$) of $\mathscr{C}$ consisting of pairwise disjoint segments as in the proof of the *Vitali Covering Lemma*, we have

$$\lambda^*\left(A \setminus \bigcup_{i \in I} [a_i, b_i]\right) = 0 \tag{8.12}$$

and

$$\sum_{i \in I} \lambda([a_i, b_i]) < \infty.$$

If I is finite, $\{[a_i, b_i]\}_{i \in I}$ is the desired subcollection.
If I is infinite, without loss of generality, we can regard that $I = \mathbb{N}$.
Since

$$\sum_{i=1}^{\infty} \lambda([a_i, b_i]) < \infty,$$

we infer that

$$\forall \varepsilon > 0 \; \exists n \in \mathbb{N} : \sum_{i=n+1}^{\infty} \lambda([a_i, b_i]) < \varepsilon. \tag{8.13}$$

Further,

$$A \setminus \bigcup_{i=1}^{n} [a_i, b_i] \subseteq \left[A \setminus \bigcup_{i=1}^{\infty} [a_i, b_i]\right] \cup \bigcup_{i=n+1}^{\infty} [a_i, b_i].$$

Exercise 8.11. Verify.

Where by the *monotonicity*, *subadditivity*, and *σ-subadditivity* of the Lebesgue outer measure λ^* (see Definition 4.2 and Remarks 4.2), considering (8.12) and (8.13), we have

$$\lambda^*\left(A \setminus \bigcup_{i=1}^{n}[a_i, b_i]\right) \le \lambda^*\left(A \setminus \bigcup_{i=1}^{\infty}[a_i, b_i]\right) + \lambda^*\left(\bigcup_{i=n+1}^{\infty}[a_i, b_i]\right)$$
$$= \lambda^*\left(\bigcup_{i=n+1}^{\infty}[a_i, b_i]\right) \le \sum_{i=n+1}^{\infty}\lambda^*([a_i, b_i]) = \sum_{i=n+1}^{\infty}\lambda([a_i, b_i]) < \varepsilon,$$

which makes $\{[a_1, b_1], \dots, [a_n, b_n]\}$ the desired subcollection and completes the proof. □

8.3 Monotone Functions

With all the Lebesgue measure and integration theory machinery at hand, we are to consider here in greater detail the familiar *monotone functions*, which are basic for our the further discourse on *functions of bounded variation* and *absolutely continuous functions*.

8.3.1 Definition and Certain Properties

Definition 8.3 (Monotone Function). A real-valued function $f : I \to \mathbb{R}$ on an nondegenerate interval $I \subseteq \mathbb{R}$, which is either increasing on I:

$$\forall, x, x' \in I, \ x < x' : \ f(x) \le f(x')$$

or decreasing on I:

$$\forall, x, x' \in I, \ x < x' : \ f(x) \ge f(x'),$$

is called *monotone on* I.

Remark 8.3. If the inequalities in the prior definition are strict, the function $f : I \to \mathbb{R}$ is called *strictly increasing* or *strictly decreasing* on I, respectively.

Exercise 8.12. Give some examples of monotone and nonmonotone functions.

Theorem 8.5 (Discontinuities of Monotone Functions). *A monotone function $f : I \to \mathbb{R}$ on an nondegenerate interval $I \subseteq \mathbb{R}$ is continuous on I, except at countably many points of I, at which the function f has jump discontinuities.*

Exercise 8.13. Prove.

Hint. Without loss of generality, consider the case of an *increasing* function $f : I \to \mathbb{R}$; show that there can exist only *jump discontinuities*, i. e.,

(i) $f(x-) < f(x+)$ if $x \in I$ is an *interior point* of the interval I,
(ii) $f(a) < f(a+)$ if $a \in \mathbb{R}$ is the left endpoint of the interval I, or
(iii) $f(b-) < f(b)$ if $b \in \mathbb{R}$ is the right endpoint of the interval I,

where all the left and right limits are *finite*; show that, the associated *open intervals* of the form

$$(f(x-), f(x+)) \quad (x \in I \text{ is an interior point of jump discontinuity}),$$

if any, are *pairwise disjoint*.

By the *Characterization of Riemann Integrability* (Theorem 6.10), we immediately obtain the following.

Corollary 8.1 (Riemann Integrability of a Monotone Function). *A monotone function $f : [a,b] \to \mathbb{R}$ $(-\infty < a < b < \infty)$ is Riemann integrable on $[a,b]$, and hence, also Lebesgue integrable on $[a,b]$.*

8.3.2 Total Jump and Jump Function

Lemma 8.1 (Jump Estimate). *Let $f : [a,b] \to \mathbb{R}$ $(-\infty < a < b < \infty)$ be an increasing function and let $x_i \in (a,b)$, $i = 1,\dots,n$, $(n \in \mathbb{N})$ be arbitrary distinct points. Then*

$$[f(a+) - f(a)] + \sum_{i=1}^{n} [f(x_i+) - f(x_i-)] + [f(b) - f(b-)] \le f(b) - f(a). \tag{8.14}$$

Proof. Without loss of generality, suppose that

$$a < x_1 < \cdots < x_n < b$$

and let $x_0 := a$, $x_{n+1} := b$.

Choosing arbitrary points $y_i \in (a,b)$, $i = 0,1,\dots,n$ such that

$$x_i < y_i < x_{i+1},\ i = 0,1,\dots,n,$$

in view of the fact that the function f *increases* on $[a,b]$, we have

$$\begin{aligned} &f(a+) - f(a) \le f(y_0) - f(a), \\ &f(x_i+) - f(x_i-) \le f(y_i) - f(y_{i-1}),\ i = 1,\dots,n, \\ &f(b) - f(b-) \le f(b) - f(y_n). \end{aligned}$$

Adding up, we arrive at estimate (8.14).

Exercise 8.14. Verify. □

The following statement provides an estimate for the *total jump* of an increasing function $f : [a,b] \to \mathbb{R}$ $(-\infty < a < b < \infty)$ over the interval $[a,b]$.

Theorem 8.6 (Total Jump Theorem). *Let $f : [a,b] \to \mathbb{R}$ $(-\infty < a < b < \infty)$ be an increasing function and $\{x_i\}_{i\in I} \subset (a,b)$, where $I \subseteq \mathbb{N}$, be the countable set of its interior points of discontinuity. Then*

$$[f(a+) - f(a)] + \sum_{i\in I}[f(x_i+) - f(x_i-)] + [f(b) - f(b-)] \le f(b) - f(a), \tag{8.15}$$

the left-hand side being the total jump of the function f over the interval $[a,b]$.

Proof. If $I = \emptyset$, i. e., f is continuous on (a,b), in view of the fact that f increases on $[a,b]$, (8.15) turns into the estimate

$$[f(a+) - f(a)] + [f(b) - f(b-)] = [f(b) - f(a)] - [f(b-) - f(a+)] \le f(b) - f(a).$$

Exercise 8.15. Explain.

If the set I and is nonempty and *finite*, i. e., without loss of generality, we can regard that $I = \{1,\dots,n\}$ for some $n \in \mathbb{N}$, the statement follows immediately from the *Jump Estimate Lemma* (Lemma 8.1).

If the set I is infinite, i. e., without loss of generality, we can regard that $I = \mathbb{N}$, as follows from the *Jump Estimate Lemma* (Lemma 8.1), for each $n \in \mathbb{N}$,

$$[f(a+) - f(a)] + \sum_{i=1}^{n}[f(x_i+) - f(x_i-)] + [f(b) - f(b-)] \le f(b) - f(a).$$

Where, passing to the limit as $n \to \infty$, we arrive at the estimate

$$[f(a+) - f(a)] + \sum_{i=1}^{\infty}[f(x_i+) - f(x_i-)] + [f(b) - f(b-)] \le f(b) - f(a),$$

which completes the proof. □

Definition 8.4 (Jump Function). Let $f : [a,b] \to \mathbb{R}$ $(-\infty < a < b < \infty)$ be an increasing function and $\{x_i\}_{i\in I} \subset (a,b)$, where $I \subseteq \mathbb{N}$, be the countable set of its interior points of discontinuity. The increasing step function

$$s(x) := \begin{cases} 0, & x = a, \\ [f(a+) - f(a)] + \sum_{i\in I: x_i<x}[f(x_i+) - f(x_i-)] + [f(x) - f(x-)], & a < x \le b, \end{cases}$$

is called the *jump function* of the function f.

Remarks 8.4.
- The jump function $s : [a,b] \to [0,\infty)$ is *well-defined* and *increasing*.
- For an arbitrary $a \le x \le b$, the value the jump function $s(x)$ is the *total jump* of the original function f over the subinterval $[a,x]$.

This, in particular, implies that, if an increasing function $f : [a,b] \to \mathbb{R}$ is *continuous* on $[a,b]$, $s(x) = 0$, $x \in [a,b]$.

- For arbitrary $a \le x < x' \le b$,

$$s(x') - s(x) = [f(x+) - f(x)] + \sum_{i \in I:\, x < x_i < x'} [f(x_i+) - f(x_i-)] + [f(x') - f(x'-)]$$

  is the *total jump* of the function f over the subinterval $[x, x']$.

Exercise 8.16.
(a) Verify.
(b) Find the jump functions for the following increasing functions
  (i) $f(x) := \operatorname{sgn}(x)$, $x \in [-1,1]$, and
  (ii) $f(x) := \begin{cases} x, & -1 \le x < 0, \\ x^2 + 1, & 0 \le x < 2, \\ 5, & 2 \le x \le 3. \end{cases}$

Using the jump function $s : [a,b] \to \mathbb{R}$ of an increasing function $f : [a,b] \to \mathbb{R}$, one can produce a continuous increasing function on $[a,b]$.

Theorem 8.7 (Difference Function Theorem). *Let $f : [a,b] \to \mathbb{R}$ $(-\infty < a < b < \infty)$ be an increasing function. The difference function*

$$g(x) := f(x) - s(x), \; a \le x \le b,$$

is increasing and continuous on $[a,b]$.

Proof. Let $a \le x < x' \le b$ be arbitrary. Applying the *Total Jump Theorem* (Theorem 8.6) (see Remarks 8.4), to the function f on the subinterval $[x, x']$, we have

$$s(x') - s(x) \le f(x') - f(x). \tag{8.16}$$

Exercise 8.17. Verify.

Where we infer that

$$g(x) := f(x) - s(x) \le f(x') - s(x') =: g(x'),$$

i. e., the difference function g is *increasing* on $[a,b]$.

Further, passing to the limit as $x' \to x+$ in (8.16), we arrive at

$$s(x+) - s(x) \le f(x+) - f(x). \tag{8.17}$$

On the other hand, as follows from the definition of s (see Remarks 8.4),

$$f(x+) - f(x) \le s(x') - s(x).$$

Exercise 8.18. Explain.

Whence, passing to the limit as $x' \to x+$, we arrive at

$$f(x+) - f(x) \le s(x+) - s(x). \tag{8.18}$$

Inequalities (8.17) and (8.18) jointly imply that

$$g(x+) := f(x+) - f(x) = s(x+) - s(x) =: g(x),\ x \in [a, b). \tag{8.19}$$

Similarly, it can be shown that

$$g(x'-) = g(x'),\ x \in (a, b]. \tag{8.20}$$

Exercise 8.19. Show.

Equalities (8.19) and (8.20) jointly imply that the difference function g is continuous on $[a, b]$.

Exercise 8.20. Explain. □

Remarks 8.5.
- Thus, subtraction the jump s of an increasing function f from it removes all its discontinuities, if any.
- In particular, if an increasing function $f : [a, b] \to \mathbb{R}$ is *continuous* on $[a, b]$, $s(x) = 0$, $x \in [a, b]$, (see Remarks 8.4), and hence, in this case, $g(x) := f(x) - s(x) = f(x)$, $x \in [a, b]$.

Exercise 8.21. Find the difference functions for the increasing functions from Exercise 8.16 (b).

8.3.3 Derivative Numbers of Increasing Functions

Now, let us take a closer look at the derivative numbers of increasing functions.

Theorem 8.8 (Derivative Numbers of Increasing Functions). *Let $f : [a, b] \to \mathbb{R}$ $(-\infty < a < b < \infty)$ be an increasing function. Then:*
(1) *all the derivative numbers of the function f are nonnegative;*
(2) *if the function f is strictly increasing and*

$$\exists p \ge 0\ \exists E \subseteq [a, b]\ \forall x \in E\ \exists\ \mathrm{Df}(x) :\ \mathrm{Df}(x) \le p,$$

then

$$\lambda^*(f(E)) \le p\lambda^*(E);$$

(3) *if the function f is strictly increasing and*

$$\exists q \geq 0 \, \exists E \subseteq [a,b] \, \forall x \in E \, \exists \, \mathrm{Df}(x) : \; \mathrm{Df}(x) \geq q,$$

then

$$\lambda^*(f(E)) \geq q\lambda^*(E),$$

where, in parts (2) *and* (3), λ^* *is the Lebesgue outer measure.*

Remark 8.6. Parts (2) and (3) of the prior theorem can be viewed as generalizations of estimates easily derived from the classical *Mean Value Theorem* for any $E = [c,d] \subseteq [a,b]$, in which case

$$\lambda^*(E) = d - c \quad \text{and} \quad \lambda^*(f(E)) = f(d) - f(c).$$

Proof.

(1) Part (1) immediately follows from the fact that f is increasing.

Exercise 8.22. Explain.

(2) Since $E \subseteq [a,b]$, which, by the *monotonicity* of the Lebesgue outer measure λ^* (see Definition 4.2), implies that

$$\lambda^*(E) \leq \lambda^*([a,b]) = \lambda([a,b]) = b - a < \infty,$$

and hence, by *Approximation of Sets with Finite Outer Measure Corollary* (Corollary 4.1) (Section 4.8, Problem 14), for any $\varepsilon > 0$, there exists an *open* set $G \subseteq \mathbb{R}$ such that

$$E \subseteq G \quad \text{and} \quad \lambda(G) < \lambda^*(E) + \varepsilon. \tag{8.21}$$

Let $x \in E$ be arbitrary. Then there is a sequence $(x_n)_{n\in\mathbb{N}} \subset [a,b] \setminus \{x\}$ such that

$$\lim_{n\to\infty} x_n = x \quad \text{and} \quad \lim_{n\to\infty} \frac{f(x_n) - f(x)}{x_n - x} = \mathrm{Df}(x) \leq p.$$

Since the G is open and $x \in G$, for all sufficiently large $n \in \mathbb{N}$, the *segment* $d_n(x)$ consisting of all points *between* x and x_n is contained in G:

$$d_n(x) \subset G. \tag{8.22}$$

Exercise 8.23. Explain.

Further, for any $c > p$, in view of the fact that f is *strictly increasing* (see Definition 8.3 and Remark 8.3), the estimate

$$0 < \frac{f(x_n) - f(x)}{x_n - x} < c \tag{8.23}$$

also holds for all sufficiently large $n \in \mathbb{N}$.

Exercise 8.24. Explain.

Without loss of generality, we can regard that both inclusion (8.22) and estimate (8.23) hold for all $n \in \mathbb{N}$.
For each $n \in \mathbb{N}$, let $\Delta_n(x)$ be the segment consisting of all points *between* $f(x)$ and $f(x_n)$. Observe that, in view of the fact that f is *strictly increasing*, all such segments are *nondegenerate* (i. e., have positive length).
Further,

$$\forall x \in E, n \in \mathbb{N} : \lambda(d_n(x)) = |x_n - x| \quad \text{and} \quad \lambda(\Delta_n(x)) = |f(x_n) - f(x)|,$$

and hence, by (8.23),

$$\forall c > p, x \in E, n \in \mathbb{N} : \lambda(\Delta_n(x)) < c\lambda(d_n(x)), \tag{8.24}$$

which, since

$$\forall x \in E : \lambda(d_n(x)) \to 0, \ n \to \infty,$$

implies that, for each $x \in E$, the segments $\Delta_n(x)$ are arbitrarily short for all sufficiently large $n \in \mathbb{N}$.
Considering this and the fact that

$$\forall x \in E, n \in \mathbb{N} : f(x) \in \Delta_n(x),$$

we infer that the segment collection

$$\mathscr{C} := \{\Delta_n(x) \mid x \in E, n \in \mathbb{N}\}$$

is a *Vitali cover* for the image set $f(E)$ (see Examples 8.2).
Since f is *strictly increasing*,

$$f(E) \subseteq [f(a), f(b)],$$

by the *monotonicity* of the Lebesgue outer measure λ^* (see Definition 4.2)

$$\lambda^*(f(E)) \le \lambda^*([f(a), f(b)]) = \lambda([f(a), f(b)]) = f(b) - f(a) < \infty. \tag{8.25}$$

Therefore, by the *Vitali Covering Lemma* (Theorem 8.3), $\mathscr{C}$ has a countable subcollection $\{\Delta_{n(i)}(x_i)\}_{i\in I}$ ($I \subseteq \mathbb{N}$) consisting of *pairwise disjoint* segments such that

$$\lambda\left(f(E) \setminus \bigcup_{i\in I} \Delta_{n(i)}(x_i)\right) = 0.$$

Where, since

$$f(E) \subseteq \left[f(E) \setminus \bigcup_{i\in I} \Delta_{n(i)}(x_i)\right] \cup \bigcup_{i\in I} \Delta_{n(i)}(x_i),$$

by the *monotonicity* and *σ-subadditivity* of the Lebesgue outer measure λ^* (see Definition 4.2) and (8.24), we infer that

$$\begin{aligned}\forall\, c > p:\ \lambda^*(f(E)) &\le \lambda^*\left(f(E) \setminus \bigcup_{i\in I} \Delta_{n(i)}(x_i)\right) + \lambda^*\left(\bigcup_{i\in I} \Delta_{n(i)}(x_i)\right)\\ &= \lambda^*\left(\bigcup_{i\in I} \Delta_{n(i)}(x_i)\right) \le \sum_{i\in I} \lambda^*\left(\Delta_{n(i)}(x_i)\right) = \sum_{i\in I} \lambda\left(\Delta_{n(i)}(x_i)\right)\\ &< c \sum_{i\in I} \lambda(d_{n(i)}(x_i)).\end{aligned}$$

Since f is *strictly increasing*, along with the segments $\Delta_{n(i)}(x_i)$, $i \in I$, *pairwise disjoint* are the segments $d_{n(i)}(x_i)$, $i \in I$.

Exercise 8.25. Explain.

Further, since, by the *σ-additivity* of the Lebesgue measure λ,

$$\sum_{i\in I} \lambda(d_{n(i)}(x_i)) = \lambda\left(\bigcup_{i\in I} d_{n(i)}(x_i)\right),$$

we have

$$\forall\, c > p:\ \lambda^*(f(E)) < c\lambda\left(\bigcup_{i\in I} d_{n(i)}(x_i)\right).$$

Considering that

$$\bigcup_{i\in I} d_{n(i)}(x_i) \subseteq G,$$

by the *monotonicity* of the Lebesgue outer measure λ^* (see Definition 4.2) and (8.21), we infer that

$$\forall\, c > p, \varepsilon > 0:\ \lambda^*(f(E)) < c\lambda\left(\bigcup_{i\in I} d_{n(i)}(x_i)\right) \le c\lambda(G) < c\left[\lambda^*(E) + \varepsilon\right].$$

Where, passing to the limit as $c \to p+$ and $\varepsilon \to 0+$, we arrive at the desired estimate

$$\lambda^*(f(E)) \le p\lambda^*(E).$$

(3) Observe that, for $q = 0$, the statement is trivially true.
Suppose that $q > 0$. Since f is *strictly increasing*, as is shown in the proof of part (2) (see (8.25)), $\lambda^*(f(E)) < \infty$, which, by *Approximation of Sets with Finite Outer Measure Corollary* (Corollary 4.1) (Section 4.8, Problem 14), implies that, for any $\varepsilon > 0$, there exists an *open* set $G \subseteq \mathbb{R}$ such that

$$f(E) \subseteq G \quad \text{and} \quad \lambda(G) < \lambda^*(f(E)) + \varepsilon. \tag{8.26}$$

Let $x \in E$ be arbitrary. Then there is a sequence $(x_n)_{n\in\mathbb{N}} \subset [a,b] \setminus \{x\}$ such that

$$\lim_{n\to\infty} x_n = x \quad \text{and} \quad \lim_{n\to\infty} \frac{f(x_n) - f(x)}{x_n - x} = \mathrm{D}f(x) \geq q,$$

and hence, for any $0 < c < q$, without loss of generality, we can regard the estimate

$$\frac{f(x_n) - f(x)}{x_n - x} > c \tag{8.27}$$

to hold for all $n \in \mathbb{N}$.
Then for the nondegenerate segments $d_n(x)$ and $\Delta_n(x)$, $x \in E$, $n \in \mathbb{N}$, defined as in the proof of part (2), we have

$$\forall\, 0 < c < q, x \in E, n \in \mathbb{N} : \lambda(\Delta_n(x)) > c\lambda(d_n(x)). \tag{8.28}$$

Let

$$C := \{x \in E \,|\, f \text{ is } \textit{continuous} \text{ at } x\}.$$

By the *Discontinuities of Monotone Functions Theorem* (Theorem 8.5), we infer that the set $E \setminus C$ is *countable*, which implies that

$$\lambda^*(E \setminus C) = \lambda(E \setminus C) = 0$$

(see Examples 4.9), and hence, by the *monotonicity* of the Lebesgue outer measure λ^* (see Definition 4.2),

$$\lambda^*(C) \leq \lambda^*(E) \leq \lambda^*(E \setminus C) + \lambda^*(C) = \lambda^*(C),$$

i. e.,

$$\lambda^*(E) = \lambda^*(C).$$

For each $x \in C$, by the *continuity* of f at x, for all sufficiently large $n \in \mathbb{N}$,

$$\Delta_n(x) \subseteq G.$$

Exercise 8.26. Explain.

Without loss of generality, we can regard that the prior inclusion holds for all $n \in \mathbb{N}$.
As is easily seen, the segment collection

$$\mathscr{C} := \{d_n(x) \,|\, x \in C, n \in \mathbb{N}\}$$

is a *Vitali cover* for the set C (see Examples 8.2).

Exercise 8.27. Explain.

Hence, by the *Vitali Covering Lemma* (Theorem 8.3), $\mathscr{C}$ has a countable subcollection $\{d_{n(i)}(x_i)\}_{i\in I}$ ($I \subseteq \mathbb{N}$) consisting of *pairwise disjoint* segments such that

$$\lambda^*\left(C \setminus \bigcup_{i\in I} d_{n(i)}(x_i)\right) = 0.$$

Where, since

$$C \subseteq \left[C \setminus \bigcup_{i\in I} d_{n(i)}(x_i)\right] \cup \bigcup_{i\in I} d_{n(i)}(x_i),$$

by the *monotonicity* and *σ-subadditivity* of the Lebesgue outer measure λ^* (see Definition 4.2) and (8.28), we infer that

$$\begin{aligned}\forall\, 0 < c < q : \lambda^*(E) = \lambda^*(C) &\le \lambda^*\left(C \setminus \bigcup_{i\in I} d_{n(i)}(x_i)\right) + \lambda^*\left(\bigcup_{i\in I} d_{n(i)}(x_i)\right)\\ &= \lambda^*\left(\bigcup_{i\in I} d_{n(i)}(x_i)\right) \le \sum_{i\in I} \lambda^*\left(d_{n(i)}(x_i)\right) = \sum_{i\in I} \lambda\left(d_{n(i)}(x_i)\right)\\ &< \frac{1}{c}\sum_{i\in I} \lambda(\Delta_{n(i)}(x_i)).\end{aligned}$$

Since f is *strictly increasing* along with the segments $d_{n(i)}(x_i)$, $i \in I$, *pairwise disjoint* are the segments $\Delta_{n(i)}(x_i)$, $i \in I$.

Further, since, by the *σ-additivity* of the Lebesgue measure λ,

$$\sum_{i\in I} \lambda(\Delta_{n(i)}(x_i)) = \lambda\left(\bigcup_{i\in I} \Delta_{n(i)}(x_i)\right),$$

we have

$$\forall\, 0 < c < q : \lambda^*(E) < \frac{1}{c}\lambda\left(\bigcup_{i\in I} \Delta_{n(i)}(x_i)\right).$$

Considering that

$$\bigcup_{i\in I} \Delta_{n(i)}(x_i) \subseteq G,$$

by the *monotonicity* of the Lebesgue outer measure λ^* (see Definition 4.2) and (8.26), we infer that

$$\forall\, 0 < c < q, \varepsilon > 0 : \lambda^*(E) < \frac{1}{c}\lambda\left(\bigcup_{i\in I} \Delta_{n(i)}(x_i)\right) \le \frac{1}{c}\lambda(G) < \frac{1}{c}\left[\lambda^*(f(E)) + \varepsilon\right].$$

Where, passing to the limit as $c \to p-$ and $\varepsilon \to 0+$, we arrive at

$$\lambda^*(E) \le \frac{1}{p}\lambda^*(f(E)),$$

or equivalently, the desired estimate

$$\lambda^*(f(E)) \ge q\lambda^*(E).$$ □

The prior theorem has the following two useful corollaries.

Corollary 8.2 (Infinite Derivative Number Set). *Let $f : [a,b] \to \mathbb{R}$ $(-\infty < a < b < \infty)$ be an increasing function. Then*

$$E_\infty := \{x \in [a,b] \mid \exists Df(x) : Df(x) = \infty\}$$

is a null set relative to the Lebesgue measure, i. e., $\lambda^*(E_\infty) = \lambda(E_\infty) = 0$.

Proof. Suppose first that the function f is *strictly increasing*. Then the assumption that

$$\lambda^*(E_\infty) > 0$$

would imply, by the *Derivative Numbers of Increasing Functions Theorem* (Theorem 8.8, part (3)), that

$$\lambda^*(f(E_\infty)) = \infty,$$

Exercise 8.28. Explain.

The latter, by the *monotonicity* of the Lebesgue outer measure λ^* (see Definition 4.2), *contradicts* the fact that

$$f(E_\infty) \subseteq [f(a), f(b)] \text{ with } \lambda^*([f(a), f(b)]) = \lambda([f(a), f(b)]) = f(b) - f(a) < \infty$$

implying that the assumption is *false*, and hence,

$$\lambda(E_\infty) = \lambda^*(E_\infty) = 0.$$

If f is *not* strictly increasing, then setting

$$g(x) := f(x) + x, \ x \in [a,b],$$

we obtain a *strictly increasing* function such that

$$\frac{g(x) - g(x_0)}{x - x_0} = \frac{f(x) - f(x_0)}{x - x_0} + 1, \ x, x_0 \in [a,b], x \ne x_0,$$

and hence,

$$E_\infty := \{x \in [a,b] \mid \exists Df(x) : Df(x) = \infty\} = \{x \in [a,b] \mid \exists Dg(x) : Dg(x) = \infty\},$$

which, by the proven above, implies that

$$\lambda(E_\infty) = \lambda^*\left(\{x \in [a,b] \mid \exists Dg(x) : Dg(x) = \infty\}\right) = 0. \qquad \square$$

Corollary 8.3 (Inequality Set). *Let $f : [a,b] \to \mathbb{R}$ $(-\infty < a < b < \infty)$ be an increasing function. Then*

(1) *for any $0 < p < q < \infty$,*

$$E_{p,q} := \{x \in [a,b] \mid \exists D_1f(x), D_2f(x) : D_1f(x) < p < q < D_2f(x)\} \tag{8.29}$$

is a null set relative to the Lebesgue measure, i. e., $\lambda^*(E_{p,q}) = \lambda(E_{p,q}) = 0$;

(2)

$$N := \{x \in [a,b] \mid \exists D_1f(x), D_2f(x) : D_1f(x) \neq D_2f(x)\} \tag{8.30}$$

is a null set relative to the Lebesgue measure, i. e., $\lambda^*(N) = \lambda(N) = 0$.

Proof.

(1) Suppose first that the function f is *strictly increasing*. Then, by the *Derivative Numbers of Increasing Functions Theorem* (Theorem 8.8, parts (2) and (3)),

$$q\lambda^*(E_{p,q}) \le \lambda^*(f(E_{p,q})) \le p\lambda^*(E_{p,q}),$$

which, in view of the fact that $p < q$, implies that

$$\lambda^*(E_{p,q}) = 0,$$

and hence the set $E_{p,q}$ is Lebesgue measurable and

$$\lambda(E_{p,q}) = 0$$

(see Remarks 4.4).

Exercise 8.29. Explain.

If f is not strictly increasing, we can apply the above to the strictly increasing

$$g(x) := f(x) + x,\ x \in [a,b],$$

and replacing p and q by $p+1$ and $q+1$, respectively.

Exercise 8.30. Fill in the details.

(2) We can equivalently redefine the set N defined (8.30) by as follows:

$$N := \{x \in [a,b] \mid \exists D_1 f(x), D_2 f(x) : D_1 f(x) < D_2 f(x)\},$$

and hence, it allows the following representation as a *countable union*:

$$N = \bigcup_{p,q \in \mathbb{Q}} E_{p,q}, \tag{8.31}$$

were the sets $E_{p,q}$ are defined by (8.29).

Since, in view of the proven in part (1), each set $E_{p,q}$ in representation (8.31) is a *λ-null set*, the rest follows by the *σ-subadditivity* of the Lebesgue outer measure λ^* (see Definition 4.2) (cf. Section 9.7, Problem 1).

Exercise 8.31. Verify equality (8.31) and fill in the details. □

Remark 8.7. The two prior corollaries do not contradict the results of Examples 8.1, 2 and 3, since the functions in the aforementioned examples are *not* increasing on any segment $[a,b]$ $(-\infty < a < b < \infty)$.

8.3.4 Differentiability of Monotone Functions

The *Inequality Set Corollary* (Corollary 8.3) has the following consequential implication.

Theorem 8.9 (Differentiability of Increasing Functions). *An increasing function* $f : [a,b] \to \mathbb{R}$ $(-\infty < a < b < \infty)$ *has a (finite or infinite) derivative* $f'(x)$ *a. e. relative to the Lebesgue measure* λ *on* $[a,b]$ *and is differentiable (i. e., has a finite derivative* $f'(x)$*) a. e. relative to the Lebesgue measure* λ *on* $[a,b]$.

Proof. By the *Inequality Set Corollary* (Corollary 8.3, part (2)) and the *Uniqueness of Derivative Numbers Theorem* (Theorem 8.2), $f'(x)$ *exists* a. e. relative to the Lebesgue measure λ on $[a,b]$.

Exercise 8.32. Explain.

Further, since also, by the *Infinite Derivative Number Set* (Corollary 8.2), $f'(x) < \infty$ a. e. λ on $[a,b]$, we infer that f is *differentiable* a. e. relative to the Lebesgue measure λ on $[a,b]$. □

We immediately arrive at the subsequent generalization.

Theorem 8.10 (Lebesgue's Theorem). *A monotone function* $f : [a,b] \to \mathbb{R}$ $(-\infty < a < b < \infty)$ *has a (finite or infinite) derivative* $f'(x)$ *a. e. relative to the Lebesgue measure* λ *on* $[a,b]$ *and is differentiable (i. e., has a finite derivative* $f'(x)$*) a. e. relative to the Lebesgue measure* λ *on* $[a,b]$.

Exercise 8.33. Explain why the statement is true for a *decreasing* function $f : [a, b] \to \mathbb{R}$ $(-\infty < a < b < \infty)$.

Remark 8.8. In view of the fact that the (finite or infinite) derivative f' of a *monotone function* $f : [a, b] \to \mathbb{R}$ $(-\infty < a < b < \infty)$ exists a. e. relative to the Lebesgue measure λ on $[a, b]$, henceforth, we can regard f' to be defined on the entire interval $[a, b]$ as follows:

$$f'(x) := \begin{cases} f'(x) & \text{for those } x \in [a,b] \text{ at which } f \text{ has a derivative,} \\ 0 & \text{for those } x \in [a,b] \text{ at which } f \text{ does not have a derivative} \end{cases}$$

or, if necessary, as follows:

$$f'(x) := \begin{cases} f'(x) & \text{for those } x \in [a,b] \text{ at which } f \text{ is differentiable,} \\ 0 & \text{for those } x \in [a,b] \text{ at which } f \text{ is not differentiable,} \end{cases}$$

due to the *completeness* of the Lebesgue measure on $([a, b], \Sigma^*([a, b]))$ (see Section 4.7.1), obtaining in both cases a Lebesgue measurable function $f' : [a, b] \to [0, \infty]$ or $f' : [a, b] \to [0, \infty)$, respectively (see Examples 5.3 and *Borel Measurability of Derivative Proposition* (Proposition 5.7) (Section 5.12, Problem 9)).

8.3.5 Total Change Estimate

Theorem 8.11 (Total Change Estimate). *Let $f : [a, b] \to \mathbb{R}$ $(-\infty < a < b < \infty)$ be an monotone function on $[a, b]$. Then $f' \in L([a, b], \lambda)$:*

(1) *if the function f is increasing,*

$$\int_{[a,b]} f'(x)\, d\lambda(x) \le f(b) - f(a); \tag{8.32}$$

(2) *if the function f is decreasing,*

$$\int_{[a,b]} f'(x)\, d\lambda(x)\, dx \ge f(a) - f(b). \tag{8.33}$$

Proof. The case of a decreasing function f immediately follows from the case of an increasing function.

Exercise 8.34. Explain.

Hence, let us suppose that $f : [a, b] \to \mathbb{R}$ $(-\infty < a < b < \infty)$ be an *increasing function* on $[a, b]$.

Let us extend the function of f to an *increasing function* on the segment $[a, b + 1]$ as follows:

$$f(x) := \begin{cases} f(x), & a \le x \le b, \\ f(b), & b < x \le b+1. \end{cases}$$

Then, for all $x \in [a, b]$ at which f has a derivative, except, possibly, at b,

$$f'(x) = \lim_{n\to\infty} n\,[f(x+1/n) - f(x)]\,,$$

which, by *Lebesgue's Theorem* (Theorem 8.10), is a. e. relative to the Lebesgue measure λ on $[a, b]$.

Extending the domain of the derivative function f' to the entire segment $[a, b]$ as in Remark 8.8 and considering that the Lebesgue measure λ is *complete* (see Section 4.7.1), by the *Measurability of Limit A. E. Theorem* (Theorem 5.9), the derivative function $f' : [a, b] \to [0, \infty]$ is Lebesgue measurable as the limit a. e. λ on $[a, b]$ of the sequence

$$g_n(x) := n\,[f(x+1/n) - f(x)]\,,\ n \in \mathbb{N}, x \in [a, b], \tag{8.34}$$

of Borel, and hence, Lebesgue measurable functions (see Examples 5.3 and cf. *Borel Measurability of Derivative Proposition* (Proposition 5.7) (Section 5.12, Problem 9)).

Since the derivative function $f' : [a, b] \to [0, \infty]$ is *nonnegative*, its (finite or infinite) Lebesgue integral

$$\int_{[a,b]} f'(x)\,d\lambda(x) \in [0, \infty]$$

exists (see Definition 6.2).

Since the extended function f is *increasing* on $[a, b+1]$, the functions g_n defined in (8.34) are *nonnegative*, and hence, by *Fatou's Lemma* (Theorem 6.6) and the *Properties of Lebesgue Integral. II* (Theorem 6.3, part (3)),

$$\int_{[a,b]} f'(x)\,d\lambda(x) \le \varliminf_{n\to\infty} \int_{[a,b]} g_n(x)\,d\lambda(x) = \varliminf_{n\to\infty} n \int_{[a,b]} [f(x+1/n) - f(x)]\,d\lambda(x); \tag{8.35}$$

also, for each $n \in \mathbb{N}$, by the *Riemann Integrability of a Monotone Function Corollary* (Corollary 8.1), the latter integral can be understood in the Riemann sense, and hence, by the properties of the Riemann integral (see, e. g., [13]),

$$\begin{aligned} \forall\, n \in \mathbb{N} : \int_{[a,b]} [f(x+1/n) - f(x)]\,d\lambda(x) &= \int_a^b [f(x+1/n) - f(x)]\,dx \\ &= \int_a^b f(x+1/n)\,dx - \int_a^b f(x)\,dx = \int_{a+1/n}^{b+1/n} f(x)\,dx - \int_a^b f(x)\,dx \end{aligned}$$

$$= \int_b^{b+1/n} f(x)\,dx - \int_a^{a+1/n} f(x)\,dx \le \int_b^{b+1/n} f(b)\,dx - \int_a^{a+1/n} f(a)\,dx = \frac{1}{n}[f(b) - f(a)]$$
$$\le f(b) - f(a). \tag{8.36}$$

Exercise 8.35. Verify each step.

From (8.35) and (8.36), we infer that

$$\int_{[a,b]} f'(x)\,d\lambda(x) \le f(b) - f(a) < \infty,$$

which, in particular, implies that $f' \in L([a,b],\lambda)$ (see Definition 6.4). □

Remark 8.9. As the following example shows, the derivative of a *nonmonotone* function $f : [a,b] \to \mathbb{R}$ $(-\infty < a < b < \infty)$, which is even continuous, need not be Lebesgue integrable on $[a,b]$.

Example 8.3. The function

$$f(x) := \begin{cases} x^2 \sin(x^{-2}), & 0 < x \le 1, \\ 0, & x = 0, \end{cases}$$

is differentiable but not monotone on $[0,1]$ and its derivative function f', which is only discontinuous at 0, is *not* Lebesgue integrable on $[0,1]$.

Exercise 8.36. Verify.

Remark 8.10. The momentous example of the *Cantor function* considered in the subsequent section shows that the inequality in (8.32) and (8.33) can be strict, which means that, the *total change formula*:

$$\int_{[a,b]} f'(x)\,d\lambda(x) = f(b) - f(a),$$

which is in place for any continuously differentiable on $[a,b]$ function $f : [a,b] \to \mathbb{R}$ ($f \in C^1[a,b]$), i. e., when f is differentiable on $[a,b]$ and the derivative function f' is *continuous* on $[a,b]$, *does not* stretch to monotone functions.

8.3.6 The Cantor Function

The following celebrated example is of the *Cantor function* $c : [0,1] \to [0,1]$, which is increasing and continuous on $[0,1]$ and is not a constant, but

$$c'(x) = 0 \text{ a. e. } \lambda \text{ on } [0,1], \tag{8.37}$$

and hence,

$$\int_{[0,1]} c'(x)\,d\lambda(x) = 0 < 1 = c(1) - c(0). \tag{8.38}$$

Let C be the *Cantor set*. Its complementary intervals can be naturally grouped as follows:

- group 1 consists of $(1/3, 2/3)$,
- group 2 consists of two intervals $(1/9, 2/9)$ and $(7/9, 8/9)$;
- group 3 consists of four intervals $(1/27, 2/27)$, $(7/27, 8/27)$, $(19/27, 20/27)$, $(25/27, 26/27)$; etc.

The nth group consists of 2^{n-1} intervals of length 3^{-n}.

Let

$$c(x) := 1/2,\ x \in (1/3, 2/3),\ c(x) := 1/4,\ x \in (1/9, 2/9),\ c(x) := 3/4,\ x \in (7/9, 8/9).$$

On the four intervals of group 3, the function f is consecutively equal

$$\frac{1}{8}, \frac{3}{8}, \frac{5}{8}, \frac{7}{8}.$$

On the 2^{n-1} intervals of group n, the function f is consecutively equal

$$\frac{1}{2^n}, \frac{3}{2^n}, \frac{5}{2^n}, \dots, \frac{2^n - 1}{2^n}.$$

Thus, c is defined on the *open Cantor set* $G := [0,1] \setminus C$, is *constant* on each of its constituent intervals, and is *increasing* on G.

Exercise 8.37. Verify.

Let us define c on the Cantor set C setting $c(0) := 0$, $c(1) := 1$, and, for any $x_0 \in C \cap (0,1)$,

$$c(x_0) := \sup\{c(x) \mid x \in G, x < x_0\}$$

Now, c is defined and *increasing* on the whole interval $[0,1]$.

Exercise 8.38. Verify.

Furthermore, $c : [0,1] \to [0,1]$ is *continuous* on $[0,1]$. Indeed if $x_0 \in [0,1]$ is a point of discontinuity of c, then necessarily $x \in C$ and, since c is increasing at least one of the intervals $(c(x_0-), c(x_0))$ or $(c(x_0), c(x_0+))$ does not contain any value of c, which *contradicts* the fact that the set $c(G)$ is *dense* in $[0,1]$.

Thus, $c : [0,1] \to [0,1]$ is an *increasing continuous function* with

$$c'(x) = 0,\ x \in G,$$

and hence, (8.37) and (8.38) hold.

Remark 8.11. The Cantor function's example also demonstrates that a continuous function $f : [a, b] \to \mathbb{R}$ $(-\infty < a < b < \infty)$, whose derivative is equal to 0 a. e. relative to the Lebesgue measure λ on $[a, b]$, need not be constant.

8.4 Functions of Bounded Variation

In this section, we are to study the *functions of bounded variation*, which encompass the monotone ones, and to prove, in particular, the *Jordan Decomposition Theorem for Functions* (Theorem 8.17) showing that any real-valued such a function can be represented as the difference of two increasing functions.

8.4.1 Definition, Examples, Properties

Definition 8.5 (Variation, Total Variation). Let f be a real- or complex-valued function on a segment $[a, b]$ $(-\infty < a < b < \infty)$ and

$$P := \{a = x_0 < x_1 < \cdots < x_n = b\} \quad (n \in \mathbb{N})$$

be a partition of $[a, b]$.

The *variation of f on $[a, b]$ relative to P* is the nonnegative number

$$V_a^b(f, P) := \sum_{k=1}^{n} |f(x_k) - f(x_{k-1})| \in [0, \infty).$$

The *total variation of f on $[a, b]$* is the (possibly infinite) nonnegative value

$$V_a^b(f) := \sup\left\{V_a^b(f, P) \,\middle|\, P \text{ is a } \textit{partition} \text{ of } [a, b]\right\} \in [0, \infty].$$

Remark 8.12. Naturally, we can define $V_a^a(f) := 0$.

Exercise 8.39. What is the total variation of a *step function* (see Definition 8.4)?

Definition 8.6 (Functions of Bounded Variation). A real- or complex-valued function on f on a segment $[a, b]$ $(-\infty < a < b < \infty)$ with

$$V_a^b(f) < \infty,$$

is said to be of *bounded variation on* $[a, b]$.

The set of all such functions is denoted by $\mathrm{BV}[a, b]$.

More generally, a real- or complex-valued function f on an nondegenerate interval $I \subseteq \mathbb{R}$ is said to be *of bounded variation on I* if, for an arbitrary segment $[a, b] \subseteq I$, f is of bounded variation on $[a, b]$.

Theorem 8.12 (Monotone is of Bounded Variation).

(1) *An increasing function* $f : [a,b] \to \mathbb{R}$ *on a segment* $[a,b]$ $(-\infty < a < b < \infty)$ *is of bounded variation on* $[a,b]$ *with*

$$V_a^b(f) = f(b) - f(a).$$

(2) *A decreasing function* $f : [a,b] \to \mathbb{R}$ *on a segment* $[a,b]$ $(-\infty < a < b < \infty)$ *is of bounded variation on* $[a,b]$ *with*

$$V_a^b(f) = f(a) - f(b).$$

Exercise 8.40. Prove.

Examples 8.4.

1. In particular, the *Cantor function* $c : [0,1] \to [0,1]$ (see Section 8.3.6) is of bounded variation on $[0,1]$ and

$$V_0^1(c) = c(1) - c(0) = 1 - 0 = 1.$$

2. Generally,

$$V_a^b(f) \neq |f(b) - f(a)|.$$

For example, for the function

$$f(x) := \begin{cases} 0, & x = 0, \\ 1, & 0 < x \le 1/2, \\ 0, & 1/2 < x \le 1, \end{cases}$$
$$V_0^1(f) = 2 > |f(1) - f(0)| = 0.$$

3. The *bounded* on $[0,1]$ function

$$f(x) := \begin{cases} \cos\frac{\pi}{2x}, & 0 < x \le 1 \\ 0, & x = 0, \end{cases}$$

is *not* of bounded variation on $[0,1]$. Indeed, for any $n \in \mathbb{N}$ and the partition

$$P_n := \{0, 1/(2n), 1/(2n-1), \dots, 1/3, 1/2, 1\},$$

we have

$$V_0^1(f, P_n) = \underbrace{2 + \cdots + 2}_{n \text{ terms}} = 2n \to \infty,\ n \to \infty,$$

and hence,

$$V_0^1(f) = \infty.$$

4. The *continuous* on $[0,1]$ function

$$f(x) := \begin{cases} x\cos\frac{\pi}{2x}, & 0 < x \le 1 \\ 0, & x = 0, \end{cases}$$

is *not* of bounded variation on $[0,1]$. Indeed, for any $n \in \mathbb{N}$ and the partition

$$P_n := \{0, 1/(2n), 1/(2n-1), \dots, 1/3, 1/2, 1\},$$

we have

$$V_0^1(f, P_n) = \frac{2}{2n} + \dots + \frac{2}{2} = \sum_{k=1}^{n} \frac{1}{k} \to \infty,\ n \to \infty,$$

and hence,

$$V_0^1(f) = \infty,$$

which also implies that the graph of f is *not* a rectifiable curve (see the *Characterization of Continuous Functions of Bounded Variation* (Proposition 8.2) Section 8.6, Problem 3).

5. A *Lipschitz*[1] *continuous function* $f : [a,b] \to \mathbb{F}$ $(-\infty < a < b < \infty$, $\mathbb{F} = \mathbb{R}$ or $\mathbb{F} = \mathbb{C})$, i. e., such that

$$\exists L > 0\ \forall x, x' \in [a,b] : |f(x') - f(x)| \le L|x' - x|. \tag{8.39}$$

We denote the set of all such functions by $\mathrm{LC}[a,b]$.
In particular, a *continuously differentiable function* $f \in C^1[a,b]$, i. e., with a continuous derivative function $f' \in C[a,b]$, is Lipschitz continuous, and hence, of bounded variation on $[a,b]$. However, as the example of the absolute value function

$$f(x) := |x|,\ x \in [-1,1],$$

shows the converse is not true, i. e., a Lipschitz continuous function need not be continuously differentiable.
Thus, we have the following chain of *strict* inclusions:

$$C^1[a,b] \subset \mathrm{LC}[a,b] \subset \mathrm{BV}[a,b]. \tag{8.40}$$

Exercise 8.41.

(a) Show that $f : [a,b] \to \mathbb{C}$ $(-\infty < a < b < \infty)$ is of bounded variation on $[a,b]$ *iff* $\mathrm{Re} f, \mathrm{Im} f : [a,b] \to \mathbb{R}$ are of bounded variation on $[a,b]$.

1 Sigismund Lipschitz (1832–1903).

Hint. Use the estimate

$$\max\left[|\operatorname{Re} f(x)|, |\operatorname{Im} f(x)|\right] \le |f(x)| \le |\operatorname{Re} f(x)| + |\operatorname{Im} f(x)|, \; x \in [a, b]. \tag{8.41}$$

(b) Verify the prior examples.

Hint. In 4, for $f : [a, b] \to \mathbb{R}$, apply the *Mean Value Theorem*; for $f : [a, b] \to \mathbb{C}$, separate the real and imaginary parts and apply inequalities 8.41.

Theorem 8.13 (Characterization of Constant Functions). *A real- or complex-valued function f on a segment $[a, b]$ $(-\infty < a < b < \infty)$ is constant iff $V_a^b(f) = 0$.*

Exercise 8.42. Prove.

Theorem 8.14 (Of Bounded Variation is Bounded). *Every function f of bounded variation on a segment $[a, b]$ $(-\infty < a < b < \infty)$ is bounded on $[a, b]$ and*

$$\sup_{a \le x \le b} |f(x)| \le |f(a)| + V_a^b(f). \tag{8.42}$$

Proof. Let $f \in BV[a, b]$ be arbitrary. Then, for any $x \in [a, b]$,

$$|f(x)| - |f(a)| \le |f(x) - f(a)| + |f(b) - f(x)| \le V_a^b(f).$$

Where

$$|f(x)| \le |f(a)| + V_a^b(f), \; x \in [a, b],$$

which implies that f is bounded on $[a, b]$ and estimate (8.42) holds. □

Remark 8.13. As Examples 8.4 show, the converse statement is not true, i. e., a bounded on a segment $[a, b]$ $(-\infty < a < b < \infty)$ function (even a continuous one) need not be of bounded variation on $[a, b]$.

Theorem 8.15 (Combinations of Functions of Bounded Variation). *The sum, difference, and product of real- or complex-valued functions of bounded variation on a segment $[a, b]$ $(-\infty < a < b < \infty)$ are of bounded variation on $[a, b]$. More precisely,*

(1) *for any $f, g \in BV[a, b]$,*

$$V_a^b(f \pm g) \le V_a^b(f) + V_a^b(g);$$

(2) *for any $f, g \in BV[a, b]$, $fg \in BV[a, b]$ and*

$$V_a^b(fg) \le \sup_{[a,b]} |g| V_a^b(f) + \sup_{[a,b]} |f| V_a^b(g).$$

In particular, for an arbitrary $f \in BV[a, b]$ and any scalar c, $cf \in BV[a, b]$ and

$$V_a^b(cf) = |c| V_a^b(f).$$

Exercise 8.43. Prove.

Hint. In the case of *product*, apply the prior theorem.

Remarks 8.14.

- Thus, for any $-\infty < a < b < \infty$, the set $BV[a,b]$ of functions of bounded variation on $[a,b]$ is a (real or complex) *vector space* (moreover, an *algebra*) relative to the pointwise operations (see, e. g., [14]).
- Concerning the *quotient* of functions of bounded variation, see Section 8.6, Problem 4.

8.4.2 Additivity of Total Variation, Total Variation Function

Lemma 8.2 (Partition Lemma). *Let $f \in BV[a,b]$ $(-\infty < a < b < \infty)$. For any partitions P and P' of $[a,b]$ such that P' is finer than P, i. e., $P \subseteq P'$,*

$$V_a^b(f,P) \le V_a^b(f,P').$$

Exercise 8.44. Prove (cf. Remark 6.18).

Theorem 8.16 (Additivity of Total Variation). *For a real- or complex-valued function on f on a segment $[a,b]$ $(-\infty < a < b < \infty)$ to be of bounded variation on $[a,b]$ it is necessary that, for any $c \in (a,b)$, and sufficient that, for some $c \in (a,b)$, $f \in BV[a,c] \cap BV[c,b]$, in which case*

$$V_a^b(f) = V_a^c(f) + V_c^b(f).$$

Proof. Let us fix a $c \in (a,b)$.

By the prior *Partition Lemma* (Lemma 8.2), in the definition of $V_a^b(f)$, the supremum can be taken over all partitions P containing the point c, every such a partition both generating and being generated by partitions P_1 and P_2 of $[a,c]$ and $[c,b]$, respectively, and

$$V_a^b(f,P) = V_a^c(f,P_1) + V_c^b(f,P_2).$$

Exercise 8.45. Explain.

Taking the supremum over all such partitions, we arrive at

$$V_a^b(f) = V_a^c(f) + V_c^b(f),$$

which also proves the *necessity*, in which case $c \in (a,b)$ can be arbitrary, and the *sufficiency*.

Exercise 8.46. Explain. □

Remark 8.15. Hence, for any function $f \in \mathrm{BV}[a,b]$ $(-\infty < a < b < \infty)$, the *total variation function of f*

$$v(x) := V_a^x(f),\ x \in [a,b], \tag{8.43}$$

is *well-defined* and *increasing* on $[a,b]$.

Exercise 8.47. Verify.

8.4.3 Jordan Decomposition Theorem

Now, we are ready to prove the following fundamental statement.

Theorem 8.17 (Jordan Decomposition Theorem for Functions). *A real-valued function $f : [a,b] \to \mathbb{R}$ is of bounded variation on a segment $[a,b]$ $(-\infty < a < b < \infty)$ iff it can be represented as the difference of two increasing functions on $[a,b]$, such a representation being called a Jordan decomposition of the function f.*

Proof. *"If"* part immediately follows from the *Monotone is of Bounded Variation Theorem* (Theorem 8.12) and the *Combinations of Functions of Bounded Variation Theorem* (Theorem 8.15).

"Only if" part. Let a function $f : [a,b] \to \mathbb{R}$ $(-\infty < a < b < \infty)$ be of bounded variation on $[a,b]$. Then the function

$$g(x) := v(x) - f(x),\ x \in [a,b],$$

where $v(x)$, $x \in [a,b]$, is the *total variation function of f* (see (8.43)), is *increasing* on $[a,b]$. Indeed, for any $a \le x < y \le b$, by the definition of the total variation and the prior theorem,

$$\begin{aligned}\forall\, a \le x < x' \le b :\ f(x') - f(x) &\le |f(x') - f(x)| \le V_x^{x'}(f)\\ &= V_a^{x'}(f) - V_a^x(f) =: v(x') - v(x),\end{aligned}$$

and hence,

$$\forall\, a \le x < x' \le b :\ g(x) := v(x) - f(x) \le v(x') - f(x') =: g(x').$$

Also, clearly,

$$f(x) = v(x) - g(x),\ x \in [a,b]. \qquad \square$$

Remark 8.16. A *Jordan decomposition* of a real-valued function $f \in \mathrm{BV}[a,b]$ $(-\infty < a < b < \infty)$ is *not* unique. Indeed, if

$$f(x) = f_1(x) - f_2(x),\ x \in [a,b],$$

is a Jordan decomposition of f, then, for any increasing function $h : [a, b] \to \mathbb{R}$, e. g., $h(x) := x$, $x \in [a, b]$, so is

$$f(x) = [f_1(x) + h(x)] - [f_2(x) + h(x)], \ x \in [a, b],$$

8.4.4 Derivative of a Function of Bounded Variation

Corollary 8.4 (Derivative of a Function of Bounded Variation). *A real- or complex-valued function f of bounded variation on a segment $[a, b]$ $(-\infty < a < b < \infty)$ is differentiable a. e. relative to the Lebesgue measure λ on $[a, b]$ and $f' \in L([a, b], \lambda)$.*

Proof. For a real-valued $f : [a, b] \to \mathbb{R}$, the statement follows immediately from *Jordan Decomposition Theorem for Functions* (Theorem 8.17), *Lebesgue's Theorem* (Theorem 8.10) (see also Remark 8.8), the *Total Change Estimate Theorem* (Theorem 8.11), and the fact that, due to the *linearity* of the Lebesgue integral, $L([a, b], \lambda)$ is a *vector space* (see Remarks 6.11).

Exercise 8.48. Explain.

For a complex-valued $f : [a, b] \to \mathbb{C}$, the statement follows from the real case by separating the real and imaginary parts and the fact that, due to the *linearity* of the Lebesgue integral, $L([a, b], \lambda)$ is a *vector space* (see Remarks 6.11).

Exercise 8.49. Explain. □

Remarks 8.17.

- Thus, by the prior theorem, any real- or complex-valued function f on a segment $[a, b]$ $(-\infty < a < b < \infty)$, whose derivative function f' is *not* Lebesgue integrable on $[a, b]$, is *not* of bounded variation on $[a, b]$. In particular, the function from Example 8.3 is *not* of bounded variation on $[0, 1]$.
- As the following example shows, for a function $f \in BV[a, b]$ $(-\infty < a < b < \infty)$, the derivative function f' need not be bounded on $[a, b]$, which in particular, implies that a function of bounded variation need not be Lipschitz continuous (see (8.39)).

Example 8.5. By the *Monotone is of Bounded Variation Theorem* (Theorem 8.12), the *increasing* function

$$f(x) := \sqrt{x}, \ x \in [0, 1],$$

is of *bounded variation* on $[0, 1]$ and its derivative function

$$f'(x) := \begin{cases} \frac{1}{2\sqrt{x}}, & x \in (0, 1], \\ 0, & x = 0, \end{cases}$$

(see Remark 8.8) is Lebesgue integrable on $[0,1]$ by the *Improper Integral of Type 2 Theorem* (Theorem 6.12) (see also Remarks 6.23) but *unbounded* (cf. Example 8.7).

Exercise 8.50. Explain.

Remark 8.18. For the function f from the prior example, $f' \in R[x,1]$ for any $0 < x < 1$ and

$$\int_0^1 |f'(t)|\,dt := \lim_{x\to 0+}\int_x^1 \frac{1}{2}t^{-1/2}\,dt = \lim_{x\to 0+} x^{1/2}\Big|_{0+}^{1} = \frac{1}{2} < \infty,$$

and the *total change formula*

$$f(x) - f(0) = \int_0^x f'(t)\,dt,\ x \in [0,1],$$

holds.

The function f from the prior example is more than just continuous and of bounded variation on $[0,1]$, it is *absolutely continuous* on $[0,1]$. Such functions are introduced and studied in the subsequent section.

8.5 Absolutely Continuous Functions

The prior exercise motivates the following discourse of the concept of an *absolutely continuous* function $f : [a,b] \to \mathbb{F}$ on a segment $[a,b]$ $(-\infty < a < b < \infty)$, for which $f' \in L([a,b],\lambda)$ and the *total change formula*

$$f(x) - f(a) = \int_{[a,x]} f'(t)\,d\lambda(t),\ x \in [a,b],$$

holds.

8.5.1 Definition, Examples, Properties

Definition 8.7 (Absolutely Continuous Functions). A real- or complex-valued function f on a segment $[a,b]$ $(-\infty < a < b < \infty)$ is called *absolutely continuous on* $[a,b]$ if, for any $\varepsilon > 0$, there exists a $\delta > 0$ such that, for an arbitrary finite collection $\{(a_i,b_i)\}_{i=1}^n$ $(n \in \mathbb{N})$ of pairwise disjoint subintervals of $[a,b]$ such that

$$\sum_{i=1}^n (b_i - a_i) < \delta,$$

$$\sum_{i=1}^n |f(b_i) - f(a_i)| < \varepsilon.$$

The set of all such functions is denoted by $AC[a, b]$.

More generally, a real- or complex-valued function f on an nondegenerate interval $I \subseteq \mathbb{R}$ is said to be *absolutely continuous on* I if, for an arbitrary segment $[a, b] \subseteq I$, f is absolutely continuous on $[a, b]$.

Remarks 8.19.

- As follows from inequalities (8.41), a complex-valued function $f : [a, b] \to \mathbb{C}$ $(-\infty < a < b < \infty)$ is absolutely continuous on $[a, b]$ *iff* its real and imaginary parts $\operatorname{Re} f, \operatorname{Im} f : [a, b] \to \mathbb{R}$ are absolutely continuous on $[a, b]$.
- If $f \in AC[a, b]$, then

$$\forall\, [c, d] \subseteq [a, b] : f \in AC[c, d].$$

- Each real- or complex-valued function f absolutely continuous on a segment $[a, b]$ $(-\infty < a < b < \infty)$ is also *uniformly continuous* on $[a, b]$, i. e.,

$$\forall\, \varepsilon > 0\, \exists\, \delta > 0\, \forall\, x, x' \in [a, b],\ |x' - x| < \delta : |f(x') - f(x)| < \varepsilon,$$

which, considering that $[a, b]$ is a *compact set* in $\mathbb{R}$ (see Example 1.18), is equivalent to the conventional continuity of f on $[a, b]$ (see, e. g., [14])

However, as the following example shows, the converse is not true, i. e., a continuous on a segment $[a, b]$ function need not be absolutely continuous on $[a, b]$.

Exercise 8.51. Verify.

Example 8.6. The *Cantor function* $c : [0, 1] \to [0, 1]$ (see Section 8.3.6) is increasing and continuous but *not* absolutely continuous on $[0, 1]$. Indeed, for the pairwise disjoint collection $\{[a_i, b_i]\}_{i=1}^{2^n}$ of 2^n $(n \in \mathbb{N})$ subsegments of $[0, 1]$ obtained at the nth stage of constructing the Cantor set, we have

$$\sum_{i=1}^{2^n} (b_i - a_i) = \left(\frac{2}{3}\right)^n \quad \text{(can be made arbitrarily small)},$$

but

$$\sum_{i=1}^{2^n} [c(b_i) - c(a_i)] = c(1) - c(0) = 1$$

Exercise 8.52. Explain.

Remark 8.20. Thus, absolute continuity is stronger than uniform continuity. The subsequent statement and remark show that Lipschitz continuity is yet stronger than absolute continuity.

Proposition 8.1 (Lipschitz Continuous is Absolutely Continuous). *Each real- or complex-valued function f Lipschitz continuous on a segment $[a, b]$ $(-\infty < a < b < \infty)$ is also absolutely continuous on $[a, b]$.*

Exercise 8.53. Prove.

Remark 8.21. However, as the following example shows, the converse is not true, i. e., a real- or complex-valued function absolutely continuous on a segment $[a,b]$ $(-\infty < a < b < \infty)$ need not be Lipschitz continuous on $[a,b]$, and hence, we have the following chain of *strict* inclusions:

$$C^1[a,b] \subset \mathrm{LC}[a,b] \subset \mathrm{AC}[a,b] \tag{8.44}$$

(see (8.40)).

Example 8.7. The *increasing* and *continuous* function

$$f(x) := \sqrt{x},\ x \in [0,1],$$

is *absolutely continuous* but *not* Lipschitz continuous on $[0,1]$ (cf. Example 8.5).

Exercise 8.54. Verify.

Hint. Use the result of Section 8.6, Problem 7 (b).

Theorem 8.18 (Combinations of Absolutely Continuous Functions). *The sum, difference, and product of real- or complex-valued functions absolutely continuous on a segment $[a,b]$ $(-\infty < a < b < \infty)$ are absolutely continuous on $[a,b]$ and so is the quotient, provided the divisor does not take the zero value on $[a,b]$.*

Exercise 8.55. Prove.

Remarks 8.22.
- Thus, for any $-\infty < a < b < \infty$, the set $\mathrm{AC}[a,b]$ of absolutely continuous on $[a,b]$ functions is a (real or complex) *vector space* (moreover, an *algebra*) relative to the pointwise operations (see, e. g., [14]).
- The composition of two absolutely continuous functions need not be absolutely continuous (see, e. g., [19]).

8.5.2 Characterization of Absolute Continuity

The following statement, being similar to the *Jordan Decomposition Theorem for Functions* (Theorem 8.17), elucidates the nature of absolutely continuous functions.

Theorem 8.19 (Characterization of Absolute Continuity). *A real-valued function $f : [a,b] \to \mathbb{R}$ is absolutely continuous on a segment $[a,b]$ $(-\infty < a < b < \infty)$ iff it can be represented as the difference of two increasing and absolutely continuous functions on $[a,b]$, and hence, is of bounded variation on $[a,b]$.*

Proof. "*If*" part immediately follows from the *Combinations of Absolutely Continuous Functions Theorem* (Theorem 8.18).

"Only if" part. Suppose that a real-valued function $f : [a,b] \to \mathbb{R}$ is absolutely continuous on a segment $[a,b]$ $(-\infty < a < b < \infty)$.

Let us first show that f is necessarily of *bounded variation* on $[a,b]$, i. e.,

$$f \in BV[a,b].$$

Indeed, there exists a $\delta > 0$, which meets the absolute continuity condition for f on $[a,b]$ with $\varepsilon = 1$ (see Definition 8.7).

Hence, for any partition $P := \{x_0, \dots, x_m\}$ $(m \in \mathbb{N})$ of $[a,b]$ with

$$\max_{1\le k\le m}(x_k - x_{k-1}) < \delta$$

we have

$$V_{x_{k-1}}^{x_k}(f) \le 1,\ i = 1, \dots, m,$$

and hence, $f \in BV[x_{i-1}, x_i]$, $i = 1, \dots, m$.

Exercise 8.56. Explain.

Therefore, by the *Additivity of Total Variation Theorem* (Theorem 8.16), $f \in BV[a,b]$ and

$$V_a^b(f) = \sum_{i=1}^{m} V_{x_{k-1}}^{x_k}(f) \le m.$$

Hence, the *total variation function*

$$v(x) := V_a^x(f),\ x \in [a,b],$$

is well-defined and *increasing* on $[a,b]$. As is shown in the proof of the *Jordan Decomposition Theorem for Functions* (Theorem 8.17), the function $v - f$ is also *increasing* $[a,b]$ and, clearly,

$$f(x) = v(x) - [v(x) - f(x)],\ x \in [a,b].$$

By the *Combinations of Absolutely Continuous Functions Theorem* (Theorem 8.18), it suffices to show now that the total variation function v is absolutely continuous on $[a,b]$.

By the absolute continuity of f on $[a,b]$, for any $\varepsilon > 0$, there exists a $\delta > 0$ such that, for every finite collection $\{(a_i, b_i)\}_{i=1}^n$ $(n \in \mathbb{N})$ of pairwise disjoint subintervals of $[a,b]$ with

$$\sum_{i=1}^{n}(b_i - a_i) < \delta,$$

we have

$$\sum_{i=1}^{n} |f(b_i) - f(a_i)| < \varepsilon.$$

Then, for arbitrary partitions P_i of $[a_i, b_i]$, $i = 1, \dots, n$,

$$\sum_{i=1}^{n} V_{a_i}^{b_i}(f, P_i) < \varepsilon.$$

Exercise 8.57. Explain.

Where, taking supremum over all partitions P_i, $i = 1, \dots, n$, we arrive at

$$\sum_{i=1}^{n} V_{a_i}^{b_i}(f) \le \varepsilon.$$

Since, by the *Additivity of Total Variation Theorem* (Theorem 8.16),

$$V_{a_i}^{b_i}(f) = V_a^{b_i}(f) - V_a^{a_i}(f) =: v(b_i) - v(a_i),\ i = 1, \dots, n,$$

the latter is equivalent to

$$\sum_{i=1}^{n} [v(b_i) - v(a_i)] \le \varepsilon,$$

which implies that the total variation function is absolutely continuous on $[a, b]$, i. e.,

$$v \in \mathrm{AC}[a, b]$$

and completes the proof. □

Remarks 8.23.

- In the proof of the prior characterization, it is shown, in particular that the *total variation function* v (see (8.43)) of an arbitrary real-valued function $f : [a, b] \to \mathbb{R}$ is absolutely continuous on a segment $[a, b]$ $(-\infty < a < b < \infty)$ is also absolutely continuous on $[a, b]$.
- The decomposition of a real-valued function $f : [a, b] \to \mathbb{R}$ is absolutely continuous on a segment $[a, b]$ $(-\infty < a < b < \infty)$ into the difference of two increasing and absolutely continuous functions on $[a, b]$ obtained via the prior characterization is its *Jordan decomposition* (see the *Jordan Decomposition Theorem for Functions* (Theorem 8.17)), which is not unique. Indeed, if

  $$f(x) = f_1(x) - f_2(x),\ x \in [a, b],$$

  is such a decomposition of f, then, for any increasing and absolutely continuous function $h : [a, b] \to \mathbb{R}$, e. g., $h(x) := x$, $x \in [a, b]$, so is

  $$f(x) = [f_1(x) + h(x)] - [f_2(x) + h(x)],\ x \in [a, b].$$

Now, we can prove the following characterization.

Theorem 8.20 (Total Variation Characterization of Absolute Continuity). *A real- or complex-valued function $f \in \mathrm{BV}[a,b]$ ($-\infty < a < b < \infty$) is absolutely continuous on $[a,b]$ iff its total variation function*

$$v(x) := V_a^x(f),\ a \le x \le b,$$

is absolutely continuous on $[a,b]$.

Exercise 8.58. Prove.

Hint. For the *"only if"* part, separate real and imaginary parts and apply the prior characterization. For the *"if"* part, use the estimate

$$|f(x') - f(x)| \le V_x^{x'}(f),\ a < x < x' < b.$$

Corollary 8.5 (Absolutely Continuous is of Bounded Variation). *An complex-valued function $f : [a,b] \to \mathbb{C}$ absolutely continuous function on a segment ($-\infty < a < b < \infty$) is of bounded variation on $[a,b]$.*

Proof. The statement follows from the *Characterization of Absolute Continuity* (Theorem 8.19) and the *Combinations of Absolutely Continuous Functions Theorem* (Theorem 8.18) by separating the real and imaginary parts (see Remarks 8.19). □

Remark 8.24. As the example of the Cantor function shows, the converse is not true, i. e., a real- or complex-valued function of bounded variation, even monotone and continuous, on a segment $[a,b]$ ($-\infty < a < b < \infty$) need not be absolutely continuous on $[a,b]$, and hence, we have the following chain of *strict* inclusions:

$$C^1[a,b] \subset \mathrm{LC}[a,b] \subset \mathrm{AC}[a,b] \subset \mathrm{BV}[a,b]$$

(see (8.44)).

Moreover, the *total variation function* (see (8.43)) of a function in each of these classes also belongs to the same class (see Section 8.6, Problem 8).

8.5.3 Derivative of an Absolutely Continuous Function

The *Derivative of a Function of Bounded Variation Corollary* (Corollary 8.4) has the following

Corollary 8.6 (Derivative of an Absolutely Continuous Function). *A real- or complex-valued absolutely continuous function f on a segment $[a,b]$ ($-\infty < a < b < \infty$) is differentiable a. e. relative to the Lebesgue measure λ on $[a,b]$ and $f' \in L([a,b],\lambda)$.*

8.5.4 Singular Functions

Definition 8.8 (Singular Function). A real- or complex-valued function of f of bounded variation on a segment $[a, b]$ $(-\infty < a < b < \infty)$ is said to be *singular on* $[a, b]$ if

$$f'(x) = 0 \text{ a. e. } \lambda \text{ on } [a, b].$$

Remark 8.25. Each constant function is singular. However, as the example of the *Cantor function* (see Section 8.3.6), which is a *singular function* on $[0, 1]$, demonstrates, a singular function need not be constant.

By the following statement, the only singular absolutely continuous functions are constants.

Theorem 8.21 (Characterization of Absolutely Continuous Singularity). *A real- or complex-valued absolutely continuous function f on a segment $[a, b]$ $(-\infty < a < b < \infty)$ is singular on $[a, b]$ iff f is a constant function on $[a, b]$.*

Proof. "*If*" part is trivial.

"*Only if*" part. Suppose that $f \in AC[a, b]$ and

$$f'(x) = 0 \text{ a. e. } \lambda \text{ on } [a, b]$$

and let

$$E := \{x \in (a, b) \,|\, f'(x) = 0\}$$

and $\varepsilon > 0$ be arbitrary.

Then, for each $x \in E$,

$$\frac{|f(x + h) - f(x)|}{h} < \varepsilon, \tag{8.45}$$

for all sufficiently small $h > 0$.

Thus, for any $\varepsilon > 0$,

$$\mathscr{C}_\varepsilon := \{[x, x + h] \subset (a, b) \,|\, x \in E, h > 0 \text{ satisfies (8.45)}\} \tag{8.46}$$

is a *Vitali cover* of E (see Definition 8.2).

Since $\lambda(E) = b - a < \infty$, by the *Vitali Covering Lemma* (*Alternative Version*) (Theorem 8.4), for any $\delta > 0$, there exists a finite subcollection of $\mathscr{C}$

$$\{[x_1, x_1 + h_1], \dots, [x_n, x_n + h_n]\} \quad (n \in \mathbb{N})$$

consisting of pairwise disjoint subsegments of (a, b) such that

$$\lambda\left(E \setminus \bigcup_{i=1}^{n} [x_i, x_i + h_i]\right) < \delta.$$

Without loss of generality, we can regard that

$$a < x_1 < x_2 < \cdots < x_n < b,$$

and hence, the intervals

$$[a, x_1), (x_1 + h_1, x_2), \dots, (x_{n-1} + h_{n-1}, x_n), (x_n + h_n, b]$$

remain after removing the segments $d_1, \dots, d_n$, from $[a, b]$.

Since

$$E \subseteq \left[E \setminus \bigcup_{i=1}^{n} [x_i, x_i + h_i] \right] \cup \bigcup_{i=1}^{n} [x_i, x_i + h_i],$$

by the *subadditivity* and *additivity* of Lebesgue measure λ (Theorem 3.1),

$$\begin{aligned} b - a = \lambda(E) &\leq \lambda\left(\bigcup_{i=1}^{n} [x_i, x_i + h_i] \right) + \lambda\left(E \setminus \bigcup_{i=1}^{n} [x_i, x_i + h_i] \right) \\ &< \sum_{i=1}^{n} \lambda([x_i, x_i + h_i]) + \delta = \sum_{i=1}^{n} h_i + \delta, \end{aligned}$$

and hence,

$$\begin{aligned} &(x_1 - a) + \sum_{i=1}^{n} [x_i - (x_{i-1} + h_{i-1})] + [b - (x_n + h_n)] \\ &\quad = (b - a) - \sum_{i=1}^{n} h_i < \delta. \end{aligned}$$

By the absolute continuity of f, we can regard $\delta > 0$ to be small enough so that

$$|f(x_1) - f(a)| + \sum_{i=1}^{n} |f(x_i) - f(x_{i-1} + h_{i-1})| + |f(b) - f(x_n + h_n)| < \varepsilon. \tag{8.47}$$

Recall that, by the definition of the segments $[x_i, x_i + h_i]$, $i = 1, \dots, n$, (see (8.45), (8.46)),

$$\sum_{i=1}^{n} |f(x_i + h_i) - f(x_i)| < \sum_{i=1}^{n} \varepsilon h_i < \varepsilon(b - a). \tag{8.48}$$

Adding (8.47) and (8.48), we obtain:

$$\begin{aligned} |f(b) - f(a)| &\leq |f(x_1) - f(a)| + \sum_{i=1}^{n} |f(x_i) - f(x_{i-1} + h_{i-1})| + |f(b) - f(x_n + h_n)| \\ &\quad + \sum_{i=1}^{n} |f(x_i + h_i) - f(x_i)| < \varepsilon(1 + b - a). \end{aligned}$$

Where, passing to the limit as $\varepsilon \to 0+$, we infer that $f(b) = f(a)$.

Applying the same argument on $[a, c]$ for an arbitrary $c \in (a, b]$, we infer that

$$f(c) = f(a), \; c \in (a, b],$$

which implies that f is constant. □

Remark 8.26. The prior theorem allows to immediately disqualify the Cantor function $c : [0, 1] \to [0, 1]$ from being absolutely continuous on $[0, 1]$.

8.5.5 Antiderivative and Total Change Formula

The following fundamental statement generalizes the part of the *Fundamental Theorem of Calculus*, which asserts that every continuous on an interval function has an antiderivative, extending it to the Lebesgue integrable functions.

Theorem 8.22 (Antiderivative of a Lebesgue Integrable Function). *Let f be a real- or complex-valued function on a segment $[a, b]$ $(-\infty < a < b < \infty)$ such that $f \in L([a, b], \lambda)$. Then the function*

$$F(x) := \int_{[a,x]} f(t)\, d\lambda(t), \; x \in [a, b], \tag{8.49}$$

is absolutely continuous on $[a, b]$ and

$$F'(x) = f(x) \text{ a. e. } \lambda \text{ on } [a, b].$$

Proof. By the *Properties of Lebesgue Integral. I* (Theorem 6.1, part (7)), the function F is well-defined by (8.49) and, by the *Absolute Continuity of Lebesgue Integral Corollary* (Corollary 9.1) (see Section 9.7, Problem 7), F is *absolutely continuous* on $[a, b]$. By the *Derivative of an Absolutely Continuous Function* (Corollary 8.6), F is differentiable a. e. λ on $[a, b]$ and $F' \in L([a, b], \lambda)$.

Therefore, since, due to the *linearity* of the Lebesgue integral, $L([a, b], \lambda)$ is a *vector space* (see Remarks 6.11), $F' - f \in L([a, b], \lambda)$ and it remains to show that

$$F'(x) - f(x) = 0 \text{ a. e. } \lambda \text{ on } [a, b]$$

(for proof, see, e. g., [19, 22]). □

Examples 8.8.

1. By the *Characterization of Riemann Integrability* (Theorem 6.10), the *sign function*

$$\operatorname{sgn}(x) := \begin{cases} 1, & x > 0, \\ 0, & x = 0, \\ -1, & x < 0, \end{cases}$$

is Riemann, and hence, Lebesgue integrable on $[-1,1]$ and its antiderivative function on $[-1,1]$ is

$$F(x) := \int_{[-1,x]} \operatorname{sgn}(t)\,d\lambda(t) = \int_{-1}^{x} \operatorname{sgn}(t)\,dt = \begin{cases} x-1, & x>0, \\ -x-1, & -1\le x\le 0 \end{cases} = |x|-1.$$

Consistently with the prior theorem, F is differentiable on $[-1,1]\setminus\{0\}$ and

$$F'(x) = \operatorname{sgn}(x),\ x\in[-1,1]\setminus\{0\}.$$

2. The *Dirichlet function*

$$f(x) := \chi_{\mathbb{Q}}(x),\ x\in[0,1],$$

is not Riemann but Lebesgue integrable on $[0,1]$ (see Examples 6.11) and its antiderivative function on $[0,1]$ is

$$F(x) := \int_{[0,x]} \chi_{\mathbb{Q}}\,d\lambda = \lambda(\mathbb{Q}\cap[0,x]) = 0,\ x\in[0,1].$$

Consistently with the prior theorem, F is differentiable on $[0,1]$ and

$$F'(x) = 0 = f(x),\ x\in\mathbb{Q}^c\cap[0,x].$$

Now, we are to stretch the classical *total change formula* recovering a function from its derivative from continuously differentiable functions to absolutely continuous functions. In fact, the following theorem shows that the *total change formula* is a characteristic property for the absolute continuous functions in the class of functions of bounded variation.

Theorem 8.23 (Total Change Theorem). *A real- or complex-valued function $f\in \mathrm{BV}[a,b]$ $(-\infty<a<b<\infty)$ is absolutely continuous on $[a,b]$ iff the total change formula*

$$f(x)-f(a) = \int_{[a,x]} f'(t)\,d\lambda(t) \tag{8.50}$$

holds for each $x\in[a,b]$.

Proof. *"If"* part follows immediately from the *Derivative of a Function of Bounded Variation Corollary* (Corollary 8.4) and the *Antiderivative of a Lebesgue Integrable Function Theorem* (Theorem 8.22).

"Only if" part. Suppose that $f\in \mathrm{AC}[a,b]$. By the *Derivative of an Absolutely Continuous Function Corollary* (Corollary 8.6), $f'\in L([a,b],\lambda)$ and, by the *Antiderivative of a Lebesgue Integrable Function Theorem* (Theorem 8.22), the function

$$g(x) := \int_{[a,x]} f'(t)\,d\lambda(t),\ x\in[a,b],$$

is *absolutely continuous* on $[a,b]$, and

$$g'(x) - f'(x) = 0 \text{ a. e. } \lambda \text{ on } [a,b].$$

By the *Characterization of Absolutely Continuous Singularity* (Theorem 8.21), $g - f$ is a *constant function* on $[a,b]$, and hence, in view of $g(a) = 0$, we infer that

$$-f(a) = g(a) - f(a) = g(b) - f(b) = \int_{[a,b]} f'(t)\, d\lambda(t) - f(b),$$

i. e.,

$$f(b) = f(a) + \int_{[a,b]} f'(t)\, d\lambda(t).$$

Applying the same argument on the segment $[a,x]$ for an arbitrary $x \in (a,b]$ we obtain the total change formula (see (8.50)) for any $x \in [a,b]$. □

Remark 8.27. The prior theorem allows to immediately disqualify the increasing and continuous *Cantor function* $c : [0,1] \to [0,1]$ from being absolutely continuous on $[0,1]$. Indeed,

$$c(1) - c(0) = 1 \neq 0 = \int_{[0,1]} c'(x)\, d\lambda(x).$$

Example 8.9. The absolute value function

$$f(x) := |x|,\ x \in [-1,1],$$

is Lipschitz continuous, and hence, absolutely continuous on $[-1,1]$, but $f \notin C^1[-1,1]$ (see Examples 8.4 and Proposition 8.1).

Consistently with the prior theorem,

$$f'(x) = \operatorname{sgn}(x) \in R[-1,1] \subset L([-1,1],\lambda)$$

(see Remark 8.8 and inclusion (6.20)).

And, for each $x \in [-1,1]$,

$$f(x) - f(-1) = |x| - 1 = \int_{-1}^{x} \operatorname{sgn}(t)\, d\lambda(t) = \int_{[-1,x]} f'(t)\, d\lambda(t)$$

(see Examples 8.8).

Now, we can characterize absolutely continuous functions in the class of increasing functions.

Theorem 8.24 (Characterization of Increasing Absolutely Continuous Functions). *An increasing function $f : [a, b] \to \mathbb{R}$ $(-\infty < a < b < \infty)$ is absolutely continuous on $[a, b]$ iff*

$$\int_a^b f'(x)\,dx = f(b) - f(a). \tag{8.51}$$

Proof. "*Only if*" follows immediately from the *Total Change Theorem* (Theorem 8.23).

"*If*" part. Suppose $f : [a, b] \to \mathbb{R}$ is an increasing function and (8.51) holds.

Since f is increasing, by the *Total Change Estimate Theorem* (Theorem 8.11) $f' \in L([a, b], \lambda)$ and the function

$$h(x) := f(x) - \int_{[a,x]} f'(t)\,d\lambda(t),\ x \in [a, b],$$

is also *increasing* on $[a, b]$.

Exercise 8.59. Verify.

Hence, since, by (8.51), $h(b) = h(a)$, we infer that

$$h(x) \equiv c$$

on $[a, b]$ with some $c \in \mathbb{R}$.

Thus,

$$f(x) = c + \int_{[a,x]} f'(t)\,d\lambda(t),\ x \in [a, b],$$

which, by the *Antiderivative of a Lebesgue Integrable Function Theorem* (Theorem 8.22) implies that $f \in \mathrm{AC}[a, b]$ completing the proof. □

Corollary 8.7 (Characterization of Monotone Absolutely Continuous Functions). *A monotone function $f : [a, b] \to \mathbb{R}$ $(-\infty < a < b < \infty)$ is absolutely continuous on $[a, b]$ iff*

$$\int_a^b f'(x)\,dx = f(b) - f(a).$$

Exercise 8.60. Prove.

8.5.6 Lebesgue Decomposition Theorem

The following statement shows that every function of bounded variation has an absolutely continuous and a singular component.

Theorem 8.25 (Lebesgue Decomposition Theorem for Functions). *A real- or complex-valued function f of bounded variation on a segment $[a,b]$ $(-\infty < a < b < \infty)$ can be represented as the sum of two functions of bounded variation on $[a,b]$, one of which is absolutely continuous and the other is singular on $[a,b]$, such a representation being called a Lebesgue decomposition of the function f.*

Exercise 8.61. Prove.

Hint. Set

$$g(x) := \int_a^x f'(t)\,dt \quad \text{and} \quad h(x) := f(x) - \int_a^x f'(t)\,dt, \quad x \in [a,b],$$

and show that g is absolutely continuous and h is singular on $[a,b]$.

Remark 8.28. A *Lebesgue decomposition* of a real- or complex-valued function $f \in BV[a,b]$ $(-\infty < a < b < \infty)$ is *not* unique. Indeed, if

$$f(x) = f_1(x) + f_2(x),\ x \in [a,b],$$

is a Lebesgue decomposition of f, then, for any nonzero constant function $h(x) \equiv c$ on $[a,b]$ with some $c \in \mathbb{C} \setminus \{0\}$, so is

$$f(x) = [f_1(x) + h(x)] + [f_2(x) - h(x)],\ x \in [a,b].$$

Example 8.10. For the function

$$f(x) := \begin{cases} x, & x \in [-1,1] \setminus \{0\}, \\ 1, & x = 0, \end{cases}$$

which is of bounded variation on $[-1,1]$, one of its Lebesgue decompositions is

$$f(x) = g(x) + h(x),\ x \in [-1,1],$$

where the function $g(x) := x$, $x \in [-1,1]$, is absolutely continuous on $[-1,1]$ and the function

$$h(x) := \begin{cases} 0, & x \in [-1,1] \setminus \{0\}, \\ 1, & x = 0, \end{cases}$$

is singular on $[-1,1]$.

Exercise 8.62. For the functions f, g, and h from the prior example, evaluate the total variations

$$V_{-1}^{1}(g),\ V_{-1}^{1}(h),\ \text{and}\ V_{-1}^{1}(f).$$

8.6 Problems

1. Prove that the *Dirichlet function*

$$f(x) := \chi_{\mathbb{Q}}(x)$$

is not of bounded variation on any interval $[a,b]$.

2. Show that the function

$$f(x) := \begin{cases} x^2 \cos\frac{\pi}{2x}, & 0 < x \le 1 \\ 0, & x = 0, \end{cases}$$

is of bounded variation on $[0,1]$ (cf. Examples 8.4).

3. Prove

Proposition 8.2 (Characterization of Continuous Functions of Bounded Variation). *A real-valued continuous function f on a segment $[a,b]$ $(-\infty < a < b < \infty)$ is of bounded variation on $[a,b]$ iff its graph is a rectifiable curve, i. e.,*

$$\sup\left\{\sum_{i=1}^{n} \sqrt{|x_i - x_{i-1}|^2 + |f(x_i) - f(x_{i-1})|^2} \,\middle|\, P \text{ is a partition of } [a,b]\right\} < \infty.$$

4. (a) Prove

Proposition 8.3 (Quotient of Functions of Bounded Variation). *Let $f, g \in \mathrm{BV}[a,b]$ $(-\infty < a < b < \infty)$ and*

$$\inf_{[a,b]} |g| > 0.$$

Then $\frac{f}{g} \in \mathrm{BV}[a,b]$ and

$$V_a^b\left(\frac{f}{g}\right) \le \frac{1}{\inf_{[a,b]} |g|} V_a^b(f) + \frac{\sup_{[a,b]} |f|}{(\inf_{[a,b]} |g|)^2} V_a^b(g).$$

In particular, $\frac{1}{g} \in \mathrm{BV}[a,b]$ and

$$V_a^b(fg) \le \frac{1}{(\inf_{[a,b]} |g|)^2} V_a^b(g).$$

(b) Give an example showing that the condition

$$\inf_{[a,b]} |g| > 0$$

in the prior proposition is essential and cannot be dropped.

5. Prove that, for a real-valued function $f \in BV[a,b]$ $(-\infty < a < b < \infty)$,

$$f(x) = [f(x) + v(x)] - v(x),\ x \in [a,b],$$

is a *Jordan decomposition* of f.

6. For $f(x) := \sin x$, $x \in [0, 2\pi]$,
   (a) show that $f \in BV[0, 2\pi]$;
   (b) find the *total variation function* for f;
   (c) find a *Jordan decomposition* of f.

7. Let $f \in C[a,b]$ $(-\infty < a < b < \infty)$ be such that

$$\forall\, c \in (a,b) : f \in AC[c,b].$$

Prove that
   (a) f need not be absolutely continuous on $[a,b]$;
   (b) f is absolutely continuous on $[a,b]$ provided it is *increasing* on $[a,b]$.

8. Show that the total variation function of a real- or complex-valued Lipschitz continuous function on a segment $[a,b]$ $(-\infty < a < b < \infty)$ is also Lipschitz continuous on $[a,b]$.

9. Let a real-valued function $f : [a,b] \to \mathbb{R}$ $(-\infty < a < b < \infty)$ be of bounded variation on a segment $[a,b]$ $(-\infty < a < b < \infty)$ and

$$v(x) := V_a^x(f),\ x \in [a,b],$$

be its *total variation function*. Prove that
   (a) $|f'(x)| \le v'(x)$ a. e. λ on $[a,b]$;
   (b) $\int_a^b |f'(x)|\, dx \le V_a^b(f)$;
   (c) the latter turns into equality *iff* $f \in AC[a,b]$.

10. Prove

Theorem 8.26 (Integration by Parts Formula). *Let f and g be real- or complex-valued absolutely continuous functions on a segment $[a,b]$ $(-\infty < a < b < \infty)$, then*

$$\int_{[a,b]} f(x)g'(x)\, d\lambda(x) = f(b)g(b) - f(a)g(a) - \int_{[a,b]} f'(x)g(x)\, d\lambda(x).$$

11. * Prove

Proposition 8.4 (BV $[a,b]$ is a Banach space). *For any $-\infty < a < b < \infty$, the vector space $BV[a,b]$ of functions of bounded variation on $[a,b]$ is a (real or complex) Banach space relative to the norm*

$$BV[a,b] \ni f \mapsto \|f\| := |f(a)| + V_a^b(f).$$

12. Show that the vector space $\mathrm{BV}[a,b]$ is *incomplete* relative to the supremum norm

$$\mathrm{BV}[a,b] \ni f \mapsto \|f\| := \sup_{a \le x \le b} |f(x)|.$$

Hint. Give an example of a fundamental but divergent sequence based on Examples 8.4, 4.

9 Signed Measures

In this chapter, we generalize the notion of measure by allowing its values to be negative to the notion of *signed measure* (or a *charge*) and prove a number of consequential statements, including the *Hahn Decomposition Theorem* (Theorem 9.2), the *Jordan Decomposition Theorem* (Theorem 9.3), and the *Radon–Nikodym Theorem* (Theorem 9.5), elucidating the nature of such signed measures.

9.1 Definition and Examples

Definition 9.1 (Signed Measure). Let (X, Σ) be a measurable space. A set function $\nu : \Sigma \to (-\infty, \infty]$ is called a *signed measure* (or a *charge*) on Σ if:

(1) $\nu(\emptyset) = 0$;
(2) ν is σ-additive.

Remarks 9.1.
- Thus, a signed measure ν taking only nonnegative values, i. e., $\nu : \Sigma \to [0, \infty]$, is merely a measure. To emphasize this fact, we can contextually refer to such as a *positive measure*.
- One can similarly define a signed measure $\nu : \Sigma \to [-\infty, \infty)$ on a measurable space (X, Σ) satisfying conditions (1), (2) of the prior definition (as soon as *not both* infinite values $\pm\infty$ are assumed). However, since this case is merely sign opposite to the defined one, without loss of generality, our discourse can be restricted to signed measures taking values in $(-\infty, \infty]$.

Definition 9.2 (Signed Measure Space). Let (X, Σ) be measurable space with a signed measure ν on the σ-algebra Σ. The triple (X, Σ, ν) is called a *signed measure space*.

Examples 9.1.
1. Let (X, Σ) be a measurable space. If ν_+ and ν_- are (positive) measures on the σ-algebra Σ, with ν_- being *finite*, then

$$\nu(A) := \nu_+(A) - \nu_-(A), \ A \in \Sigma, \tag{9.1}$$

is a signed measure on Σ, which is finite (σ-finite) *iff* ν_+ is finite (σ-finite), respectively.

In particular, if λ_G and λ_H are the *Lebesgue–Stieltjes measures* on the Borel measurable space $(\mathbb{R}, \mathscr{B}(\mathbb{R}))$ associated with an increasing right-continuous functions $G, H : \mathbb{R} \to \mathbb{R}$, respectively, with λ_H being *finite*, i. e.,

$$\lambda_H(\mathbb{R}) = H(\infty) - H(-\infty) < \infty$$

https://doi.org/10.1515/9783110600995-009

(see Section 4.7.1 and Examples 4.9), then, by the *Jordan Decomposition Theorem for Functions* (Theorem 8.17),

$$F(x) := G(x) - H(x),\ x \in \mathbb{R}, \tag{9.2}$$

is a real-valued right-continuous function *of bounded variation* on $\mathbb{R}$ (see Definition 8.6) and

$$\nu_F(A) := \lambda_G(A) - \lambda_H(A),\ A \in \mathscr{B}(\mathbb{R}), \tag{9.3}$$

is a σ-finite signed measure on $\mathscr{B}(\mathbb{R})$, which is finite *iff* λ_G is finite, i. e.,

$$\lambda_G(\mathbb{R}) = G(\infty) - G(-\infty) < \infty.$$

For the next two examples, let (X, Σ, μ) be a measure space.

2. If $f : X \to \overline{\mathbb{R}}$ and $f \in L(X, \mu)$, i. e.,

$$\int_X f\, d\mu < \infty$$

(see Definition 6.4), by the *σ-Additivity of Lebesgue Integral* (Corollary 6.1), the set function

$$\nu_f(A) := \int_A f\, d\mu := \int_A f_+\, d\mu - \int_A f_-\, d\mu,\ A \in \Sigma, \tag{9.4}$$

(see Definition 6.3), where

$$f_+(x) := \max(f(x), 0) \text{ and } f_-(x) := -\min(f(x), 0),\ x \in X, \tag{9.5}$$

are the *positive* and *antinegative parts of* f, respectively, (see *Positive and Antinegative Parts of a Function Corollary* (Corollary 5.4)) with $f_+, f_- \in L(X, \mu)$, is a *finite* signed measure on Σ, with

$$\nu_+(A) := \int_A f_+\, d\mu \text{ and } \nu_-(A) := \int_A f_-\, d\mu - \int_A f_-\, d\mu,\ A \in \Sigma, \tag{9.6}$$

being *finite* (positive) measures on the σ-algebra Σ.

3. If $f : X \to \overline{\mathbb{R}}$ is a Σ-measurable function and we only require that

$$\int_X f_-\, d\mu < \infty,$$

then

$$\nu_f(A) = \nu_+(A) - \nu_+(A),\ A \in \Sigma,$$

(see (9.4)–(9.6)) is a signed measure on the σ-algebra Σ, which may take the value ∞.

In particular, for an arbitrary Σ-measurable nonnegative function $f : X \to [0, \infty]$, $f_- \equiv 0$ and we have the positive measure

$$\mu_f(A) := \int_A f\,d\mu,\ A \in \Sigma,$$

(see the *σ-Additivity of Lebesgue Integral* (Theorem 6.2)).

Exercise 9.1. Verify.

Remark 9.2. As is proved in the *Jordan Decomposition Theorem* (Theorem 9.3), all signed measures are of the form given by (9.1).

9.2 Elementary Properties

Theorem 9.1 (Properties of Signed Measure). *Let (X, Σ, ν) be a signed measure space.*

(1) *ν is additive on Σ.*

(2) *If for $A, B \in \Sigma$ with $B \subseteq A$ and $\nu(B) < \infty$, then*

$$\nu(A \setminus B) = \nu(A) - \nu(B).$$

(3) *If $A \in \Sigma$ with $\nu(A) < \infty$, then*

$$\forall\, B \subseteq A,\ B \in \Sigma :\ \nu(B) < \infty.$$

(4) *If $B \in \Sigma$ with $\nu(B) = \infty$, then*

$$\forall\, A \supseteq B,\ A \in \Sigma :\ \nu(A) = \infty.$$

(5) *ν is continuous, i. e.,*
   (a) *for any increasing sequence $(A_n)_{n\in\mathbb{N}}$ in Σ,*

$$\nu\Big(\lim_{n\to\infty} A_n\Big) = \lim_{n\to\infty} \nu(A_n) \quad \textit{(continuity from below)},$$

   where $\lim_{n\to\infty} A_n = \bigcup_{n=1}^{\infty} A_n$;
   (b) *for any decreasing sequence $(A_n)_{n\in\mathbb{N}}$ in Σ with $\mu(A_N) < \infty$ for some $N \in \mathbb{N}$,*

$$\nu\Big(\lim_{n\to\infty} A_n\Big) = \lim_{n\to\infty} \nu(A_n) \quad \textit{(continuity from above)},$$

   where $\lim_{n\to\infty} A_n = \bigcap_{n=1}^{\infty} A_n$.

Proof.

(1)

Exercise 9.2. Prove (see Exercise 3.1).

(2) Follows from the *additivity* of ν.

Exercise 9.3. Prove (see the proof of the *Properties of Measure* (Theorem 3.1)).

(3) By the *additivity* of ν,

$$\forall B \subseteq A,\ B \in \Sigma :$$

and in view of the $\nu(A \setminus B) > -\infty$, we infer that

$$\nu(B) = \nu(A) - \nu(A \setminus B) < \infty.$$

(4) Part (4) is equivalent *contrapositive* of (3).

(5)

Exercise 9.4. Prove by modifying the proof of the *Continuity of Measure Theorem* (Theorem 3.2) using the above properties. □

Remark 9.3. The requirement of $\mu(A_n)$, $n \in \mathbb{N}$, being *eventually finite* in the *continuity of signed measure from above* is essential and cannot be dropped.

Exercise 9.5. Explain why $\mu(A_n)$, $n \in \mathbb{N}$, is *eventually finite*. Give a corresponding example.

9.3 Hahn Decomposition

Here, we are to prove the famed *Hahn Decomposition Theorem*[1] (Theorem 9.2).

9.3.1 Positive, Negative, and Null Set

To proceed, we need to define the following notions that are quite natural in the context.

Definition 9.3 (Positive, Negative, and Null Set). Let (X, Σ, ν) be a signed measure space.

- A set $A \in \Sigma$ is called a *positive set* relative to ν or a *ν-positive set* if

$$\forall B \in \Sigma,\ B \subseteq A :\ \nu(B) \geq 0.$$

1 Hans Hahn (1879–1934).

- A set $A \in \Sigma$ is called a *negative set* relative to ν or a ν-*negative set* if

$$\forall B \in \Sigma,\ B \subseteq A : \ \nu(B) \le 0.$$

- A set $A \in \Sigma$ is called a *null set* relative to ν or a ν-*null set* if

$$\forall B \in \Sigma,\ B \subseteq A : \ \nu(B) = 0.$$

Remarks 9.4.
- A set is simultaneously ν-positive and ν-negative *iff* it is a ν-null set.
- Any Σ-measurable subset of a ν-positive/negative/null set is also a ν-positive/negative/null set, respectively.
- If ν is a *positive measure* on Σ, the above definition of a ν-null set is consistent with Definition 4.5 (see Remark 4.5).

Exercise 9.6. Verify.

Example 9.2. Let (X, Σ, μ) be a measure space and $f : X \to \overline{\mathbb{R}}$ a Σ-measurable function with

$$\int_X f_- \, d\mu < \infty$$

(see Examples 9.1), then

$$X_+ := \{x \in X \,|\, f(x) \ge 0\}, \ X_- := \{x \in X \,|\, f(x) \le 0\}, \text{ and } X_0 := \{x \in X \,|\, f(x) = 0\}$$

are, respectively, a *positive*, a *negative*, and a *null set* relative to the signed measure

$$\nu_f(A) := \int_A f \, d\mu, \ A \in \Sigma,$$

on Σ (see Examples 9.1).

Exercise 9.7. Verify.

9.3.2 Negative Subset Lemma

We also need the following lemma.

Lemma 9.1 (Negative Subset Lemma). *Let (X, Σ, ν) be a signed measure space. For any set $A \in \Sigma$ with $\nu(A) < \infty$, there exists a ν-negative subset B of A such that $\nu(B) \le \nu(A)$.*

Proof. Suppose that a set $A \in \Sigma$ with $\nu(A) < \infty$ is arbitrary.

Let us show that

$$\forall \varepsilon > 0 \, \exists A_\varepsilon \in \Sigma, \, A_\varepsilon \subseteq A \text{ with } \nu(A_\varepsilon) \le \nu(A) \, \forall B \in \Sigma, \, B \subseteq A_\varepsilon : \, \nu(B) \le \varepsilon. \tag{9.7}$$

Indeed, assume the opposite, i. e., that

$$\exists \varepsilon > 0 \, \forall A_\varepsilon \in \Sigma, \, A_\varepsilon \subseteq A \text{ with } \nu(A_\varepsilon) \le \nu(A) \, \exists B \in \Sigma, \, B \subseteq A_\varepsilon : \, \nu(B) > \varepsilon. \tag{9.8}$$

The latter being true, in particular, if we choose A_ε to be A itself, and hence,

$$\exists B_1 \in \Sigma, \, B_1 \subseteq A : \, \nu(B_1) > \varepsilon.$$

Furthermore, by the *Properties of Signed Measure* (Theorem 9.1),

$$\nu(B_1) < \infty$$

and

$$\nu(A \setminus B_1) = \nu(A) - \nu(B_1) < \nu(A) - \varepsilon < \nu(A)$$

Therefore, (9.8) holds if we choose A_ε to be $A \setminus B_1$, which implies that

$$\exists B_2 \in \Sigma, \, B_2 \subseteq A \setminus B_1 : \, \nu(B_2) > \varepsilon.$$

Continuing inductively, we obtain a *pairwise disjoint* sequence $(B_n)_{n\in\mathbb{N}}$ of Σ-measurable subsets of A such that

$$\forall n \in \mathbb{N} : \, \nu(B_n) > \varepsilon.$$

Then, for $\Sigma \ni B := \bigcup_{n=1}^\infty B_n \subseteq A$, by the *σ-additivity* of ν,

$$\nu(B) = \sum_{n=1}^\infty \nu(B_n) = \infty,$$

which, by the *Properties of Signed Measure* (Theorem 9.1), *contradicts* the fact that $\nu(A) < \infty$. Thus, (9.7) holds, indeed.

For each $n \in \mathbb{N}$, let

$$A_n := A_{1/n} \subseteq A,$$

where $A_{1/n}$ is a Σ measurable subset of A from (9.7) corresponding to $\varepsilon = 1/n$.

Without loss of generality, we can regard that the set sequence $(A_n)_{n\in\mathbb{N}}$ is *decreasing*, i. e.,

$$A_{n+1} \subseteq A_n, \; n \in \mathbb{N}.$$

Exercise 9.8. Explain.

Then

$$\Sigma \ni B := \bigcap_{n=1}^{\infty} A_n \subseteq A$$

and, by the *continuity of v from above* (Theorem 9.1), in view of (9.7),

$$\nu(B) = \lim_{n\to\infty} \nu(A_n) \le \nu(A).$$

Further, for an arbitrary $\Sigma \ni C \subseteq B$, by the *continuity of v from above* (Theorem 9.1),

$$\nu(C) = \nu(C \cap B) = \nu\left(\bigcap_{n=1}^{\infty} C \cap A_n\right) = \lim_{n\to\infty} \nu(C \cap A_n) \le \lim_{n\to\infty} (1/n) = 0,$$

which, showing that B is a *ν-negative set*, completes the proof. □

9.3.3 Hahn Decomposition Theorem

Theorem 9.2 (Hahn Decomposition Theorem). *Let (X, Σ, ν) be a signed measure space.*

(1) *There exists a complementary pair of a ν-negative set X_- and a ν-positive set $X_+ = X_-^c$ called a Hahn decomposition of X relative to ν.*

(2) *Given a Hahn decomposition X_-, X_+ of X, a pair of complementary sets $X'_- \in \Sigma$, $X'_+ = (X'_-)^c \in \Sigma$ is a Hahn decomposition of X relative to ν iff*

$$X'_- \triangle X_- = X'_+ \triangle X_+$$

is a ν-null set.

Proof.

(1) Let

$$L := \inf\{\nu(A) \,|\, A \in \Sigma\}.$$

Then

$$-\infty \le L \le 0.$$

Exercise 9.9. Explain.

There exists a set sequence $(A_n)_{n\in\mathbb{N}}$ in Σ such that

$$\infty > \nu(A_n) \to L,\ n \to \infty. \tag{9.9}$$

By the *Negative Subset Lemma* (Lemma 9.1), without loss of generality, we can regard that, for each $n \in \mathbb{N}$ the set A_n is *ν-negative*.

Exercise 9.10. Explain.

Then the set

$$X_- := \bigcup_{n=1}^{\infty} A_n$$

is also *negative*. Indeed, setting

$$B_n := A_n \setminus \bigcup_{k=1}^{n-1} A_k \in \Sigma,\ n \in \mathbb{N}, \quad (A_0 := \emptyset),$$

we obtain a *pairwise disjoint* sequence of *negative* sets with

$$\bigcup_{n=1}^{\infty} B_n = \bigcup_{n=1}^{\infty} A_n = X_-.$$

Exercise 9.11. Explain.

Hence, by the *σ-additivity* of ν,

$$\forall\, C \in \Sigma,\ C \subseteq X_- : \ \nu(C) = \nu(C \cap X_-) = \nu\left(\bigcap_{n=1}^{\infty} C \cap B_n\right) = \sum_{n=1}^{\infty} \nu(C \cap B_n) \le 0.$$

Furthermore, for each $n \in \mathbb{N}$, by the *additivity* of ν (Theorem 9.1) and the *negativity* of X_-,

$$L \le \nu(X_-) = \nu(A_n) + \nu(X_- \setminus A_n) \le \nu(A_n).$$

Where, passing to the limit as $n \to \infty$ in view of (9.9), we infer that

$$L = \nu(X_-),$$

which, in particular, is implies that

$$-\infty < L \le 0.$$

The set $X_+ := X_-^c$ is *positive*. Indeed, assume that this is not true, i. e.,

$$\exists\, B \in \Sigma,\ B \subseteq X_+ : \ \nu(B) < 0.$$

Hence, by the *additivity* of ν (Theorem 9.1),

$$\nu(X_- \cup B) = \nu(X_-) + \nu(B) < L := \inf\{\nu(A) \,|\, A \in \Sigma\},$$

which is a *contradiction*.

(2) *"Only if"* part. Suppose that a ν-negative set X'_- and a ν-positive set X'_+ form a Hahn decomposition of X relative to ν.
Then

$$X'_- \setminus X_- = X_+ \setminus X'_+. \tag{9.10}$$

Exercise 9.12. Verify.

This implies that the set $X'_- \setminus X_- = X_+ \setminus X'_+$ is both ν-negative and ν-positive, and hence (see Remarks 9.4), is a ν-*null set*.

Exercise 9.13. Explain.

Similarly,

$$X_- \setminus X'_- = X'_+ \setminus X_+ \tag{9.11}$$

and $X_- \setminus X'_- = X'_+ \setminus X_+$ is a ν-*null set*.
Thus, by (9.10) and (9.11)

$$X'_- \triangle X_- = (X'_- \setminus X_-) \cup (X_- \setminus X'_-) = (X'_+ \setminus X_+) \cup (X_+ \setminus X'_+) = X'_+ \triangle X_+$$

and $X'_- \triangle X_- = X'_+ \triangle X_+$ is a ν-*null set* being the union of two null sets (see Section 9.7, Problem 1).

Exercise 9.14. Prove the *"if"* part. □

Remark 9.5. As follows in the proof of the *Hahn Decomposition Theorem* (Theorem 9.2), given a signed measure space (X, Σ, ν), for a Hahn decomposition X_-, X_+ of X relative to ν,

$$\mu(X_-) = \inf\{\nu(A) \,|\, A \in \Sigma\} \in (-\infty, 0].$$

That is,

$$\mu(X_-) = \min\{\nu(A) \,|\, A \in \Sigma\}.$$

Also,

$$\mu(X_+) = \max\{\nu(A) \,|\, A \in \Sigma\} \in [0, \infty].$$

Indeed, by the *additivity* of ν (Theorem 9.1) and the ν-*negativity* of X_- and ν-*positivity* of X_+, for each $A \in \Sigma$,

$$\nu(A) = \nu(A \cap X_+) + \nu(A \cap X_-) \le \nu(A \cap X_+) \le \nu(A \cap X_+) + \nu(A^c \cap X_+) = \nu(X_+).$$

Example 9.3. For a measure space (X, Σ, μ) and a Σ-measurable function $f : X \to \overline{\mathbb{R}}$ with

$$\int_X f_- \, d\mu < \infty,$$

the pairs of Σ-measurable sets

$$X_- := \{x \in X \mid f(x) < 0\}, \; X'_- := \{x \in X \mid f(x) \le 0\}$$

and

$$X_+ := \{x \in X \mid f(x) \ge 0\}, \; X'_+ := \{x \in X \mid f(x) > 0\}$$

are, respectively, *negative* and *positive* relative to the signed measure

$$\nu(A) := \int_A f \, d\mu, \; A \in \Sigma$$

(see Examples 9.1 and Example 9.2), the pairs X_-, X_+ and X'_-, X'_+ being Hahn decompositions of X relative to ν.

Exercise 9.15. Verify.

9.4 Jordan Decomposition Theorem

The *Jordan Decomposition Theorem* (Theorem 9.3) to be proved in this section is a generalization of the *Jordan Decomposition Theorem for Functions* (Theorem 8.17).

9.4.1 Mutual Singularity of Measures

To proceed, we need to introduce the notion of *mutual singularity of measures.*

Definition 9.4 (Mutual Singularity of Measures). Two (positive) measures μ and ν on a measurable space (X, Σ) are called *mutually singular* if there exists a set $E \in \Sigma$ for which

$$\mu(E) = \nu(E^c) = 0.$$

Notation. $\mu \perp \nu$.

Remark 9.6. The relationship of being mutually singular is *symmetric.*

Examples 9.4.

1. For the Lebesgue measurable space $([0,1], \Sigma^*([0,1]))$, the measures

$$\mu(A) := \lambda(A \cap [0,1/2)), \; \nu(A) := \lambda(A \cap [1/2,1]), \; A \in \Sigma^*([0,1])$$

   are *mutually singular* with $\mu([1/2,1]) = \nu([0,1/2)) = 0$.

2. On the Lebesgue measurable space $(\mathbb{R}, \Sigma^*)$, the Lebesgue measure λ and the Lebesgue–Stieltjes measure λ_F generated by the increasing continuous function

$$F(x) := \begin{cases} 0, & x < 0, \\ c(x), & 0 \le x \le 1, \\ 1, & x > 1, \end{cases}$$

where $c : [0,1] \to [0,1]$ is the *Cantor function* (see Section 8.3.6), are *mutually singular* with $\lambda(C) = \lambda_F(C^c) = 0$. The latter follows from the fact that, for any interval $(a,b] \subseteq C^c$,

$$\lambda_F((a,b]) := F(b) - F(a) = 0.$$

Exercise 9.16. Verify.

9.4.2 Jordan Decomposition Theorem

Theorem 9.3 (Jordan Decomposition Theorem). *Let (X, Σ, ν) be a signed measure space.*

(1) *There exist a positive measure ν_+ and a positive finite measure ν_- on the σ-algebra Σ, which are mutually singular and such that*

$$\nu(A) = \nu_+(A) - \nu_-(A), \ A \in \Sigma.$$

(2) *The measure ν_+ is finite (σ-finite) provided the signed measure ν is finite (σ-finite), respectively.*

(3) *Such a decomposition of the signed measure ν, called its Jordan decomposition, is unique.*

Proof.

(1) Let $X_-, X_+ \in \Sigma$ be a Hahn decomposition of X relative to ν and

$$\nu_+(A) := \nu(A \cap X_+), \ \nu_-(A) := -\nu(A \cap X_-), \quad A \in \Sigma.$$

(i) As follows from part (2) of the *Hahn Decomposition Theorem* (Theorem 9.2) (see Section 9.7, Problem 2), the set functions ν_+ and ν_- are well-defined, i. e., are independent of the underlying Hahn decomposition;

(ii) ν_+ and ν_- are *mutually singular* positive measures on Σ, with ν_- being *finite*;

(iii) $\nu(A) = \nu_+(A) - \nu_-(A)$, $A \in \Sigma$.

Exercise 9.17. Prove (i)–(iii).

(2) The measure ν_+ is finite (σ-finite) provided the signed measure ν is finite (σ-finite), respectively.

Exercise 9.18. Prove.

(3) Suppose that $\nu = \nu'_+ - \nu'_-$ is another such a decomposition of ν. The measures ν'_+ and ν'_-, being *mutually singular*, there are complementary sets $X'_- \in \Sigma$, $X'_+ = (X'_-)^c \in \Sigma$ such that

$$\nu'_+(X'_-) = \nu'_-(X'_+) = 0.$$

Then, by the *monotonicity* of ν'_- (Theorem 3.1),

$$\forall A \in \Sigma,\ A \subseteq X'_+ : \ 0 \le \nu'_-(A) \le \nu'_-(X'_+) = 0,$$

which implies that

$$\forall A \in \Sigma,\ A \subseteq X'_+ : \ \nu(A) = \nu'_+(A) - \nu'_-(A) = \nu'_+(A) - 0 = \nu'_+(A) \ge 0,$$

and hence, X'_+ is a *ν-positive* set.
Similarly, it can be shown that X'_- is a *ν-negative* set.

Exercise 9.19. Show.

Hence, $X'_-, X'_+ \in \Sigma$ is a *Hahn decomposition of X relative to ν*, which, as follows from part (2) of the *Hahn Decomposition Theorem* (Theorem 9.2) (see Section 9.7, Problem 2), implies that

$$\forall A \in \Sigma : \nu'_+(A) := \nu(A \cap X'_+) = \nu(A \cap X_+) =: \nu_+(A) \text{ and}$$
$$\nu'_-(A) := \nu(A \cap X'_-) = \nu(A \cap X_-) =: \nu_-(A)$$

completing the proof of the uniqueness part and the entire statement. □

Example 9.5. For a measure space (X, Σ, μ) and a Σ-measurable function $f : X \to \overline{\mathbb{R}}$ with

$$\int_X f_- \, d\mu < \infty,$$

the signed measure

$$\nu_f(A) := \int_A f \, d\mu, \ A \in \Sigma,$$

on the σ-algebra Σ has the following *Jordan decomposition*:

$$\nu = \nu_+ - \nu_-,$$

where

$$\nu_+(A) := \int_A f_+ \, d\mu, \ \nu_-(A) := \int_A f_- \, d\mu, \ A \in \Sigma,$$

(see Examples 9.1, Example 9.2, and Example 9.3).

9.4.3 Total Variation of a Signed Measure

Definition 9.5 (Total Variation of a Signed Measure). Let (X, Σ, ν) be a signed measure space and $\nu = \nu_+ - \nu_-$ be the Jordan decomposition of the signed measure ν. The measure

$$|\nu| = \nu_+ + \nu_-$$

on the σ-algebra Σ is called the *total variation measure of* ν and the value $|\nu|(X)$ is called the *total variation of* ν *on* X.

Example 9.6. For a measure space (X, Σ, μ) and a Σ-measurable function $f : X \to \overline{\mathbb{R}}$ with

$$\int_X f_- \, d\mu < \infty,$$

the signed measure

$$\nu_f(A) := \int_A f \, d\mu, \ A \in \Sigma,$$

on the σ-algebra Σ has the following *total variation measure*:

$$|\nu_f|(A) := \int_A f_+ \, d\mu + \int_A f_- \, d\mu = \int_A |f| \, d\mu, \ A \in \Sigma,$$

where

$$\nu_+(A) := \int_A f_+ \, d\mu, \ \nu_-(A) := \int_A f_- \, d\mu, \ A \in \Sigma,$$

(see Example 9.5).

9.5 Radon–Nikodym Theorem

The celebrated theorem *Radon–Nikodym Theorem* (Theorem 9.5) to be proved in this section furnishes, under certain conditions, a representation of a signed measure absolutely continuous relative to a measure as the Lebesgue integral relative to the latter.

9.5.1 Absolute Continuity of Signed Measure

To proceed, we need to introduce the notion of *absolute continuity* of a signed measure.

Definition 9.6 (Absolute Continuity of Signed Measure). A signed measure $\nu : \Sigma \to (-\infty, \infty]$ on a measure space (X, Σ, μ) is called *absolutely continuous relative to the (positive) measure* μ if

$$\forall A \in \Sigma : \mu(A) = 0 \Rightarrow \nu(A) = 0.$$

Notation. $\nu \ll \mu$.

Examples 9.7.
1. On an arbitrary measurable space (X, Σ), the *zero measure* is absolutely continuous relative to any measure μ and only the zero measure is absolutely continuous relative to itself.
2. For an arbitrary signed measure $\nu : \Sigma \to (-\infty, \infty]$ on a measure space (X, Σ, μ) with the *Jordan decomposition* $\nu = \nu_+ - \nu_-$ (see the *Jordan Decomposition Theorem* (Theorem 9.3))
$$\nu \ll |\nu|, \ \nu_+ \ll |\nu|, \text{ and } \nu_- \ll |\nu|,$$
where $|\nu| := \nu_+ + \nu_-$ the *total variation measure of* ν on Σ (see Definition 9.5).
3. For a measure space (X, Σ, μ) and a Σ-measurable function $f : X \to \overline{\mathbb{R}}$ with
$$\int_X f_- \, d\mu < \infty$$
(see Examples 9.1) the signed measure
$$\nu_f(A) := \int_A f \, d\mu, \ A \in \Sigma,$$
is absolutely continuous relative to the measure μ.
In particular, for an arbitrary Σ-measurable nonnegative function $f : X \to [0, \infty]$, $f_- \equiv 0$ and the positive measure
$$\mu_f(A) := \int_A f \, d\mu, \ A \in \Sigma,$$
is absolutely continuous relative to the measure μ.
4. Two *mutually singular* nonzero (positive) measures μ and ν on a measurable space (X, Σ) are *not* absolutely continuous relative to each other.
In particular, on the Lebesgue measurable space $(\mathbb{R}, \Sigma^*)$, the Lebesgue measure λ and the Lebesgue–Stieltjes measure λ_F generated by the increasing continuous function

$$F(x) := \begin{cases} 0, & x < 0, \\ c(x), & 0 \le x \le 1, \\ 1, & x > 1, \end{cases}$$

where $c : [0,1] \to [0,1]$ is the *Cantor function*, (see Examples 9.4) are *not* absolutely continuous relative to each other.

5. For two *mutually singular* nonzero (positive) measures μ and ν on a measurable space (X, Σ),

$$\nu \ll \mu \Leftrightarrow \nu \text{ is the } \textit{zero measure} \text{ on } \Sigma.$$

Exercise 9.20.
(a) Verify.
(b) Let μ_1 and μ_2 be measures on a measurable space (X, Σ) and $\mu := \mu_1 + \mu_2$. Prove that $\mu_i \ll \mu$, $i = 1, 2$ (see Section 3.3, Problem 1).

Theorem 9.4 (Characterizations of Absolute Continuity). *Let $\nu : \Sigma \to (-\infty, \infty]$ be a signed measure on a measure space (X, Σ, μ) with the Jordan decomposition $\nu = \nu_+ - \nu_-$. The following statements are equivalent:*
(1) $\nu \ll \mu$.
(2) $\nu_+ \ll \mu$ *and* $\nu_- \ll \mu$.
(3) $|\nu| \ll \mu$.

Proof. Let us prove the following closed chain of implications:

$$(1) \Rightarrow (2) \Rightarrow (3) \Rightarrow (1).$$

(1) $\Rightarrow$ (2). Suppose $\nu \ll \mu$ and let sets $X_-, X_+ \in \Sigma$ be a Hahn decomposition of X relative to ν (see the *Hahn Decomposition Theorem* (Theorem 9.2)). Then

$$\forall A \in \Sigma \text{ with } \mu(A) = 0 : \ \nu_+(A) := \nu(A \cap X_+) = 0,$$

since by the *monotonicity* of μ (Theorem 3.1), $\mu(A \cap X_+) = 0$. Hence, $\nu_+ \ll \mu$.

Similarly, one can show that $\nu_- \ll \mu$.

Exercise 9.21. Show.

(2) $\Rightarrow$ (3). Suppose $\nu_+ \ll \mu$ and $\nu_- \ll \mu$. Then

$$\forall A \in \Sigma \text{ with } \mu(A) = 0 : \ |\nu|(A) := \nu_+(A) + \nu_-(A) = 0 + 0 = 0,$$

and hence, $|\nu| \ll \mu$.

(3) $\Rightarrow$ (1). Suppose $|\nu| \ll \mu$. Then

$$\forall A \in \Sigma \text{ with } \mu(A) = 0 : \ |\nu|(A) := \nu_+(A) + \nu_-(A) = 0,$$

and hence, $\nu_+(A) = \nu_-(A) = 0$, which implies that

$$\nu(A) := \nu_+(A) - \nu_-(A) = 0,$$

i. e., $\nu \ll \mu$. □

9.5.2 Radon–Nikodym Theorem

Theorem 9.5 (Radon–Nikodym Theorem). *Let $\nu : \Sigma \to (-\infty, \infty]$ be a σ-finite signed measure on a σ-finite measure space (X, Σ, μ). If $\nu \ll \mu$, then there exists a Σ-measurable function $f : X \to \mathbb{R}$ such that*

$$\nu(A) = \int_A f\, d\mu, \ A \in \Sigma.$$

Provided ν is a positive measure, the function f in the above representation is nonnegative, i. e., $f : X \to [0, \infty)$.

If g is another such a function, then $f(x) = g(x)$ a. e. μ on X.[2]

Proof. First, suppose that μ and ν are *finite* (positive) measures on Σ and let

$$\mathcal{F} := \left\{ g : X \to [0, \infty) \,\middle|\, g \text{ is } \Sigma\text{-measurable}, \forall A \in \Sigma : \int_A g\, d\mu \le \nu(A) \right\}. \tag{9.12}$$

The set $\mathcal{F}$ is not empty since $g(x) \equiv 0 \in \mathcal{F}$.

If $g_1, g_2 \in \mathcal{F}$, then $\max(g_1, g_2) \in \mathcal{F}$. Indeed, let

$$B := \{x \in X \,|\, g_1(x) \ge g_2(x)\} \in \Sigma.$$

Remark 9.7. Explain why $B \in \Sigma$.

Then, for each $A \in \Sigma$, by the *additivity* of the Lebesgue integral (the *Properties of Lebesgue Integral. II* (Theorem 6.3, part (1))) and ν (Theorem 3.1), we have

$$\begin{aligned}\int_A \max(g_1, g_2)\, d\mu &= \int_{A \cap B} \max(g_1, g_2)\, d\mu + \int_{A \cap B^c} \max(g_1, g_2)\, d\mu \\ &= \int_{A \cap B} g_1\, d\mu + \int_{A \cap B^c} g_2\, d\mu \le \nu(A \cap B) + \nu(A \cap B^c) = \nu(A).\end{aligned}$$

Hence, inductively, if $g_i \in \mathcal{F}$, $i = 1, \ldots, n$ ($n \in \mathbb{N}$), then, in view of the *properties of measurable functions* (Theorem 5.3), $\max(g_1, \ldots, g_n) \in \mathcal{F}$.

2 Johann Radon (1887–1956), Otto Nikodym (1887–1974).

Let

$$S := \sup_{g \in \mathscr{F}} \int_X g \, d\mu. \tag{9.13}$$

By the definition of $\mathscr{F}$ (see (9.12)),

$$0 \le S \le \nu(X) < \infty.$$

There exists a sequence $(g_n)_{n\in\mathbb{N}}$ in $\mathscr{F}$, in which we can regard $g_1(x) \equiv 0$, such that

$$\int_X g_n \, d\mu \to S, \ n \to \infty. \tag{9.14}$$

Then the sequence

$$f_n := \max(g_1, \dots, g_n) \in \mathscr{F}, \ n \in \mathbb{N},$$

is increasing:

$$0 = g_1(x) \le f_n(x) \le f_{n+1}(x), \ n \in \mathbb{N}, x \in X,$$

and hence, by the *Sequences of Measurable Functions Theorem* (Theorem 5.4), the limit function

$$f(x) := \lim_{n\to\infty} f_n(x) \in [0, \infty], \ x \in X,$$

is Σ-measurable. Further, by the *Monotone Convergence Theorem* (Theorem 6.4),

$$\forall A \in \Sigma : \ \lim_{n\to\infty} \int_A f_n \, d\mu = \int_A f \, d\mu.$$

Since, for each $n \in \mathbb{N}$, by the *Properties of Lebesgue Integral. I* (Theorem 6.1, part (3)),

$$\int_X g_n \, d\mu \le \int_X f_n \, d\mu \le S,$$

passing to the limit as $n \to \infty$ in view of (9.14), we arrive at

$$\int_X f \, d\mu = S. \tag{9.15}$$

Also, since $f_n \in \mathscr{F}$, $n \in \mathbb{N}$, in view of (9.12),

$$\forall A \in \Sigma : \ \int_A f \, d\mu = \lim_{n\to\infty} \int_A f_n \, d\mu \le \nu(A). \tag{9.16}$$

Therefore, $f \in L(X,\mu)$, which, by the *Properties of Lebesgue Integral. II* (Theorem 6.3, part (8)), implies that f is *finite a. e.* μ *on* X. Thus, we can redefine f to be 0 on the μ-null set

$$\{x \in X \mid f(x) = \infty\}$$

obtaining a Σ-measurable function $f : X \to [0,\infty)$, which, by the *Properties of Lebesgue Integral. II* (Theorem 6.3, part (3)), satisfies (9.15) and (9.16), the latter implying that $f \in \mathscr{F}$.

To prove that f is the desired function, let us show that

$$\forall A \in \Sigma: \ \varphi(A) := \nu(A) - \int_A f\,d\mu = 0. \tag{9.17}$$

Since ν is a finite measure on Σ and $f \in \mathscr{F}$, by the *σ-Additivity of the Lebesgue Integral* (Theorem 6.2), we conclude that φ is also a *finite measure* on Σ. Moreover, $\varphi \ll \mu$.

Exercise 9.22. Verify.

Assume that (9.17) is not true, i. e.,

$$\exists B \in \Sigma: \ \varphi(B) > 0,$$

and hence, since $\varphi \ll \mu$, $\mu(B) > 0$.

Exercise 9.23. Explain.

Then

$$\exists \beta > 0: \ \varphi(B) > \beta\mu(B). \tag{9.18}$$

On Σ, consider the *finite signed measure*

$$\psi(A) := \varphi(A) - \beta\mu(A), \ A \in \Sigma. \tag{9.19}$$

Since $\varphi \ll \mu$ and by (9.18),

$$\psi \ll \mu \quad \text{and} \quad \psi(B) > 0.$$

Let $X_-, X_+ \in \Sigma$ be a Hahn decomposition of X relative to ψ, with X_- being ψ-negative and X_+ being ψ-positive. Then, by the *additivity* of ψ (Theorem 9.1) in view of the ψ-positivity/negativity of the sets X_+ and X_-, respectively, we have

$$0 < \psi(B) = \psi(B \cap X_+) + \psi(B \cap X_-) \le \psi(B \cap X_+) \le \psi(B \cap X_+) + \psi(B^c \cap X_+) = \psi(X_+),$$

and hence, since $\psi \ll \mu$,

$$\mu(X_+) > 0. \tag{9.20}$$

The function $f + \beta\chi_{X_+}$ belongs to the collection $\mathscr{F}$ since, for each $A \in \Sigma$,

$$\begin{aligned}
\int_A (f + \beta\chi_{X_+})\,d\mu &= \int_{A\cap X_-} (f + \beta\chi_{X_+})\,d\mu + \int_{A\cap X_+} (f + \beta\chi_{X_+})\,d\mu \\
&= \int_{A\cap X_-} f\,d\mu + \int_{A\cap X_+} (f + \beta)\,d\mu \\
&\le \nu(A\cap X_-) + \int_{A\cap X_+} f\,d\mu + \beta\mu(A\cap X_+) && \text{by (9.17) and (9.19);} \\
&= \nu(A\cap X_+) + \nu(A\cap X_-) - \psi(A\cap X_+) && \text{since } \psi(A\cap X_+) \ge 0; \\
&\le \nu(A\cap X_+) + \nu(A\cap X_-) = \nu(A).
\end{aligned}$$

However, by (9.15) and (9.20),

$$\int_X (f + \beta\chi_{X_+})\,d\mu = \int_X f\,d\mu + \beta\mu(X_+) = S + \beta\mu(X_+) > S,$$

which *contradicts* the definition of S (see (9.13)).

Thus, (9.17) holds, and hence,

$$\nu(A) = \int_A f\,d\mu,\ A \in \Sigma.$$

Now, suppose that μ and ν are *σ-finite* (positive) measures on Σ. Then there exists a *pairwise disjoint* sequence $(X_n)_{n\in\mathbb{N}}$ in Σ such that

$$\bigcup_{n=1}^{\infty} X_n = X \quad \text{and} \quad \mu(X_n), \nu(X_n) < \infty,\ n \in \mathbb{N}.$$

Exercise 9.24. Explain.

By the proven above, for the *finite* (positive) measures μ and ν on the trace σ-algebra $\Sigma \cap X_n$ for each $n \in \mathbb{N}$, we infer that, for each $n \in \mathbb{N}$, there exists a $\Sigma \cap X_n$-measurable function $f_n : X_n \to [0, \infty)$ such that

$$\forall A \in \Sigma : \ \nu(A \cap X_n) = \int_{A\cap X_n} f_n\,d\mu.$$

Then, for each $n \in \Sigma$, the function

$$\tilde{f}_n(x) := \begin{cases} f_n(x), & x \in X_n, \\ 0, & x \in X_n^c \end{cases}$$

is nonnegative, finite-valued, and Σ-measurable, and hence, by the *Series of Measurable Functions Corollary* (Corollary 5.5), the nonnegative and finite-valued function

$$f(x) := \sum_{n=1}^{\infty} \tilde{f}_n(x),\ x \in X,$$

is also Σ-measurable. Moreover, for each $A \in \Sigma$, by the *Integrating Series Proposition* (Proposition 6.1) and *σ-additivity* of ν,

$$\int_A f\,d\mu = \sum_{n=1}^{\infty} \int_A \tilde{f}_n\,d\mu = \sum_{n=1}^{\infty} \int_{A\cap X_n} f_n\,d\mu = \sum_{n=1}^{\infty} \nu(A \cap X_n) = \nu(A).$$

If ν is a *σ-finite* signed measure, by the *Jordan Decomposition Theorem* (Theorem 9.3),

$$\nu = \nu_+ - \nu_-$$

where ν_+ is a σ-finite measure, ν_- is a finite measure on Σ.

By the *Characterizations of Absolute Continuity* (Theorem 9.4), $\nu_+ \ll \mu$ and $\nu_- \ll \mu$. The rest follows by applying the above to the measures ν_+ and ν_-.

Exercise 9.25. Fill in the details.

To prove the *uniqueness* part, first, suppose that the signed measure ν is *finite*, which implies that $f, g \in L(X, \mu)$, and hence, by the *Characterization of Equality A. E.* (Proposition 6.3) (Section 6.11, Problem 5), $f(x) = g(x)$ a. e. μ on X.

Now, suppose that the signed measure ν is *σ-finite*. Then there exists a sequence $(X_n)_{n\in\mathbb{N}}$ in Σ such that

$$\bigcup_{n=1}^{\infty} X_n = X \quad \text{and} \quad \nu(X_n) < \infty,\ n \in \mathbb{N}.$$

Applying the above to μ and the *finite* signed measure ν on the trace σ-algebra $\Sigma \cap X_n$ for each $n \in \mathbb{N}$, we infer that $\{x \in X_n \,|\, f(x) \neq g(x)\}$, $n \in \mathbb{N}$, is a *μ-null set*, and hence,

$$\{x \in X \,|\, f(x) \neq g(x)\} = \bigcup_{n=1}^{\infty} \{x \in X_n \,|\, f(x) \neq g(x)\}$$

is also μ-null set.

Exercise 9.26. Verify. □

9.5.3 Radon–Nikodym Derivative

Definition 9.7 (Radon–Nikodym Derivative). Let $\nu : \Sigma \to (-\infty, \infty]$ be a signed measure on a measure space (X, Σ, μ). A Σ-measurable function $f : X \to \overline{\mathbb{R}}$ such that

$$\nu(A) = \int_A f\, d\mu, \; A \in \Sigma,$$

if exists, is called the *density* or *Radon–Nikodym derivative of* ν *relative to* μ.

Notation. $f = \frac{d\nu}{d\mu}$ or sometimes $d\nu = f d\mu$.

Remarks 9.8.
- If there exists $f = \frac{d\nu}{d\mu}$, it is not unique. Indeed, for any Σ-measurable function $g : X \to \overline{\mathbb{R}}$ such that $g = f \pmod{\mu}$, $g = \frac{d\nu}{d\mu}$.
- Under the conditions of the *Radon–Nikodym Theorem*, a finite-valued Radon–Nikodym derivative of ν relative to μ, $f = \frac{d\nu}{d\mu}$, exists and is unique up to a. e. μ on X.

9.6 Lebesgue Decomposition Theorem

The following definition generalizes mutual singularity of measures (see Definition 9.4).

Definition 9.8 (Singularity of Signed Measure). A signed measure $\nu : \Sigma \to (-\infty, \infty]$ on a measure space (X, Σ, μ) is said to be *singular relative to the measure* μ if there exists a set $E \in \Sigma$ for which

$$\mu(E) = 0 \quad \text{and} \quad \forall B \in \Sigma,\ B \subseteq E^c : \nu(B) = 0.$$

Notation. $\nu \perp \mu$.

The subsequent statement is a generalization of *Lebesgue Decomposition Theorem for Functions* (Theorem 8.25).

Theorem 9.6 (Lebesgue Decomposition Theorem). *Let* $\nu : \Sigma \to (-\infty, \infty]$ *be a σ-finite signed measure on a σ-finite measure space* (X, Σ, μ). *Then there is a unique representation*

$$\nu(A) = \nu_1(A) + \nu_2(A), \; A \in \Sigma,$$

with signed measures $\nu_i : \Sigma \to (-\infty, \infty]$, $i = 1, 2$, *such that*

$$\nu_1 \ll \mu \quad \textit{and} \quad \nu_2 \perp \mu,$$

which is called the Lebesgue decomposition of ν *relative to* μ.

For proof, see, e. g., [1, 7].

9.7 Problems

1. Let (X, Σ, ν) be a signed measure space. Prove that a countable union of ν-positive/negative/null sets is a ν-positive/negative/null set, respectively.
2. Let (X, Σ, ν) be a signed measure space. Prove that, if $X_-, X_+ \in \Sigma$ and $X'_-, X'_+ \in \Sigma$ are Hahn decompositions of X relative to ν, then

$$\forall A \in \Sigma : \ \nu(A \cap X_-) = \nu(A \cap X'_-), \ \nu(A \cap X_+) = \nu(A \cap X'_+).$$

3. Let (X, Σ, μ) be a measure space, $f, g : X \to [0, \infty)$ be Σ-measurable functions such that

$$f(x)g(x) = 0 \text{ a. e. } \mu \text{ on } X,$$

and

$$\mu(A) := \int_A f \, d\mu, \ \nu(A) := \int_A g \, d\mu, \ A \in \Sigma.$$

Prove that $\mu \perp \nu$.

4. * Let $\nu : \Sigma \to (-\infty, \infty]$ be a signed measure on a measurable space and (X, Σ) with the total variation measure $|\nu|$. Prove that, for any $A \in \Sigma$,

$$|\nu|(A) := \sup\left\{ \sum_{i=1}^{n} |\nu(B_i)| \ \middle| \ n \in \mathbb{N}, \ B_i \in \Sigma, \ i = 1, \ldots, n, \text{ are pairwise disjoint and } \bigcup_{i=1}^{n} B_i = A \right\}.$$

5. Prove

Proposition 9.1 (Null Set Characterization). *Let $\nu : \Sigma \to (-\infty, \infty]$ be a signed measure on a measurable space and (X, Σ) with the total variation measure $|\nu|$. A set $A \in \Sigma$ is a ν-null set iff it is a $|\nu|$-null (i. e., $|\nu|(A) = 0$).*

Hint. Use the *Jordan decomposition of ν* (see the *Jordan Decomposition Theorem* (Theorem 9.3)).

6. Let $\nu : \Sigma \to (-\infty, \infty]$ be a signed measure on a measurable space (X, Σ) with the *Jordan decomposition* $\nu = \nu_+ - \nu_-$ and the *total variation measure* $|\nu| := \nu_+ + \nu_-$. Show that

$$\int_A f \, d\nu := \int_A f \, d\nu_+ - \int_A f \, d\nu_-, \ A \in \Sigma,$$

is a signed measure on the σ-algebra Σ and

$$\left| \int_A f \, d\nu \right| \le \int_A |f| \, d|\nu|, \ A \in \Sigma.$$

7. (a) Prove

Proposition 9.2 (Characterization of Absolute Continuity). *Let μ and ν be (positive) measures on a measurable space (X, Σ). For $\nu \ll \mu$ it is sufficient and, provided $\nu(X) < \infty$, necessary that*

$$\nu \ll \mu \Leftrightarrow \forall \varepsilon > 0 \, \exists \delta > 0 \, \forall A \in \Sigma \text{ with } \mu(A) < \delta : \nu(A) < \varepsilon.$$

Corollary 9.1 (Absolute Continuity of Lebesgue Integral). *Let (X, Σ, μ) be a measure space and $f : X \to \overline{\mathbb{R}}$ be a Σ-measurable function. If $f \in L(X, \mu)$, then*

$$\forall \varepsilon > 0 \, \exists \delta > 0 \, \forall A \in \Sigma \text{ with } \mu(A) < \delta : \int_A |f(x)| \, d\mu(x) < \varepsilon.$$

(b) Give an example showing that the condition of the finiteness of the measure ν is essential for thenecessity part of Proposition 9.2 and cannot be dropped.

Hint. For (a), to prove the *necessity* part of Proposition 9.2 *by contrapositive* or *by contradiction*, apply the *Lower Limit/Upper Limit Proposition* (Proposition 3.5) (Section 3.3, Problem 10).

8. Let $\nu : \Sigma \to (-\infty, \infty]$ be a signed measure on a measurable space (X, Σ) with the *Jordan decomposition* $\nu = \nu_+ - \nu_+$ and the *total variation measure* $|\nu| := \nu_+ + \nu_-$. Show that

$$\frac{d\nu_+}{d|\nu|} = \chi_{X_+}, \quad \frac{d\nu_-}{d|\nu|} = \chi_{X_-}, \quad \frac{d\nu}{d|\nu|} = \chi_{X_+} - \chi_{X_-}.$$

9. Let μ and ν be *mutually singular* measures on a measurable space (X, Σ). Prove that $\nu \ll \mu + \nu$ and find $\frac{d\nu}{d(\mu+\nu)}$.
10. Let μ_1 and μ_2 be σ-finite measures and ν be a σ-finite signed measure on a measurable space (X, Σ) such that $\nu \ll \mu_2$ and $\mu_2 \ll \mu_1$. Prove that $\nu \ll \mu_1$ and

$$\frac{d\nu}{d\mu_1} = \frac{d\nu}{d\mu_2} \cdot \frac{d\mu_2}{d\mu_1} \pmod{\mu}.$$

A The Axiom of Choice and Equivalents

A.1 The Axiom of Choice

To choose one sock from each of infinitely many pairs of socks requires the axiom of choice, but for shoes the axiom is not needed.
Bertrand Russell

Here, we give a concise discourse on the celebrated *Axiom of Choice*, its equivalents, and *ordered sets*.

A.1.1 The Axiom of Choice

Expository Reference to a Set by Cantor. *By a set X, we understand "a collection into a whole of definite, well-distinguished objects, called the elements of X, of our perception or of our thought."*

Axiom of Choice (1904). *For each nonempty collection $\mathscr{F}$ of nonempty sets, there is a function $f : \mathscr{F} \to \bigcup_{X\in\mathscr{F}} X$ such that*[1]

$$\mathscr{F} \ni X \mapsto f(X) \in X.$$

Or equivalently.

For each nonempty collection $\{X_i\}_{i\in I}$ of nonempty sets, there is a function $f : I \to \bigcup_{i\in I} X_i$ such that

$$I \ni i \mapsto f(i) \in X_i.$$

The function f is called a choice function on $\mathscr{F}$, respectively, on I.

See, e. g., [8, 6, 9, 16].

A.1.2 Controversy

The *Axiom of Choice* allows one to prove the following counterintuitive statements.

Theorem A.1 (Vitali Theorem (1905)). *There exists a set in $\mathbb{R}$, which is not Lebesgue measurable.*[2]

See Theorem 4.7.

1 Due Ernst Zermelo (1871–1953).

2 Giuseppe Vitali (1875–1932).

https://doi.org/10.1515/9783110600995-010

Theorem A.2 (Banach–Tarski Paradox (1924)). *Given a solid ball in 3-dimensional space, there exists a decomposition of the ball into a finite number of disjoint pieces, which can be reassembled, using only rotations and translations, into two identical copies of the original ball. The pieces involved are nonmeasurable, i. e., one cannot meaningfully assign volumes to them.*[3]

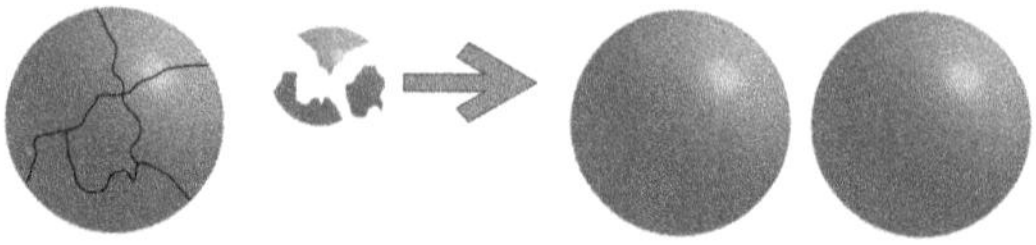

A.1.3 Timeline

- 1904: Ernst Zermelo formulates the *Axiom of Choice* in order to prove the *Well-Ordering Principle* (see Theorem A.7).
- 1939: Kurt Gödel[4] proves that, if the other standard set-theoretic *Zermelo–Fraenkel*[5] *Axioms* (see, e. g., [8, 6, 9, 16]) are consistent, they do not disprove the *Axiom of Choice.*
- 1963: Paul Cohen[6] completes the picture by showing that, if the other standard *Zermelo–Fraenkel Axioms* are consistent, they do not yield a proof of the *Axiom of Choice*, i. e., the *Axiom of Choice* is *independent.*

A.2 Ordered Sets

Here, we introduce and study various types of *order* on a set.

Definition A.1 (Partially Ordered Set). A *partially ordered set* is a nonempty set X with a *binary relation* $\leq$ of *partial order*, which satisfies the following *partial order axioms*:
1. For any $x \in X$, $x \leq x$. *Reflexivity*
2. For any $x, y \in X$, if $x \leq y$ and $y \leq x$, then $x = y$. *Antisymmetry*
3. For any $x, y, z \in X$, if $x \leq y$ and $y \leq z$, then $x \leq z$. *Transitivity*

If $x \leq y$, we say that x is a *predecessor* of y and that y is a *successor* of x.

Notation. $(X, \leq)$.

3 Alfred Tarski (1901–1983).
4 Kurt Gödel (1906–1978).
5 Abraham Fraenkel (1891–1965).
6 Paul Cohen (1934–2007).

Examples A.1.
1. An arbitrary nonempty set X is partially ordered by the *equality* (*coincidence*) relation $=$.
2. The set $\mathbb{R}$ is partially ordered by the usual order $\leq$.
3. The *power set* $\mathscr{P}(X)$ of a nonempty set X (see Section 1.1.1) is partially ordered by the set-theoretic inclusion $\subseteq$.

Remark A.1. Elements x, y of a partially ordered set $(X, \leq)$ are called *comparable* if $x \leq y$ or $y \leq x$. In a partially ordered set, incomparable elements may occur.

Exercise A.1.
(a) Verify the prior examples and remark.
(b) Give two more examples.
(c) Give an example of a partially ordered set $(X, \leq)$, in which no two distinct elements are comparable.

Remarks A.2.
- If $\leq$ is partial order on X, then the relation $\geq$ defined as follows:

$$x \geq y \Leftrightarrow y \leq x$$

  is also a partial order on X.
- If $x \leq y$ and $x \neq y$, we write $x < y$ or $y > x$.

Definition A.2 (Upper and Lower Bounds). Let Y be a nonempty subset of a *partially ordered set* $(X, \leq)$.
- An element $x \in X$ is called an *upper bound* of Y if

$$\forall y \in Y : y \leq x.$$

- An element $x \in X$ is called a *lower bound* of Y if

$$\forall y \in Y : x \leq y.$$

Remark A.3. Upper/lower bounds of a set Y need not exist.

Exercise A.2. Give corresponding examples.

Definition A.3 (Maximal and Minimal Elements). Let Y be a nonempty subset of a *partially ordered set* $(X, \leq)$.
- An element $x \in Y$ is called a *maximal element* of Y if

$$\nexists y \in Y : x < y,$$

  i. e., x has no successors in Y.

- An element $x \in Y$ is called a *minimal element* of Y if

$$\nexists y \in Y : y < x,$$

i. e., x has no predecessors in Y.

Remarks A.4.
- If an element $x \in Y$ is not comparable with all other elements of Y, it is automatically both maximal and minimal element of Y.
- A *maximal element* of Y need not be greater than all other elements in Y.
- A *minimal element* of Y need not be less than all other elements in Y.
- Maximal and minimal elements of Y need not exist nor be unique.

Exercise A.3. Give corresponding examples.

Definition A.4 (Greatest and Least Elements). Let Y be a nonempty subset of a *partially ordered set* $(X, \leq)$.
- An element $x \in Y$ is called the *greatest element* (also the *last element*) of Y if it is an *upper bound* of Y, i. e.,

$$\forall y \in Y : y \leq x.$$

- An element $x \in Y$ is called the *least element* (also the *first element*) of Y if it is a *lower bound* of Y, i. e.,

$$\forall y \in Y : x \leq y.$$

Remark A.5. The greatest/least element of a set need not exist.

Exercise A.4. Give corresponding examples.

Exercise A.5. Let $\mathscr{P}(X)$ be the power set of a set X consisting of more than one element and partially ordered by the set-theoretic inclusion $\subseteq$ and

$$\mathscr{Y} := \{A \in \mathscr{P}(X) \mid A \neq \emptyset, X\}.$$

(a) What are the *lower* and *upper bounds* of $\mathscr{Y}$?
(b) What are the *minimal* and *maximal elements* of $\mathscr{Y}$?
(c) What are the *least* and *greatest elements* of $\mathscr{Y}$?

Proposition A.1 (Properties of the Greatest and Least Elements). *Let Y be a nonempty subset of a partially ordered set $(X, \leq)$.*
(1) *If the greatest/least element of Y exists, it is unique.*
(2) *If the greatest/least element of Y exists, it is also the unique maximal/minimal element of Y.*

Exercise A.6.
(a) Prove.
(b) Give an example showing that a maximal/minimal element of Y need not be its greatest/least element.

Definition A.5 (Least Upper and Greatest Lower Bounds). Let Y be a nonempty subset of a *partially ordered set* $(X, \leq)$.
- If the set U of all upper bounds of Y is nonempty and has the least element u, u is called the *least upper bound* (the *supremum*) of Y and we write $u = \sup Y$.
- If the set L of all lower bounds of Y is nonempty and has the greatest element l, l is called the *greatest lower bound* (the *infimum*) of Y and we write $l = \inf Y$.

Remark A.6. For a subset Y in a partially ordered set $(X, \leq)$, $\sup Y$ and $\inf Y$ need not exist and, when they do, need not belong to Y.

Exercise A.7. Give corresponding examples.

Definition A.6 (Totally Ordered Set). A *totally ordered set* (also a *linearly ordered set* or a *chain*) is a partially ordered set $(X, \leq)$, in which any two elements are comparable, i. e.,

$$\forall\, x, y \in X : \; x \leq y \text{ or } y \leq x.$$

Exercise A.8.
(a) When is the *power set* $\mathscr{P}(X)$ of a nonempty set X partially ordered by the set-theoretic inclusion $\subseteq$ a chain?
(b) Show that, for a nonempty subset Y of a *totally ordered set* $(X, \leq)$, a *maximal/minimal element* of Y, when exists, is the *greatest/least element* of Y.

Definition A.7 (Well-Ordered Set). A *well-ordered set* is a totally ordered set $(X, \leq)$, in which every nonempty subset has the first element.

Examples A.2.
1. $(\mathbb{N}, \leq)$, as well as its arbitrary nonempty subset, is well ordered.
2. $(\mathbb{Z}, \leq)$ is totally ordered, but *not* well ordered.

Remarks A.7.
- As follows from *Zermelo's Well-Ordering Principle* (Theorem A.7), every nonempty set can be well ordered.
- Each well-ordered set $(X, \leq)$ is similar to $(\mathbb{N}, \leq)$ in the sense that each nonempty subset $Y \subseteq X$ has the *first element* and each element $x \in X$, except for the last one, if any, has the *unique immediate successor* (the *unique next element*) $s(x) \in X$, i. e., there is a *unique* element $s(x) \in X$ such that

$$x < s(x) \text{ and } \nexists\, y \in X : \; x < y < s(x).$$

This fact affords the possibility of inductive proofs (*transfinite induction*) and constructions over such sets similar to those over $\mathbb{N}$.

Exercise A.9.
(a) Verify the latter.
(b) Prove that, if a nonempty subset Y of a well-ordered set $(X, \leq)$ has an upper bound, it has the *least upper bound.*
(c) Prove that the usual order $\leq$ of the real line $\mathbb{R}$ restricted to any *uncountable* subset $Y \subseteq \mathbb{R}$ is a total order, but not a well order on Y.

A.3 Equivalents

The axiom of choice is obviously true, the well-ordering principle obviously false, and who can tell about Zorn's lemma?
Jerry Bona

Here, we are to prove the equivalence of the *Axiom of Choice* to following three fundamental set-theoretic principles.

Theorem A.3 (Hausdorff Maximal Principle). *In a partially ordered set, there exists a maximal chain.*

Theorem A.4 (Hausdorff Maximal Principle (Precise Version)). *In a partially ordered set, every chain is contained in a maximal chain.*

Theorem A.5 (Zorn's Lemma). *In a partially ordered set, whose every chain has an upper bound, there is a maximal element.*[7]

Theorem A.6 (Zorn's Lemma (Precise Version)). *For each element x in a partially ordered set $(X, \leq)$, whose every chain has an upper bound, there is a maximal element u in $(X, \leq)$ such that $x \leq u$.*

Theorem A.7 (Zermelo's Well-Ordering Principle). *Every nonempty set can be well ordered.*

Proof. We are to prove the following closed chain of implications:

$$AC \Rightarrow HMP \Rightarrow ZL \Rightarrow ZWOP \Rightarrow AC,$$

where the abbreviations *AC*, *HMP*, *ZL*, and *ZWOP* stand for the *Axiom of Choice*, the *Hausdorff Maximal Principle*, *Zorn's Lemma*, and *Zermelo's Well-Ordering Principle*, respectively.

7 Max Zorn (1906–1993).

AC ⇒ *HMP*
Assume the *Axiom of Choice* and let C be an arbitrary chain in a partially ordered set $(X, \leq)$ and let $\mathscr{C}$ be the collection of all chains in X containing C partially ordered by the set-theoretic inclusion $\subseteq$.

Observe that $\mathscr{C} \neq \emptyset$ since, obviously, $C \in \mathscr{C}$.

Our goal is to prove that $(\mathscr{C}, \subseteq)$ has a *maximal element* U.

Let f be a *choice function* assigning to every nonempty subset A of X one of its elements $f(A)$.

For each $A \in \mathscr{C}$, let $\hat{A}$ be the set of all those elements in X whose *adjunction* to A produces a chain in $\mathscr{C}$:

$$\hat{A} := \{x \in X \mid A \cup \{x\} \in \mathscr{C}\}.$$

Clearly, $A \subseteq \hat{A}$, the equality holding *iff* A is a maximal element in $(\mathscr{C}, \subseteq)$.

Consider a function $g : \mathscr{C} \mapsto \mathscr{C}$ defined as follows:

$$\mathscr{C} \ni A \mapsto g(A) := \begin{cases} A \cup \left\{f(\hat{A} \setminus A)\right\} & \text{if } A \subset \hat{A}, \\ A & \text{if } A = \hat{A}. \end{cases}$$

Observe that, for each $A \in \mathscr{C}$, the set $g(A)$ differs from A by *at most one element*.

Thus, to prove that $U \in \mathscr{C}$ is a maximal element in $(\mathscr{C}, \subseteq)$, one needs to show that $U = \hat{U}$, i. e., that $g(U) = U$.

Let us introduce the following temporary definition.

Definition A.8 (Tower). We call a subcollection $\mathscr{I}$ of $\mathscr{C}$ a *tower* if it satisfies the conditions:
(1) $C \in \mathscr{I}$.
(2) If $A \in \mathscr{I}$, then $g(A) \in \mathscr{I}$.
(3) If $\mathscr{D}$ is a chain in $\mathscr{I}$, then $\bigcup_{A \in \mathscr{D}} A \in \mathscr{I}$.

Observe that *towers* exist since, as is easily seen, $\mathscr{C}$ is a *tower* itself.

Furthermore, the intersection of all towers $\mathscr{I}_0$ is also a tower and is, in fact, the *smallest tower*, the *nonemptiness* of $\mathscr{I}_0$ being ensured by condition (1) of the above definition.

Let us show that $\mathscr{I}_0$ is a *chain* in $(\mathscr{C}, \subseteq)$.

We call a set $B \in \mathscr{I}_0$ *comparable* if, for each $A \in \mathscr{I}_0$ either $A \subseteq B$ or $B \subseteq A$. Thus, proving that $\mathscr{I}_0$ is a chain amounts to showing that all its elements are comparable.

Observe that there is at least one comparable set in $\mathscr{I}_0$, which is C, since it is contained in any other set in $\mathscr{I}_0$.

Let B be an arbitrary comparable set in $\mathscr{I}_0$. Suppose that $A \in \mathscr{I}_0$ and A is a *proper subset* of B. Then $g(A) \subseteq B$. Indeed, since B is comparable, either $g(A) \subseteq B$, or $B \subset g(A)$. In the latter case, A is a proper subset of a proper subset of $g(A)$, which contradicts the fact that $g(A) \setminus A$ is *at most a singleton*.

Further, consider the collection $\mathscr{U}(B)$ of all those $A \in \mathscr{I}_0$ for which either $A \subseteq g(B)$, or $g(B) \subseteq A$.

It is not difficult to make sure that $\mathscr{U}(B)$ is a *tower* (the verification of the least trivial condition (2) uses the argument of the preceding paragraph).

Since $\mathscr{U}(B)$ is a tower and $\mathscr{U}(B) \subseteq \mathscr{I}_0$, $\mathscr{U}(B) = \mathscr{I}_0$ necessarily.

All these considerations imply that, for each comparable set $B \in \mathscr{I}_0$, the set $g(B)$ is also comparable. The latter jointly with the facts that the set C is comparable and that, by condition (3), the union of the sets of a chain of comparable sets is also comparable imply that all comparable sets of $\mathscr{I}_0$ constitute a tower, and hence, they exhaust the entire $\mathscr{I}_0$.

Since $\mathscr{I}_0$ is a chain in $\mathscr{I}$, by condition (3), the set

$$U := \bigcup_{A \in \mathscr{I}_0} A \in \mathscr{I} \subseteq \mathscr{C}$$

and, obviously, U contains every set of $\mathscr{I}_0$. In particular, by condition (2) applied to the tower $\mathscr{I}_0$, $g(U) \subseteq U$, which implies that $g(U) = U$ proving that U is a maximal chain containing C.

Hence, the *Axiom of Choice* does imply the *Hausdorff Maximal Principle.*

HMP $\Rightarrow$ *ZL*

Assume the *Hausdorff Maximal Principle* and let x be an arbitrary element in a partially ordered set $(X, \leq)$, whose every chain has an upper bound.

By the *HMP*, there is a *maximal chain* U in $(X, \leq)$ containing the trivial chain $\{x\}$, i. e., $x \in U$. By the premise of *Zorn's Lemma*, U has an *upper bound* u in $(X, \leq)$. In particular, this implies that $x \leq u$.

From the maximality of the chain U, it follows that u is a *maximal element* in $(X, \leq)$. Otherwise, there would exist such an element $v \in X$ that $u < v$, which would imply that the chain U could be extended to a larger chain $U \cup \{v\}$ contradicting the maximality of U.

Thus the *Hausdorff Maximal Principle* implies the *Zorn's Lemma*.

ZL $\Rightarrow$ *ZWOP*

Assume *Zorn's Lemma* and let X be an arbitrary nonempty set.

Let

$$\mathscr{W} := \{A \subseteq X \mid \exists \leq_A \text{ a well order on } A\}.$$

Observe that $\mathscr{W} \neq \emptyset$ since it contains all *singletons* (more generally, all *finite subsets*) of X.

Exercise A.10. Explain.

The following defines a *partial order* on $\mathscr{W}$:

$$(A, \leq_A) \preceq (B, \leq_B)$$

iff

$$A \subseteq B \text{ and } \forall\, x, y \in A : x \leq_A y \Leftrightarrow x \leq_B y \text{ and } \forall\, x \in A\ \forall\, y \in B \setminus A : x <_B y. \tag{A.1}$$

Exercise A.11. Verify that $\preceq$ is a *partial order* on $\mathscr{W}$.

For an arbitrary *chain* $\mathscr{C}$ in $(\mathscr{W}, \preceq)$, let

$$D := \bigcup_{C \in \mathscr{C}} C \subseteq X.$$

We define a *total order* $\leq_D$ on D as follows:

For any $x, y \in D$, exists $C_x, C_y \in \mathscr{C}$ such that $x \in C_x$, $y \in C_y$. Considering that $(\mathscr{C}, \preceq)$ is a chain, $C_x \subseteq C_y$ or $C_y \subseteq C_x$, and hence, $x, y \in C$, where $C := \max(C_x, C_y)$, and we define

$$x \leq_D y \Leftrightarrow x \leq_C y.$$

Exercise A.12. Verify
(a) that $\leq_D$ is *well-defined*, i. e., does not depend on the choice of $(C, \leq_C) \in \mathscr{C}$, which contains both x and y and
(b) that $(D, \leq_D)$ is a *chain*.

Thus, $(D, \leq_D)$ is a *chain*.

Let E be an arbitrary nonempty subset of D. For an $x \in E$, there is a $(C_x, \leq_{C_x}) \in \mathscr{C}$ is such that $x \in C_x$. Then $E \cap C_x$ is a nonempty subset of C_x and, since $(C_x, \leq_{C_x})$ is *well ordered*, without loss of generality, we can regard x to be the least element of $E \cap C_x$. Suppose $y \in E$ and $y <_D x$. Then, considering that $(\mathscr{C}, \preceq)$ is a *chain*, there is a $(C, \leq_C) \in \mathscr{C}$ containing both x and y with two possibilities:
1. $(C, \leq_C) \preceq (C_x, \leq_{C_x})$, in which case $y \in E \cap C_x$ implying that $y <_{C_x} x$, and hence, contradicting the choice of x.
2. $(C_x, \leq_{C_x}) \preceq (C, \leq_C)$, in which case, if $y \in C_x$, then $y \in E \cap C_x$ implying that $y <_{C_x} x$, and hence, contradicting the choice of x. If $y \in C \setminus C_x$, then, by (A.1), $x <_C y$ contradicting the fact that $y <_D x$.

The obtained contradictions show that x is the least element of E in $(D, \leq_D)$ proving that $(D, \leq_D)$ is *well ordered*.

The set $(D, \leq_D)$ is an *upper bound* of $\mathscr{C}$ in $(\mathscr{W}, \preceq)$.

Exercise A.13. Verify.

The latter, by *Zorn's Lemma*, implies that $(\mathscr{W}, \preceq)$ has a *maximal element* $(M, \leq_M)$. Then $M := X$ since, otherwise, choosing any $u \in X \setminus M$, we could extend the well order on M to a well order $\leq_N$ on $N := M \cup \{u\}$ as follows:

$$\forall\, x, y \in M : x \leq_N y \Leftrightarrow x \leq_M y \quad \text{and} \quad \forall\, x \in M : x <_N u,$$

which would imply that $(M, \leq_M) \prec (N, \leq_N)$ contradicting the maximality of $(M, \leq_M)$ in $(\mathscr{W}, \preceq)$. Thus X can be well ordered, i. e., *Zorn's Lemma* implies *Zermelo's Well-Ordering Principle.*

ZWOP $\Rightarrow$ *AC*

Exercise A.14. Prove.

Hint. For a nonempty collection of nonempty sets $\{X_i\}_{i\in I}$ consider a *well order* $\preceq$ on $X := \bigcup_{i\in I} X_i$. □

As an extra demonstration of how a typical proof by *Zorn's Lemma* works, let us prove the implication

ZL $\Rightarrow$ *AC*

Assume *Zorn's Lemma* and let $\{X_i\}_{i\in I}$ be a nonempty collection of nonempty sets. Consider the collection $\mathscr{F}$ of all possible functions $f : I \supseteq D(f) \to \bigcup_{i\in I} X_i$, such that, for any i in the *domain* $D(f)$ of f, $f(i) \in X_i$.

Such functions, obviously, exist on *singletons* (more generally, on *finite subsets*) of I.

Let us introduce a *partial order* $\preceq$ on $\mathscr{F}$ as follows:

$$f \preceq g \Leftrightarrow D(f) \subseteq D(g) \text{ and } f(x) = g(x),\ x \in D(f),$$

i. e., g is an *extension* of f.

Exercise A.15. Verify that $\preceq$ is a partial order on $\mathscr{F}$.

The premise of *Zorn's Lemma* holds in $(\mathscr{F}, \preceq)$.

Exercise A.16. Verify.

Hint. For an arbitrary chain $\mathscr{C}$ in $(\mathscr{F}, \preceq)$, show that $u \in \mathscr{F}$ with the domain

$$u(i) := g(i),\ i \in D(u) := \bigcup_{g\in\mathscr{C}} D(g),\ g \in \mathscr{C},$$

is an *upper bound* of $\mathscr{C}$ in $(\mathscr{F}, \preceq)$.

Then, by *Zorn's Lemma*, there is a *maximal choice function* f in $(\mathscr{F}, \preceq)$. This, implies that the domain $D(f)$ of f is the entire indexing set I and completes the proof.

Exercise A.17. Explain.

Bibliography

[1] R. F. Bass, *Real Analysis for Graduate Students*, ver. 2.1, Richard F. Bass, 2014.

[2] A. Ya. Dorogovtsev, *Mathematical Analysis. A Reference Handbook*, Vishcha Shkola, Kiev, 1985 (Russian).

[3] A. Ya. Dorogovtsev, *Elements of the General Theory of Measure and Integral*, Vishcha Shkola, Kiev, 1989 (Russian).

[4] B. R. Gelbaum and J. M. H. Olmsted, *Counterexamples in Analysis*, Dover Publications, Inc., Mineola, New York, 2003.

[5] C. Goffman and G. Pedrick, *First Course in Functional Analysis*, 2nd ed., Chelsea Publishing Co., New York, 1983.

[6] P. R. Halmos, *Naive Set Theory*, Undergraduate Texts in Mathematics, Springer-Verlag, New York, Heidelberg, Berlin, 1974.

[7] P. R. Halmos, *Measure Theory*, Graduate Texts in Mathematics, vol. 18, Springer-Verlag, New York, Heidelberg, Berlin, 1974.

[8] F. Hausdorff, *Set Theory*, 2nd ed., Chelsea Publishing Co., New York, 1962.

[9] T. J. Jech, *The Axiom of Choice*, Dover Publications, Inc., Mineola, New York, 2008.

[10] I. Kaplansky, *Set Theory and Metric Spaces*, Allyn and Bacon, Inc., Boston, 1972.

[11] A. G. Kurosh, *Lectures on General Algebra*, 2nd ed., Nauka, Moscow, 1973 (Russian).

[12] S. Lang, *Real and Functional Analysis*, 3rd ed., Graduate Texts in Mathematics, vol. 142, Springer-Verlag, New York, 1993.

[13] M. V. Markin, *Integration for Calculus, Analysis, and Differential Equations: Techniques, Examples, and Exercises*, World Scientific Publishing Co. Pte. Ltd., New Jersey, London, Singapore, 2019.

[14] M. V. Markin, *Elementary Functional Analysis*, De Gruyter Graduate, Walter de Gruyter GmbH, Berlin, Boston, 2018.

[15] M. V. Markin, *Elementary Operator Theory*, De Gruyter Graduate, Walter de Gruyter GmbH, Berlin, Boston, 2019, ISBN 978-3-11-060096-4.

[16] G. H. Moore, *Zermelo's Axiom of Choice: Its Origins, Development, and Influence*, Dover Publications, Inc., Mineola, New York, 2013.

[17] J. Muscat, *Functional Analysis. An Introduction to Metric Spaces, Hilbert Spaces, and Banach Algebras*, Springer International Publishing, Switzerland, 2014.

[18] J. R. Munkres, *Topology*, 2nd ed., Prentice Hall, Upper Saddle River, New Jersey, 2000.

[19] I. P. Natanson, *Theory of Functions of a Real Variable*, 3rd ed., Nauka, Moscow, 1974 (Russian).

[20] K. R. Parthasarathy, *Introduction to Probability and Measure*, The Macmillan Co. of India, Ltd., Delhi, 1977.

[21] C. W. Patty, *Foundations of Topology*, Waveland Press, Inc., Prospect Heights, Illinois, 1997.

[22] H. L. Royden and P. M. Fitzpatrick, *Real Analysis*, 4th ed., Pearson, London, 2010.

[23] W. Rudin, *Real and Complex Analysis*, 3rd ed., Graduate Texts in Mathematics, vol. 142, McGraw-Hill, Inc., New York, 1987.

[24] B. Simon, *Real Analysis. A Comprehensive Course in Analysis, Part 1*, American Mathematical Society, Providence, RI, 2015.

[25] W. A. Sutherland, *Introduction to Metric and Topological Spaces*, 2nd ed., Oxford University Press Inc., New York, 2009.

[26] R. L. Wheeden and A. Zygmund, *Measure and Integral. An Introduction to Real Analysis*, 2nd ed., CRC Press, Boca Raton, Florida, 2015.

[27] J. Yeh, *Real Analysis: Theory of Measure and Integration*, 3rd ed., World Scientific, New Jersey, London, Singapore, 2014.

https://doi.org/10.1515/9783110600995-011

Index

www.ingramcontent.com/pod-product-compliance
Lightning Source LLC
Chambersburg PA
CBHW080908220326
41598CB00034B/5509
9783110600971